The BODY Problematic

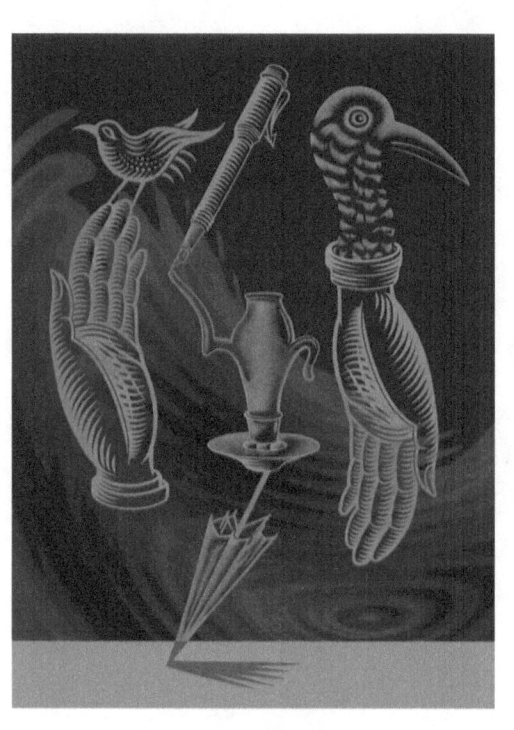

Laura Hengehold

The BODY Problematic

Political Imagination in Kant and Foucault

The Pennsylvania State University Press
University Park, Pennsylvania

LIBRARY OF CONGRESS
CATALOGING-IN-PUBLICATION DATA

Hengehold, Laura.
The body problematic : political imagination
in Kant and Foucault / Laura Hengehold.
    p.   cm.
Includes bibliographical references and index.
ISBN 978–0–271–03212–2 (pbk : alk. paper)
   1. Kant, Immanuel, 1724–1804.
   2. Foucault, Michel, 1926–1984.
   3. Body, Human.
   4. Political science.
   I. Title.
B2798.H3565 2007
193—dc22
2007009482

Copyright © 2007
The Pennsylvania State University
All rights reserved
Printed in the United States of America
Published by
The Pennsylvania State University Press,
University Park, PA 16802-1003

The Pennsylvania State University Press
is a member of the
Association of American University Presses.

It is the policy of
The Pennsylvania State University Press
to use acid-free paper. This book is printed on
stock that meets
the minimum requirements of
American National Standard
for Information Sciences—Permanence of Paper for
Printed Library Material,
ANSI Z39.48–1992.

CONTENTS

Acknowledgments vii

INTRODUCTION:
IMAGINATION AND PROBLEMATIZATION 1

PART 1
THE POLITICAL TOPOLOGY OF KANTIAN REASON 27

Drawing the Boundaries of Pure Reason 35
Transcendental and Other Topographies 44
The Quest for Unity 55
Discursivity and Materiality 67
The Virtues of Communicability 80
The Kantian Body—Missing in Action 89

PART 2
MAN AND HIS DOUBLES: TWO WAYS TO PROBLEMATIZE 117

Heterotopia and the Phenomenological World 118
In the Field of the Problematic Object 128
The Man-Form: Empirical and Transcendental 155
Materiality and Resemblance: Statements 167
Materiality and Resemblance: Bodies 181
An-aesthetic philosophy? 198

PART 3
LOCKED IN THE MARKET 211

From *Raison d'État* to *Phobie d'État* 216
Migration of Sovereignty 224
The Normal and the Normative 247
Crisis in Liberalism 269
Negative Anthropology 282
Afterword: Not Similar to Something, Just Similar 299

References 303
Index 317

FOR *Jennie*

ACKNOWLEDGMENTS

Without question, my greatest debt is to Andrew Cutrofello, for infinite patience and optimism at moments when I became lost in the materiality of language. I am unspeakably lucky to have had his friendship and mentorship during the years over which this book was composed. Thanks also to Brent Adkins, Allison Moore, Marian Staats, Craig Greenman, and Dan Price for many long nights on the phone and over drinks arguing about phenomenology, feminism, and the significance of the aesthetic; to John Sayer, for his steadfast model of the resistant aesthetic life; and to David Ingram, Peg Birmingham, and James Bernauer for helpful comments on the earliest versions of this project. I had the opportunity to work out possible implications of this project thanks to the Society for Phenomenology and Existential Philosophy, the Foucault Circle (and its indefatigable coordinator, Richard Lynch), and Eric Sean Nelson and Ben Pryor of the University of Toledo. Len Lawlor, Ladelle McWhorter, Rudolf Makkreel, Todd May, and John McCumber offered helpful criticisms and suggestions during the publication process. The manuscript has benefited greatly from their talents at finding order within the formless.

I owe a special thanks to Valentine Moulard-Leonard and to her family, who welcomed me in Paris; to Jacob Rogozinski, Eric Fassin, and the U.S. Fulbright Scholars Program for supporting my research on Foucault's unpublished lecture courses and *Anthropology* manuscript; and to L'IMEC, who generously granted me access to these materials. I am also grateful to the College of Arts and Sciences at Case Western Reserve University, who funded two follow-up trips to Paris under the W. P. Jones Faculty Development program, to the many students in Philosophy 315 who have unknowingly been the experimental subjects of this book, and to Andrea Esser and Jutta Ittner, who made possible the 2004 Max Kade workshop "Kant's Aesthetic Judgment." Alex Vikmanis and Arnaud Gerspacher contributed sage editing and formatting assistance. David Pincus taught me to appreciate affects other than resistance and to free myself, at least in part, from heteronomous imaginaries. Although no list can really do justice to the long threads of reflection that tie humans to one another (and thus I fear some names will be omitted), I would like to thank

Amanda Schaffer for her wonderfully pragmatic advice about the publishing industry, Sara Waller for moral support during the final revisions; and Dorothy Miller, Jonathan Sadowsky, Gilbert Doho, and Kunal Parker for many discussions about the state of American welfare, postcolonial politics, and the philosophy of history. A portion of Part 2 appeared under the title "'My Body, This Paper, This Fire': The Fate of Emotion in Foucault's Kantian Legacy" in *Networks: Selected Studies in Phenomenology and Existential Philosophy* 29, ed. Steven Crowell, Kelly Oliver, and Shannon Lundeen, 2003 SPEP Special Supplement, *Philosophy Today* 47, no. 5: 45–55.

INTRODUCTION: IMAGINATION AND PROBLEMATIZATION

Eleanor Bumpurs, a "270-pound, arthritic sixty-seven-year-old woman," was shot and killed by New York City police in 1984 for resisting eviction from city housing with a knife (Williams 1991, 136). Her case was one of several that crystallized African-American anger against the New York City police department and a coroner's office that was reputed to overlook police violence against poor suspects. It occurred during a period when the city of New York was struggling with a financial crisis that the banks only resolved on condition that public services be drastically cut and real estate prices be released from any controls, as they did in other cities and even debtor nations at that time (Harvey 2005, 44–48). Her daughter, who said that she had not been informed about the overdue rent, reported that her mother had suffered in the past from mental illness; housing officials acknowledged that she sometimes "saw Reagan coming through her walls" (Williams 1991, 142; Buder 1984; Raab 1984a). Family members had warned the elderly woman not to allow any strangers into the apartment, under any circumstances.

Although her death had simple physical causes, Mrs. Bumpurs would not have died if not for the way a broad range of personal and institutional expectations construed her body and her freedom. The officers sent to carry out the eviction were members of a special Emergency Service Unit supposedly trained to deal with disturbed persons. Explanations as to why they did not fire warning shots or disable her with mace when she lunged at them with her knife ranged from "she was elderly" and should have been easily subdued to "sometimes, when you're dealing with a disturbed person, [mace] makes them worse," and finally, "neighbors would have to be removed" (Buder 1984). In court testimony, no adequate reason was ever presented for firing a second, fatal shot when the first disabled Mrs. Bumpurs's knife arm (Raab 1985). Legal theorist Patricia Williams challenges us to explain:

the animus that inspired such fear and impatient contempt in a police officer that the presence of six other well-armed men could not allay his need to kill a sick old lady fighting off hallucinations with a knife. It seemed to me a fear embellished by something beyond Mrs. Bumpurs herself; something about her that filled the void between her physical, limited presence and the "immediate threat and endangerment to life" in the beholding eyes of the officer. Why was the sight of a knife-wielding woman so fearful to a shotgun-wielding policeman that he had to blow her to pieces as the only recourse, the only way to preserve his physical integrity? (1991, 144)

Williams implies that the *something* beyond Mrs. Bumpurs's physical presence existed in the policeman's imagination and occupied the metaphorical "void" between feelings of loyalty and aggression tied to his membership in this particular unit, his evaluation of the justice involved in evictions, his conception of what a "normal" versus a "disturbed" elderly woman might believe or be capable of doing when confronted with a threat to her home and personal safety, and the conditions under which he was entitled to use deadly force.

However, this void also encompassed the Housing Authority's unfulfilled expectations of communication with the Division of Social Services responsible for psychiatric evaluations and Mrs. Bumpurs's family, as well as the family's reverse expectations. One of the officers defended himself in court by casting suspicion on the motives of her family members, whom he held responsible for not paying the rent themselves (Williams 1991, 142). It was shaped by impersonal categories such as the criteria used to categorize psychiatric illness and the roster of names on which Mrs. Bumpurs entered the "imagination" of agency workers as a more or less urgent case. A psychiatrist had recently pronounced her psychotic, but no coordination with Housing Authority had taken place regarding her finances. Family members disagreed about whether she was really mentally ill (Raab 1984a); Williams also questions whether the eviction was legitimate, since papers were never served to her personally (1991, 136–37). In short, Mrs. Bumpurs's body was not *only* "physical," but had an unknown aspect involving the imagination of other bodies than her own. This also means that in some ontologically significant sense, a part of Mrs. Bumpurs *existed* in other people's bodies and codes as well as in her own flesh and language. But she was obviously unable to use this plurality to her advantage; she experienced it as a problem, and not a way to escape or take distance on her poverty.

I am also struck, however, by how often people in similar situations—the disabled, aged, poor, or those with exhausting family and work obligations—form the outer horizon of imagination for those in relatively stable or affluent circumstances. Advertisers, health officials, and banks require us to measure our self-respect against our ability to navigate their simultaneous claims. These institutions and practices project images of the successful consumer we hope to be as well as the loan-suffocated bankruptcy client "we can't imagine being." Although rich and poor are not subject to discourses and institutional surveillance in the same way, both have to make sense of expectations emerging from incongruous discourses and decide, against the backdrop of limited psychic resources and time (if not simple finances), which to ignore and which to take seriously. Often we give form to our lives not through specific hopes but by recognizing specific "formed" examples of poverty, bad luck, incompetence, or deviance that stand in for whatever unformed and nonsensical situation we hope to avoid. Crime stories on television and poverty reporting provoke their audiences to identify the "fatal flaw" explaining and tacitly justifying the suffering of others, and thus reinforce their own sense of security (unreasonably, in many cases). One wonders how a complex society that places such demands on the individual can avoid *inventing* people and situations like Mrs. Bumpurs's so that "sane" people can recognize and appreciate their own competence, however limited.

It is also worth noting that in some unspecified way, Mrs. Bumpurs did perceive her problem as political rather than merely moral or psychological. By political, I mean she thought her problem arose from power relations at a scale that involved strangers and believed any solution would require changing those power relations. We do not know how literally or metaphorically she saw "Reagan coming through her walls." But Reagan was the emblem of a sea change in American attitudes toward the poor, the civil rights movement, and the rights of private landlords and investors or entrepreneurs vis-à-vis consumers and those who depended on public services. We will never know whether, given enough time, statistics, and discussion, Mrs. Bumpurs might not have been able to phrase her complaint against the injustice she associated with Reagan without hallucinations or recourse to violence. Perhaps she could have been what I would call a mad critic—not necessarily insane, but lacking the resources and the ability to communicate with a public that could have made the conflict in her imagination *real* and therefore capable of being altered for people beyond her family.

For another example of the uneasy relationship between imagination, madness, and political criticism, consider the nineteenth-century abolitionist John Brown, who was hanged in 1859 for leading an uprising at Harpers Ferry, Virginia. He and eighteen followers (some white, some African-American) attempted to seize control of a federal arsenal in hopes of establishing a slave-free zone in the South and gradually spreading rebellion among slaves. Southern slaveholders read the uprising as evidence of an abolitionist conspiracy to invade the South, but Northern supporters (including Brown's family members) pleaded with the Virginia governor for clemency on the ground that he was insane, as did Northern Republicans who wanted to distance themselves from Brown's violent tactics and save the Union (McGlone 1995, 214–15). However, one must ask who was more violent and divisive: Brown (whose men had committed atrocities during the struggle to keep Kansas free) or the leaders of plantation society who employed vigilantes and civilian slave patrols to keep Africans under personal domination and no less violent economic exploitation?

During the preceding ten years, the Fugitive Slave Act (1850) and *Dred Scott v. Sanford* decision (1857) had deprived even free blacks of full citizenship; antimiscegenation laws discouraged whites from imagining that slaves had recognizably human emotional lives (Wallenstein 1995, 154–55). In this context, Brown's raid aroused fear less because it achieved any of its objectives (the abolitionists were defeated in a few days) than because it demonstrated cooperation among blacks and whites. "John Brown's Body" became an anthem for Northern soldiers who thought the goal of the Civil War was African-American freedom. But Brown's sanity remained a point of political and historical controversy. It was in the interest of many Northerners, even some abolitionists, to regard Brown as a "fanatic" given the hopelessness of his mission and his readiness to employ violence. His family history and his noted oscillations between audacity and indecision did suggest emotional instability. Brown insisted on his sanity, of course. But so too did the governor, who wanted to make a case for Northern aggression toward the South. Scholars have also wondered whether Brown's uprising would have seemed either so hopeless or so extreme if its aim had been liberty for a population of similarly oppressed whites (McGlone 1995, 76; Loewen 1995, 176).

John Brown actualized or made *explicit* a virtual state of war between

American whites and between African-Americans and whites—a state of war that preexisted the formation of the United States's revolutionary social contract. Because he claimed to have insight into the real, bodily antagonisms underlying political debate at the time, he was branded as a lunatic by adversaries and by some defenders. As with Mrs. Bumpurs, the simple physical aspects of his acts are indisputable, but what he thought he was doing in leading eighteen men against Harpers Ferry, and what he *did* by doing it, can only be determined by considering how his actions reinforced the actions and imaginations of other people. John McGlone suggests that historians' focus on Brown's sanity implies that the "insane" are incapable of making meaningful history. Yet retrospectively, we can also regard him as a mad critic, one who recognized some of the legal and bodily conditions excluding *slave* experience from consideration as fully human and could only communicate the violence of this experience by returning violence. Mad or not, Brown believed that he resembled African-Americans in morally significant ways and found the existence of slavery personally, indeed physically intolerable.

It is difficult to describe the kind of conflict that makes either Mrs. Bumpurs or John Brown seem or feel like mad critics—not "dogmatic," to borrow Immanuel Kant's jargon, but unable to convince anyone that their acts and words correspond to reality. Someone who has not been faced with regular conflict and dire uncertainty will probably consider it normal to leave an apartment peacefully when police arrive, if the rent has not been paid; it also seems normal for an individual to respond pragmatically rather than morally to injustices that go beyond his or her powers to redress. In a society with a history of racism, it seems normal for racial minorities to meet violent deaths or to live in economic circumstances so precarious that they are indistinguishable from mental illness—and mad to challenge the status quo. One reason a critical stance toward certain political phenomena feels "unreal" is that, in fact, the discourses and practices that claim to have authority over our bodies are far less unified than we fear or would like to believe. The human body exists as much in the imagination as in the flesh, and cannot be detached from the images and persons it resembles in significant ways. To understand and enjoy one's own body requires one to borrow from discourses and practices that correspond imperfectly to one another and to each person's unique capacities. It takes great strength to hold these discourses and practices at a distance, not to blame oneself for failing to unify them through exemplary gestures. In John Brown's case, the discourses of Christianity, law, psychiatry, or even abolitionism seem not to have exhausted the specific nature of his goals and resistances, but he spent extraordinary energy trying to bring them into line. It also takes a receptive audience to make one's own way of perceiving and describing the world *real* rather than a figment of the imagination or fleeting hallucination, like Mrs. Bumpurs's encounter with Reagan.

Every individual must struggle to maintain as much control as possible and derive as much benefit as possible from the powers associated with his or her body—powers that, moreover, can often be actualized only if one has associates, family members, a political party, or a public to observe and draw meaning from the resulting actions. There are some people, however, whose inability to communicate is a condition for others' communication and sanity. Bodies that fit a certain valorized style of individuality do not experience "individuality" as a defect or obstacle to communication. The supposedly universal subject of much science and philosophy is the subject whose circumstances require him or her to be least aware of his or her body; the able or healthy person is seldom brought up short by the fact of his or her embodiment (Leder 1990). By contrast, the disabled, women and members of national ethnic minorities, persons of so-called deviant sexuality, and those with limited financial resources are forced to attend to the potentially incongruous character of their own embodiment and find themselves confronted with "excessive" situations requiring more self-surveillance and interpretation than others may (Young 1990, 134; Freund and Fisher 1982, 79–95). Although most Western philosophers have taken the "normal" experience of reality and imagination as their starting point, there are philosophical and ethical advantages to beginning with the experience of fragmentation, as long as one does not minimize the suffering it can occasion.

Foucault's histories describe how certain bodies that threatened to create social disorder for emerging European states, such as the mentally ill, the criminal, the ill, and the sexually deviant, were confined and trained to function as the common reference point for several discourses or practices, thereby contributing to the actual increase in social order and knowledge. These bodies contain or represent certain "unthinkable" or formless qualities to the majority and enable those who are sane or well to gain a better grasp on the limits of their abilities to recognize and promote sanity or health. By marking the limits of sense and experience for a certain historical community, they play a *structural* role very similar to the one Immanuel Kant assigned to the concept of the *noumenon*—that is, of provoking reflection on the fact that human knowledge is finite and concerned only with appearances. The noumenon is a "problematic" concept that structures a field of appearances without ever corresponding to an actual object. But asylums, prisons, hospitals, markets, and the bodies that inhabit them give this concept a tangible, phenomenal form. The void linking Mrs. Bumpurs to the police officers who shot her was charged with anxiety regarding the violence and irrationality members of a racist society associate, willingly or not, with poverty and minority appearance.

Kant's claim that we can only have insight into appearances or phenomena, not things-in-themselves, has several important implications for any study of human embodiment and imagination. First, it means that even my own mind and body are accessible to me only as appearances, using concepts and perceptual patterns that are both a priori and culturally conditioned. I cannot know what my mind and body are in themselves; still less can I know the capacities or nature of other people's minds and bodies in themselves. A second point, immediately following from the first, is that although the knowledge that I *do* have of my own body and the bodies of others may be scientifically accurate and technically useful, I must use imagination to connect the right concepts with relevant observations and to identify meaningful similarities between gestures, organs, and forms. While many European thinkers define the imaginary by contrast to the real, Kant explores how imagination *contributes* to our sense of reality and our ability to act on it.

However, Kant's writing was also motivated by a struggle against dogmatism and fanaticism. Over time, he used the critical philosophy to defend reason against a variety of dogmatic opponents: empiricists and rationalists in the *Critique*s, then religious sentimentalists and culturally or historically oriented philosophers in his later works. Some scholars have suggested that Kant was even warding off his own tendencies to hypochondria or fear of mental illness. The *Schwärmer* (usually translated as "fanatic" or "enthusiast") believes he has insight into the nature of things-in-themselves, whether these are pure essences, divine commands, or mere sense data. Monique David-Ménard (1990) has argued that the *Schwärmer* and his or her ideas play the role of problematic object limiting and thereby defining the scope of critical philosophy. By refusing to think in an "uncritical" manner, that is, by insisting on human understanding and sensibility, Kant protected himself against resemblance to thinkers whom he found domineering and unstable. He also protected his sense of moral integrity and autonomy against inclinations associated with the body.

Foucault's work has been compared to Kant's on several occasions, and he acknowledged a significant debt to Kant's ideas.[1] Most studies of Kant's influence on Foucault address the conditions for possible experience or contrast Kantian ethics with Foucault's ideas about power, normalization, and care of the self. This study will focus on how each thinker tacitly or explicitly develops the ideas of embodiment and imagination. Although Kant's epistemology and ethics will play a role, I am most interested in the way Kant draws connections between imagination and feeling in his account of reflective judgment, especially aesthetic reflective judgments concerned with what is beautiful or sublime. Kant claims that the kind of feeling accompanying a judgment of beauty or sublimity does not just inform us about our own powers of thought and action, but about the extent to which those powers can be shared with others. He also distinguishes this kind of feeling from the emotions or sensations associated with the human body in biological or psychological science. In fact, "pure" feelings are cognitively significant insofar as they direct the scientist's attention to relevant forms and events, and thus enable the biological body or psychological mind to be mapped in the first place.

I argue that the feelings and imaginative acts involved in pure aesthetic judgment give us access to an aspect of the body that precedes and exceeds empirical or introspective knowledge of the body. These feelings allow us to establish morally significant identifications and repudiations within a community, as well as to create a common way of seeing and feeling among community members to which, in one way or another, all their discourses and practices refer. Often this common way of seeing and feel-

---

1. See "Michel Foucault," a dictionary entry from the early 1980s written under the pseudonym "Maurice Florence" (*AME* 459–63). Foucault's sustained thoughts about Kant can be found in three places: first, in his "thèse complementaire" on Kant's *Anthropology from a Pragmatic Point of View*, written at the same time as *History of Madness* (Foucault 1961); second, in the discussion "Man and His Doubles" found toward the end of *The Order of Things* (*OT*), which builds on many of the themes introduced in the *thèse*; and third, in his lectures and essays dealing with the theme of Enlightenment, especially "What Is Enlightenment?" (in *AME*) "What Is Revolution?" and "What Is Critique?" (in Foucault 1997). Kant also plays a prominent role in Foucault's 1966 review of Cassirer ("Une histoire restée muette," in *DE* 1:545–49) and in an interview with Giulio Preti from 1973 (in *FL*). In his 1979–80 lecture course, *Naissance de la biopolitique* (*NB*), Foucault compares Adam Smith's critique of state economic reason extensively with Kant's critique of philosophical reason. For discussions of Foucault's relationship to Kant, see especially Fimiani 1998, Cutrofello 1994, and Han 2002.

ing is structured through the confinement and rejection of people such as the mentally ill, the poor, the dying, the criminal, and the perverse. But the way marginalized bodies function as "problematic objects" for an ensemble of discourses and practices, excluded but constantly referenced, changes over time. Beginning in the eighteenth century, Europeans and their colonial descendents began to regard limit-phenomena such as madness or perversity as potential risks latent in their own bodies calling for self-motivated psychological control as well as external social control. This self-monitoring made it difficult for individuals to bear the idea of resembling the poor or perverse in "risky ways," renewing ostracism and degradation against them. By returning to bodies and pleasures that are *not empirical* and by giving moral significance to the aesthetic dimension, Foucault attempts to undo some of the disabling and dominating ways Western societies have become accustomed to investing bodies with the potential for sense and the threat of disorder or madness.

Is it possible to escape from or rearrange the void in which one's proximity or exposure to the problematic object becomes unbearable? Foucault uses the term "problematization" to mean taking an object in a field of discourse or social practice (like Mrs. Bumpurs's knife or hallucination) as a symptom of conflict or ambiguity between several ways of imaginatively structuring a social field: "Problematization doesn't mean the representation of a pre-existent object, nor the creation through discourse of an object that doesn't exist. It's the set of discursive or non-discursive practices that makes something enter into the play of the true and the false, and constitutes it as an object for thought (whether under the form of moral reflection, scientific knowledge, political analysis, etc)" (*FL* 296). Problematization is a style of thought that does not consist in producing and manipulating representations of already-existing phenomena. Rather, it is a way in which practices act on themselves, often producing new phenomena or oppositions between phenomena (Foucault 1984, 334). The goal of problematization is to create an object of thought and action about which communicable statements can be pronounced, rather than a hallucinatory idea or a situation one suffers in an unreal atmosphere. This means that sensibility and passivity (the capacity to be acted upon) may be important factors enabling one to construe a situation as more or less desirable, more or less open to change or variation. Aesthetic judgment is part of problematization because it reveals the relationship between emotional or sensory apprehension and the ability to act.

The topic of this book is embodiment and *political* imagination, not imagination in general. Many political observers are concerned by the fact that elections no longer seem to give most citizens the feeling of power and possibility they once conveyed. Others are alarmed by the increasing complexity of government, which makes citizens feel like clients in the hands of specialists. Such complexity can be easily manipulated by industries with media and legal expertise at their disposal, but seems beyond the power of the ordinary citizen to use for his or her ends. Although citizens in the older democracies are aware of past injustices and conflicts such as slavery, colonialism, genocides, and industrial exploitation of labor and the environment, their ability to imagine a future that is informed by and redresses some of the damage of those past conflicts seems to have dimmed. This affects the body insofar as the body suffers from war, imprisonment, medicine, more or less safe working conditions, and the relative health of children and aging parents. But it also affects the body's ability to step back from situations in which it may feel powerless, such as a confrontation with police, and rediscover its feelings of power.

In war, but also in many forms of peacetime discipline, the body's habits and pain reflect the invisible, imaginative expectations of community members and institutions and give them a tangible social presence they would otherwise lack (Scarry 1985). Dying, soldiers' bodies enact a conqueror's image of captured territory and surrender their own national imagination to the void. But people are constantly seeking reasons and new uses for the habits they already have, exercising the ability to exceed any given set of images through which others may recognize and control them while eluding violence. Consider a paradigmatic encounter between civilians, police, and the shadowy imaginative zone between them that Ryszard Kapuściński believed was responsible for tipping the scales against the shah's regime. In this encounter, a subtle shift in the relations between bodies imagining slightly different outcomes to their actions seemed to open the possibility of gestures that previously seemed unthinkable:

> The moment that will determine the fate of the country, the Shah, and the revolution, is the moment when one policeman walks from his post toward one man on the edge of the crowd, raises his voice, and orders the man to go home. The policeman and the man on the edge of the crowd are ordinary, anonymous people, but their meeting has historic significance.... Until now, whenever these two men approached each other, a third figure

instantly intervened between them. That third figure was fear.
... Now that fear has retreated, this perverse, hateful union has
suddenly broken up; something has been extinguished. The two
men have now grown mutually indifferent, useless to each other;
they can go their own ways. (Kapuściński, cited in Žižek 1993,
233–34)

In this incident, the Iranian civilian unexpectedly viewed his body from a vantage point that gave it far more power than if he had seen himself through the imagination of the shah's policemen. Indeed, only in that moment did the "third figure," fear, appear distinct from both bodies—like the "something" haunting Mrs. Bumpurs and provoking the officers to shoot.

Those who have certain kinds of bodies tend to live, and to believe themselves bound by, the fictions created by other groups. They see these fictions as reality and, as there may be harsh emotional and physical penalties for challenging them, conceive of imagination as the act of a body *already bound by this reality*. They are held to these fictions by the power relations vested in their bodies—the most real thing they know. But these bodies can be mixed with others and elaborated or performed according to a variety of fictions. To play a madwoman in a certain theater piece does not mean one is insane, merely that one needs another script in which the same gestures are reasonable and sane. Individuals who wish to make a joyful vision or practice recognizably *real* in the eyes of others must draw other people into a shared conviction of resemblance and power through the contrast of bodily pleasure and pain. The problem is that without practice in clinging to the invisible "real" and without witnesses who will testify to the reality and power of this "real," we are tempted to regard it as obviously "imaginary," and our imaginative tendencies as proof that we lack the realism necessary for success.

Being able to identify the right conflicts alters the being of those who feel embattled. Shifting one's frame of reference or playing two discourses off against each other may reveal incongruities that are as liberating as they are potentially frightening. Although the process takes time, reflection, and sometimes distraction, bodies can become attuned to longer or shorter segments of activity during which nothing seems at stake and "everything seems possible." By "political imagination," therefore, I do not just mean that orienting oneself in political reality requires imagination or that imagination can be exploited by states and movements, but that under certain circumstances, bodies that are neither totally governed by the imagination of others nor afraid of their own capacity to introduce disorder discover unexpected capacities for action. Making the public aware of its investment in the plurality of imaginative schemes and the scales limiting their perception and communication enables them to "think," to "step back," or to regard an institution from the standpoint of the problems it *solves* (and might solve differently). Encouraging such problematization is an important part of the artist's or political activist's work.

A few words are in order regarding the situation of this book. A "problematic" can mean the ensemble of discourses and practices that a given subject must negotiate at a specific moment in history. Using the phrase implies that these discourses and practices respond to a common problem or common set of problems, but this common structure can usually only be seen in retrospect. We do not, I believe, yet have the proper vantage point from which to recognize Mrs. Bumpurs's death as emblematic of a distinct problematic involving psychiatry, economics, criminal justice, and perhaps religion that emerged in the 1980s and 1990s. During the 1990s, an immense number of articles and books in philosophy, history, and cultural studies appeared on the subject of embodiment. I suspect that scholarly interest in the relationship between embodiment, imagination, and political power was an academic response to the kinds of economic, medical, and cultural pressures that resulted in Mrs. Bumpurs's death. In other words, scholars in these disciplines were drawn to "problematize" the body because they recognized that women's bodies, especially the bodies of poor mothers and girls, were being shaped in often violent ways by the imaginations of others and struggling against new economic obstacles for control over that imagination.

From the 1970s through the 1990s, a series of economic and political crises changed how global governance structures, national and international, drew on bodies and imagination for their stability. The oil crisis, the deregulation of currency markets, unemployment, inflation, the threat of AIDS, and the fall of the Soviet system were some of the stars in this dark constellation. Governments adopted strategies such as the privatization of national industries, reduction in social protections, and valorization of financial and information activities (as opposed to production). These tactics were supposed to reduce the state's interference in private economic and cultural life, and were recommended as a stimulus for development in the global South. In practice, they often expanded government activity favoring the business sector rather than educational or social services, and deliberately or unwittingly promoted the resurgence of religion. This "neoliberal" governmentality encouraged competitive behavior by giving individuals responsibility for preventing or surmounting risks like poverty, illness, and isolation—risks that a previous era had attempted to manage at the state level through social insurance—as well as fearful racism toward foreign or "subject" populations.

Neoliberalism captured the imagination of the wealthy, who stood to gain from making regions and companies compete with fewer social protections. But unfortunately it also depressed the imagination of movements who had begun to envision new forms of life during the 1960s and 1970s, by making alternatives seem "fanciful," "extravagant," or "irrational." The "war on drugs" and welfare reform of the 1980s and 1990s built on existing cultural associations between chaos and poverty, madness, criminality, race, and deviant sexuality. They trained Americans in a common mode of social perception that evaluates communities and forms of individuality for their potential return on emotional or financial investment. Often, the resulting vision of indefinite free trade and everyday "entrepreneurialism" reinforced old patterns of racial and class mistrust. Theorists in the 1990s looked at the evidence of neoliberalism's stress on bodies and hoped to identify a conflict in which they could assert a right to individuality without losing the resonance of collective belonging. In this situation, the political thinker's goal is to be an effective critic without feeling or being received as mad—subject to intolerable anxiety, anger, or depression, capable of imagining a future on the basis of *good* elements in a present or past that may otherwise be rife with bitterness.

What is meant here by political imagination? First, it can refer to the way individuals empirically imagine the unity or coordination of governmental and civil institutions whose activities they only encounter erratically: the post office, the school system, the police, some protestors on the corner, as well as the smattering of events and activities reported in the news. Behind these encounters, every citizen has his or her own idea of how economic health is affected by foreign diplomatic or military action, how political parties choose their candidates, how church finances or public activities are regulated by tax laws and the First Amendment, and what particular offices or institutions are appropriate targets of activist influence. Mrs. Bumpurs's imaginative picture of political reality was summed up in the image of Reagan "coming through her walls."

These large-scale interactions are the subject of scholarly analysis and classes in sociology, law, and political science. However, even teachers with practical experience in addition to scholarly knowledge are guided by imagination, and something is inevitably left out of every course or presentation. Kant calls this inevitable limitation the discursivity of human understanding: reason demands that our apprehension of particular experiences or laws ultimately form a whole, but we never experience the whole as such. Our knowledge is always limited by our level of analysis or description. If we insist on believing that political reality (or even the world) "makes sense" as a whole, then some connections or details must always be "left to the imagination."

But one can also talk about political imagination from the practical angle as a matter of what people think they can *do* to improve their own situation, individual or collective. Here political imagination consists in an estimation and articulation of *power relations*. It gives rise to questions such as Can I influence the school board? Is it possible to replace the local police captain, or at least create an ombudsman's office to handle complaints? Can I start a new school or a public watch program to replace dysfunctional institutions? Does this or that political party have the ability to persuade the country of a certain position on health care or education? Every citizen has a rough sense, whether accurate or not, of what changes are plausible at a given time with a particular set of public concerns and abilities. In 1973, for example, many Americans believed that dependence on foreign oil for energy would soon end because rising prices had created an enormous need for inexpensive, ecological alternatives. In 1989, they believed that the fall of the Soviet Union proved the impossibility of any workable socialism, quite apart from the undesirability of that particular socialism. Every election and national crisis seems to bring out a range of imaginative options that some group is willing to gamble can become a reality and that others consider delusional. The feeling of power experienced by participants is directly related to the detail and scope of imagination, although it may not be theoretically expressed; great movements and great politicians know how to do more than they can explain.

The first section of the book examines Kant's struggle to find unity in his own experience and reflection. It shows how Kant took the body's boundaries as a framework for the unity of thought in his pre-critical writings, and explains why he only associated the body with sensibility and anthropology, rather than reason and reflection, later in his career. My overall goal is to show how bodies compensate for the persistence of fractures and discontinuities in Kant's image of thought; aesthetic pleasure, in particular, blurs the boundaries between empirical and transcendental, and individual and collective, aspects of embodied experience. By contrasting individual and collective aspects of imagination in the *Critique of Pure Reason* and the *Critique of Judgment*, I show where Kantian imagination has political implications, and how these implications are structured by Kant's avoidance of *Schwärmerei* or fanaticism.

The second section of the book describes Foucault's attempt to affirm the fractured and plural image of thought revealed in certain Kantian texts. It shows how the exclusion, confinement, and normalization of bodies creates patterns of recognizable resemblance among human bodies in public and privatizes the disorder experienced by speakers and actors. Foucault regards Kant's *solutions* to metaphysical dilemmas that hampered natural science and contributed to religious fanaticism as *problems* with which we must grapple. In *The Order of Things*, Foucault describes the body as the way in which the finitude of modern thought was attributed to material conditions of life, labor, and language so that historical transformation of those conditions seemed plausible (*OT* 314). Likewise, the body is implicit in empirical knowledge associated with the human sciences and used to administer large populations in the historical period following the Enlightenment. However, the body plays a different role and represents a different problem for each discourse and practice. Although there are sensations, affects, and actions, there is *no such thing as a "body"* outside the forms of thought that take their own practices as a problem and resist the defining or limiting conditions of thought. When I speak of the "body" singular, I mean the *ensemble* of bodies, discourses, and practices that resolve their tendency to multiply or fragment by *training* individual bodies to be *visible* or *perceptible* in specific ways.

The third section of the book examines how viscerally charged conflicts between imaginative frameworks affect citizens' ability to participate in public spaces or to recognize their own agency and norm-setting ability in the forms of law, administration, and political insurgency. I use the preceding analyses to draw some conclusions regarding the recent transition from what Foucault calls a security-based art of government to neoliberalism. The security-based art of government tried to protect populations against danger and to foster biological and economic flourishing through a range of mass tactics, including collective insurance and a defensive use of racial ideology on behalf of majorities. Neoliberalism "privatizes" the risks and capacities of populations onto individuals, encouraging them to take charge of their own exposure to risk or opportunity in relative isolation or independence.

Neoliberalism exploits the body's potential in new ways by transforming every aspect of private life, such as education, marriage, childbearing, and sports, into a potentially profitable or competitive investment. The anxiety caused by participation in *so many* potentially unrelated "markets" in which one has little opportunity to evaluate or refuse competition makes it difficult and even dangerous to imagine collective transformations. Every effort to problematize or vary existing ways of life and institutions requires one to restrict imagination along certain lines—in short, not "think too far ahead," "too globally," or "too locally." But there is a great difference between restricting one's own imagination or agreeing to do so as a group for the sake of focus and coherence, and being restricted by others in such a way that one is confronted by incoherence. One can deliberately focus a lens to look at a slide or adopt a single angle of analysis for a course syllabus. But if the lens is focused by someone with different interests or quality of vision, the slide may not make any sense at all, and if a class is too detailed or too general, the student comes away unsure what she has learned or if she learned anything. What is important is that each individual balances imagination and knowledge in a different way to produce the feeling of sense and power, and yet these feelings must communicate or have a collective dimension if sense and power are to be believable, that is to say, "real."

I do not claim to have discovered the problematic governing neoliberalism, nor do I proffer specific recommendations for action. Rather, I explore the aspects of embodiment and imagination that could allow us to make *political* sense of Mrs. Bumpurs's mad criticism. This book also represents an attempt to do justice to the craziness that many citizens of modern democracies (much less the authoritarian regimes) feel when confronted with the contemporary breakdown in institutions of public security—although Foucault reminds us that these institutions were complicit with colonialism and racism during the nineteenth and most of the twentieth centuries. I hope that it remains true to the critical side of Kant's project—affirming the finitude of human understanding even to the extent of rooting egalitarianism in our *inability* to say who we resemble, rather than in a positive humanism. Citizens in Western societies will better understand the potentials or crisis of their own political imagination if they regard themselves as potentially sharing the "postcolonial" burden of incongruity between historical practices and discursive or visual codes with citizens elsewhere on the planet. Conceiving the body as a "problematic object" enables us to reflect on its function in stabilizing the multiple imaginative schemes through which others act on us and we act on ourselves. The goal is to establish a new ethical relationship to norms, one less vulnerable to hostility and conquest in the name of health and security.

Heterotopic Interval

While most philosophers and psychologists tend to think of imagination as a capacity of individual persons or of the individual "subject" of common experience, anthropologists, feminists, and postcolonial theorists frequently refer to imagination as a collective and historical phenomenon that determines who will be recognized as "man" or "human." Each set of discourses borrows certain assumptions from the others; philosophers assume that the anthropologist knows how imaginative subjects are rooted in material communities; anthropologists assume that philosophers or psychologists have demarcated imagination from other cognitive functions. The idea of *heterotopia* is one that can fruitfully be used to describe how intellectual disciplines and cultural practices overlap or borrow from one another selectively and at points of crucial ambiguity. One of my goals in this book is to make that collective meaning of imagination more concrete for philosophers by showing how the two senses overlap in Kant's work.

In Western philosophy, imagination has been regarded either as a lack or unreal *variation* on the real by some thinkers, and as an integral *part* of reality by others.[2] One tradition, extending from Aristotle and Aquinas to Edmund Husserl and Jean-Paul Sartre, regards imagination as *fantasia*, the ability to produce images of nonexistent entities *supplementing* or *negating* reality. The other tradition, to which many early modern philosophers belong (along with some twentieth-century thinkers like Henri Bergson and Walter Benjamin), regards all mental events as "images" or "ideas." For these philosophers, imagination refers not only to voluntary fictions but also to a confused or indistinct apprehension of

---

2. See Casey 1976 and Kearney 1988 for discussions of the imagination in phenomenology and twentieth-century European thought more generally.

reality. Kant acknowledges both alternatives, but he also gives imagination a role in *constituting the real as real.* We imagine because we are incapable of grasping things-in-themselves from all sides and aspects at any given moment. Far from being a defect, Kant made the inability to grasp things-in-themselves into the touchstone of *humanly verifiable* morality and science, rather than allowing human experience to be disqualified as "unreal," "mere fantasy," or confusion by contrast to divine comprehension. In this way, he created a neutral phenomenological zone from which the multiplicity of spaces and the threat of madness could be excluded.

In the *Critique of Pure Reason,* Kant describes transcendental imagination as a faculty enabling individuals to find a rule or concept for every intuition. We use imagination whenever we organize a manifold of intuited phenomena according to the basic concepts structuring experience as a temporally unified, causally interconnected, and stable "world" in which we occupy a determinate space. If the imagination were unable to organize all perceptions into a stable order corresponding to possible moments of logical judgment about objects, we would be subject to a rhapsody of mental and physical events unworthy of being called experience. Language names the products of this ontological synthesis and enables us to fix the elements of that reality. But by resembling and drawing attention to resemblances within the real, language also treats "man" the speaker in much the same way as the objects of his experience. The schematism of (human) imagination, after all, has criteria for recognizing "man" when it sees examples. Here imagination is part of the reality it helps organize.

But these acts of imagination are usually associated with individual bodies, while language is assumed to be an intrinsically collective phenomenon. Students of art history and material culture have tried in varying ways to show that rituals and artifacts train users in common perceptual and creative habits, just as language is both collective and subject to individual variation or invention. Myths, works of art, and dance direct people's bodily action, including their attention to resemblances, by reference to something invisible that is present *in* the visible, auditory, or tangible dimension. Many people in situations of relative privilege believe that their experience is *unified* by orientation toward the invisible, as Kantian reason is unified by the regulative employment of Ideas. This ideally unified experience, toward which individual ethnicities and knowledges should eventually contribute, was further supported by the idea of the nation and the stupendously productive subject of modern science.

In *Imagined Communities* (1983), Benedict Anderson describes the development of specifically modern forms of "belonging" and "exclusion" in the context of South American national revolutions. The imagined community of the nation was added to older ways of situating oneself psychically in relation to an indefinite plurality of others, such as lineage or religious community. Public spaces, so important for the tradition of deliberative democratic thought from Kant to the Frankfurt school, exist "only by virtue of imagination"—specifically, the imagination of being seen or read by strangers (Warner 2002, 8–9, 74–76). These imaginative locations organize and orient the subject in a geographical landscape claimed by warring parties. More recently, Arjun Appadurai has distinguished between multiple "scapes" of contemporary migrant consciousness, such as "ethnoscapes," "technoscapes," "financescapes," "mediascapes," and "ideoscapes," of which the Enlightenment/liberal worldview is one of the most powerful (1996, 31–36). As a result of international migration and rapid expansion in communication media, more and more citizens are aware of the extent to which their everyday lives are shaped by the imagination of others and know that not all forms of imagination involve the same attention span or point of historical reference.

However, political regimes and oppositional movements also enhance their legitimacy by emphasizing the *plurality* of imaginative and invisible spaces available to them in a cultural situation. Corrupt postcolonial states, according to Achille Mbembe, use rituals and symbols from different discourses and traditional practices to suggest their absolute domination over everyday life—exaggerating the dissonances within citizens' experience and claiming surplus power from their ability to manipulate several symbolic domains (2001, 109–15). For example, they enhance the obvious military or financial activities of government with Christian or indigenous religious symbols—representatives of the "invisible" (see also Tonda 2002, 27–28). The Mothers of the Disappeared mobilized popular awareness of invisible detainees in Argentina's political prisons by placing their own bodies, powerfully associated with maternal and religious authority in the Catholic nationalist imagination, in the public sphere as visible proxies or "doubles" for their missing children and grandchildren (Cornell 1998, 103–4; Taussig 1992, 27–28, 48–50).

Imaginative spaces and discourses such as religion, family life, the educational system, political ceremony, and warfare emanate from many points within the social body. Law, Catholic doctrine, and psychiatry do not speak about sexuality or allow their practitioners to imagine sexuality in the same way. But most citizens are required to interact within or in terms of several spaces, a practice Maria Lugones calls "world"-traveling (2003). In truth, there is no reason to talk about these spaces *except* insofar as they coexist in the actions or appearance of individual bodies. Daily encounters with police, teachers, factory bosses, or healers allow these spaces to affect one another by affecting the bodies in which they coexist. The law "learns" about psychiatry or the church when a parishioner brings a lawsuit against a priest or a therapist testifies before Congress about the effects of rehabilitation on domestic batterers.

To describe the intimate relation between ordered thought and spatial order, Foucault considers the contrasting case of a *heterotopia*, or juxtaposition of several "emplacements that are irreducible to each other and absolutely nonsuperposable" (*AME* 178). This juxtaposition contrasts with a utopia, whose imaginary order is presented in a realistic manner (*OT* xviii). Foucault's most famous example of heterotopia was the string of letters linking entries in Argentinian novelist Jorge-Luis Borges's fictional Chinese encyclopedia, supposedly contemporaneous with the *Encyclopédie* of the French Enlightenment and the artificial language project of John Wilkins, member of the English Royal Society (Borges 1964). Borges's encyclopedia attempts to do without a common imaginative *locus* for its categories. As Foucault muses, where could these animals "(i) frenzied, (j) innumerable, (k) drawn with a very fine camelhair brush"—"ever meet, except in the immaterial sound of the voice pronouncing their enumeration, or on the page transcribing it?" (*OT* xvi).

But Foucault also allows that in certain cases these "emplacements" can be brought together in a "real space" that allow some spaces to represent, contest, or reverse the others. Foucault had already experimented with this kind of heterotopia in *History of Madness*, where he describes Bosch's *Ship of Fools* as a bit of space neither in the "sane" world of commerce, government, and religion, nor in the wholly "other" world of death or the void. In his 1967 lecture "Of Other Spaces," he distinguishes between heterotopias of "crisis" such as initiations, travel, and religious experience from heterotopias of "deviance" such as asylums and prisons (*AME* 189–90). Theaters, gardens, hotel rooms, hospitals, and libraries are concrete sites where individuals simultaneously (and often only temporarily) participate in several distinct systems of social or intellectual order. Other scholars have suggested the Palais Royal of *ancien régime* Paris, in which aristocracy, revolutionaries, and the demimonde shared a common meeting ground; the Masonic lodge, an experimental zone in which social order based on class affiliation was subordinated to the secret increase of Enlightenment as well as the creation of an intellectual elite; and the early English factory, in which elements of handcraft economy were interwoven with the increasingly detailed forms of order characterizing industrial management (Hetherington 1997).

One reason for this structure's persistence in Foucault's corpus may be that it reflects a certain experience of his own childhood: as he stated in an interview, "We did not know when I was ten or eleven years old whether we would become German or remain French. When I was sixteen or seventeen, I knew only one thing: school life was an environment protected from exterior menaces, from politics" (*EST* 125). In other words, he did not know which imagined community or order of religion, commerce, and the state would ultimately appeal to his body for self-evidence and "reality." Structurally, the school has the same function as the ship, the brothel, and the letters separating categories in Borges's encyclopedia—it hovers between incompatible schemes of order but is also a potential seed crystal of order in its own right. For this reason, I use "heterotopia" to refer to the *collection* of more-or-less real imaginative schemes lacking a site that could give them a common time and place, rather than the encyclopedia, ship, or school itself, that is, the "single real place." In certain "magical" spaces such as the words of the encyclopedia or the space of a ship or classroom, these schemes can coincide, if only because the space *excludes* them all in a common way.

Foucault associates Borges's bizarre encyclopedia with the European surrealists, citing Lautréamont's toast to the beauty of a "chance encounter between a sewing machine and an umbrella on a dissecting table" in order to point out the important role played by the *table* in our ability to cognize, much less to judge aesthetically. But Borges's work, which is sometimes regarded as a bridge between Surrealism and the Latin American literary movement known as "magic realism," also reflects the phenomenal experience of life in a non–metropolitan country. In the works of authors like Cortazar, Donoso, García Márquez, and Allende, mythical and fantastic events are described in the same tone reserved for realistic ones, and the most serious historical crises are presented in the manner of fables.

The phrase "magic realism," which is also applied to non–Latin American authors such as Salman Rushdie, Ben Okri, and Leslie Marmon Silko, suggests that the "magic" is merely an aesthetic addition to realism, on which the West is already expert. But many of these authors are attempting to convey the sense of "qualified" reality or disbelief with which citizens in totalitarian *or* rapidly modernizing states are required to confront cultural products from different periods and metaphysical or moral contexts. Feminist theorists, anthropologists, and scholars in cultural studies have had to borrow an understanding of imagination that is more like a heterotopia than it is a simple variation or constitution of the "real" in order to write about worlds in which disparate religions, military forces, and points of historical reference such as tribal custom or globalization overlap.

Heterotopias are also places of *heterochrony* (AME 182–83; DE 3:581). In other words, they alert participants to the fact that even at moments of apparent continuity, their attention is claimed by a plurality of temporalities and levels of historical analysis. Foucault refers to the cemetery as a site where the eternal and the everyday are brought into jarring contiguity; Kevin Hetherington's example of the English factory likewise emphasizes the uneasy coexistence workers experienced between the speed of economic practices inherited from the era of home work and manufacture and new speeds imposed by machinery. Regimes could not use religion or ethnic history to capture their citizens' imaginations so effectively if architectural and geographical entities (such as state, church, school, hospital or healer, workplace, global Northwest or South) were not associated with spans of time and ways of living that convey the flavor of that time, such as the indigenous past, colonial past, postcolonial present, nationalist future, or global capitalist future. This means that one can perceive the juxtaposition of multiple spaces in an *event* just as in a *space* like the brothel or ship. By provoking a feeling of discontinuity between forms of corporeal and intellectual order, heterotopias inspire a type of historical reflection in which the human environment is dissociated rather than unified.

In many cases, the modern state takes responsibility for providing a unifying historical and spatial framework for the diverse spaces and schemes of disciplines, religions, and economic production of exchange. It may also participate in the active dis-integration or uneven development of social spaces and temporal rhythms using modern communication technologies and investment or consumption strategies (Harvey 2000, 122–30; Lefebvre 1991, 50–53). But the human body also plays the role of the "ship," "prison," or encyclopedia index in everyday life. It may seem bizarre to think of the body's unity as the space from which conflicting imaginary frames have been purged or brought into a temporary coexistence, but this is not surprising from a psychoanalytic point of view. Those who visit psychiatrists often suffer from being made to inhabit several powerful views of the world at the same time or to take sides with loved ones who have grown up feeling severely embattled (Davoine and Gaudillière 2004). The body can also function as an event, a site for events, or a part of some larger event because of its capacity for acting and being acted upon. There are places where the "body" opens out onto the past or onto imagination and cannot be closed or individuated. These can be sites of trauma, intense pleasure, or spiritual insight.

If Borges's encyclopedia abandons the *site* which is a necessary condition for scientific knowledge, it does reveal *symptoms* that are necessary conditions for knowing about imagination in its plurality. Maurice Merleau-Ponty (1962) gave the body a capital role in unifying phenomenological experience. According to Foucault, the body plays just as important a role in uncovering the competition and conflict between temporalities and practices that have shaped our understanding of history. The body is the most important artifact in which past and present struggles can be read, but through which these struggles can also be foreclosed. "The body manifests the stigmata of past experience and also gives rise to desires, failings, and errors. These elements may join in a body where they achieve a sudden expression, but just as often, their encounter is an engagement in which they efface each other, and pursue their insurmountable conflict" (*AME* 375). *Bodies* in the plural, in their differentiations as well as their similarities, exhibit the contradictions and failings of the discourses and practices that govern them, feed them, educate them, cure them, or send them to war.

Premodern techniques of governance applied themselves to the body, but largely in order to alter or intimidate a soul whose desirable qualities and longevity were intensely bound up with religious anticipations of a *non-bodily* existence. By contrast, the modern body is the medium through which thought takes itself as an imaginative object and the instrument by which thought resists being objectified. The idea of the soul might even be the way we think of this struggle between materially effective but imaginative elements of the body. The "body" that is an imaginative construct of political significance, and to which every living, speaking, affective body must orient itself in order to survive, is the one whose very artificiality provides a common ground for the generation of comparative data concerning real bodies, data which has been used to enslave individuals as much as to liberate or cure them. In being a medical problem, a sexual problem, a disciplinary and economic problem, and perhaps, today, even an ecological problem, the "body" enables modern societies to generate knowledge about those fields of natural functioning in a more easily communicable and rationally justifiable way than is available to societies oriented chiefly toward transcendent or otherworldly truth.

But this knowledge often conceals the emotional and political damage of conflicts, such as Mrs. Bumpurs's struggle with Reagan, that are neither "true" nor "false" in any strict sense because they refer to points of life-threatening ambiguity in a society's relations of power. Citizens of authoritarian regimes (or very troubled families) know not to speak publicly about experiences or rumors that would disrupt the functioning of other conversations and security networks on which they depend. Lawyers who cite psychiatrists (or the reverse) know that there are terms or experiences referenced by the "other discipline" that would require a lifetime of scholarship and practice to translate effectively, for no translation exists for every ramification of concepts like "custom" or "paranoia."[3] Every individual body has a different level of tolerance for nonsense of this type. But every individual body benefits from the effect of "sense." Thus the heterotopic imagination cannot be a regulative Idea in the same way as the "world" or "totality," for it cannot be the same for all citizens, although all citizens contribute to and are affected by it.[4]

Some people, like Mrs. Bumpurs, find the heterotopic imagination too close for comfort and hallucinate or suffer on its basis. Others are able to step back once they realize that there is no point in demanding that people, parties, and institutions agree in every respect on the world they share. In this book, I assume that more of the world's citizens, even in the older democracies, live with heterotopia than live with the coherent experience described as universal by philosophers and Western psychologists. Ironically, the heterotopic imagination is the norm. As I will argue

---

3. W. V. Quine expresses a similar idea when he states that theoretically laden terms are rarely amenable to translation in terms of sensory experience, but must be situated with respect to other statements in their home discourse. These discourses, in turn, can be translated into one another where they touch on related sensory phenomena, but their terms are rarely in one-to-one correspondence, making reductionism in either direction difficult (1969, 16–17, 50–51).

4. Gennochio does not believe that any of Foucault's heterotopias really escape the "interiority" of lived spatiality described by phenomenologists like Bachelard and Merleau-Ponty, although they might function as impossible "regulative ideals." "The heterotopia is ... more of an idea about space than any actual place" (1995, 43). Hetherington (1997) argues that heterotopias should indeed be understood as real (empirical) spaces in which complementary and perhaps incompatible practices of *social ordering* seize bodies at a given historical moment.

in "Negative Anthropology," however, what is "normal" in a statistical sense may also be profoundly *pathological* for certain bodies. Political imagination should create conditions under which each citizen, movement, or nation can experiment with and set its own norms, without damaging similar prospects for others.

PART 1
THE POLITICAL TOPOLOGY OF KANTIAN REASON

In 1961, Foucault submitted a translation and commentary on the genesis and structure of Kant's *Anthropology from a Pragmatic Point of View* as a "thèse complementaire" for his *doctorat d'état* at the University of Paris (Foucault 1961). The reading of Kant I offer here is tailored to addressing the kinds of issues Foucault raises in his *Introduction à l'anthropologie de Kant* and which perplexed him throughout his career. While Foucault's essay on the *Anthropology* does not figure disproportionately in my account of Kant's work, the strategy of examining "genesis and structure" is very important. Kant's claim to give a philosophical account for the totality of experience inevitably leads the reader to sort through his or her experience for aspects corresponding to, or explicable through, terms in the Kantian architectonic.[1] If, as Claude Lévi-Strauss argues, the signifier precedes and articulates the universe in its significance (1987, 60–61), then every human experience and perhaps many nonhuman experiences must fall under or be translatable into *some* Kantian concept, however distorted the translation may be. Indeed, Kant often defines terms such as sensibility, imagination, judgment, and feeling in opposition to one another and uses these terms to introduce order into the "signified" or set of representations from which coherence is expected, notably in "On the Amphiboly of Concepts of Reflection" in the *Critique of Pure Reason*. He also tries to explain the unity beneath such oppositions (for example, in the Transcendental Dialectic and the whole of the *Critique of Judgment*), a project sometimes leading to greater complexity and disunity rather than the hoped-for synthesis.

For the contemporary reader, the totality of Kant's structure can identified by the simple historical facticity of his text. On the other hand, Kant

---

1. For example, I began by looking for—and unexpectedly finding—reference to the familiar act of diagramming or sketching out an idea through gestures in the "space of the mind's eye."

did not *write* his books with a total catalogue of concepts and terms in mind. His explicit knowledge had to catch up with his feeling for the world's significance, to continue the analogy to Lévi-Strauss (1987, 61). Kant had to *develop* what readers now appropriate as a totality, responding to the intellectual and political conflicts of his particular circumstances as well as leaving room for future thoughts.[2] After reflecting on my own experience in light of Kant's structure, and reflecting on his structure as "constituted out of processes of correcting and recutting . . . patterns, regrouping, defining relationships of belonging and discovering new resources" over time, it seemed to me that Kant struggled repeatedly with the difficulty of creating a totality for reason, and that he identified this difficulty with the dangerously divisive potential of imagination or an "imaginative" understanding. It also seemed that Kant found a provisional solution to this danger in the aesthetic and anthropological capacities discussed in his *Critique of Judgment*. This is the version of Kant that Foucault drew upon and also resisted.

Because the strategy through which Kant finally mastered the "signified" of his own experience as a thinker and distanced himself from the threats of fanaticism, social disorder, and political authoritarianism is a version of the strategy that enabled disciplines like the human sciences and institutions such as medicine and the prison to flourish, I think it is worth considering this strategy in detail if we hope to exercise political imagination after Foucault. The key to Kant's strategy is positing the existence of a "problematic object," which divides the field of human thought into a manageable, knowable part and an indefinite, unknowable part. Kantian reason overcomes its bifurcations to the extent that the aesthetic appearance of human bodies and their aesthetic reactions to one another

---

2. Foucault points out that Kant was teaching and improving his anthropology course over the entire period when he was also developing and publishing his critical philosophy. He therefore reasons that Kant assumed anthropology was a structural component of the critical philosophy, one that probably showed the strains and traces of its construction (1961, 7, 120). The concepts employed in his anthropology course, after all, are the same ones found with greater theoretical development in the three *Critiques*. Foucault draws on Kant's last correspondence with Beck to argue that *anthropological* reflection enables "the major themes of the critique—relation to the object, synthesis of the multiple, universal validity of representation" to be "regrouped around the problem of communication" (18). It explains how, in practical terms, individual subjects of experience come to have a collective world (41–42). Unlike Foucault, I do not assume that Kant intended transcendental reflection and anthropological reflection to be different in kind, but to reveal different phenomena and categories of thought at work within a single structure of rational, sensible experience. Foucault's major works correspond more closely to Kant's intentions, in my opinion, than his *Anthropology* essay.

can be trained by culture and discipline. This means that insofar as Kant's efforts to reorient reason around the problematic object were adopted or repeated by other scholarly disciplines and governmental practices, they have emotional and political consequences in the anthropological sphere. Foucault explains how certain kinds of human bodies did in fact represent the unthinkability and unrepresentability of something like Kant's "problematic object" for European political authorities in the modern and contemporary periods, and how their exclusion went hand in hand with the remarkable success of social-scientific knowledge and technological development.

Kant regards reason as a basic style or "power" of thought (*Erkenntnisvermögen*) like sensibility, imagination, and judgment (*CPR* B169–70; *CJ* 55:167–68).[3] In this he follows a long tradition in philosophical psychology, although unlike many of his predecessors he distinguishes between thought in general and the contents of individual, empirical human minds. Kant is often occupied with the difficulty of differentiating basic powers of thought in relation to the totality of phenomenological, scientific, and social experience. In its theoretical employment, for example, reason is the attempt to unify the laws determining all objects of sensible experience. As part of this unifying process, reason also has a critical use—setting limits to its own legitimate exercise. In its practical employment, reason sets moral goals for human action—acts that affirm the universality and autonomy of human reason. In this second sense, reason is identified as a faculty of desire. When properly oriented toward the moral law, this desire is creative and contrasts with blind psychological inclinations. In the *Critique of Pure Reason*, finally, reason and understanding have the further task of making judgments (A132–36 / B171–75). Imagination is an aspect of the understanding, rather than sensibility, although at first Kant regarded imagination as an "independent mediator" between sensibility and understanding.[4] When he later decided that judgment should be a power of thought in its own right (in the *Critique of Judg-*

---

3. With four exceptions *all* citations to Kant refer to the volume of the Akademie edition (1908–13) in which the work appears and page numbers in the Akademie edition as provided in the margins of the English translations. One exception is the *Critique of Pure Reason*: I have complied with the standard practice of referring to this text by A/B pagination without mentioning the volume. The others are Kant 1965, Kant 1992a, and *KPW*, whose translators do not provide Akademie edition page numbers. For these, page references are to the English rather than the German versions.

4. See B152–54. The implications of his shift from the A (1781) to B (1787) editions, along with my reasons for accepting the B edition, will be addressed later.

ment), imagination was given a crucial role in revealing how pure feelings of pleasure and pain could orient the act of judgment and thereby mediate between reason and understanding.

According to most European thinkers in the seventeenth and eighteenth centuries, human experience consists of representations (ideas or *Vorstellungen*).[5] Schools of thought disagreed, however, over the ontological status of representations; were they things-in-themselves, or only surrogates for such things? Most important, how did the Creator's understanding of these phenomena differ from the human understanding, and what could this difference tell us about humanity's relation to nonhuman nature? Leibniz, Spinoza, and other rationalists referred the limitations of human understanding to a supreme intellect or authority (although Spinoza identified God with nature as a whole, causing many contemporaries to regard him as an atheist). Locke, Hume, and other empiricists assumed that representational objects were produced through sensation and subsequent reflection on sensory experience, but the sources of sensation remained mysterious and less interesting than the mind's efforts to order the results. A philosopher's view on the relationship between representations and their ultimate source had political implications in an era when European states were transforming themselves from religiously affiliated principalities into absolutist monarchies supported by administrators—and, in the case of England, to a religious commonwealth governed briefly by the people. Atheism, for example, implied that a thinker rejected divine royal authority or considered it nothing more than the brute exercise of power. As will be discussed in Part 2, Foucault's *Order of Things* is largely dedicated to examining the system of representation Kant reorganized at the level of philosophy, though the seeds of this transformation were sown in other disciplines, especially in philology.

Kant's most significant achievement was to draw a sharp divide between questions concerning the possibility of phenomenal objects in human *experience* and questions concerning the possibility of things-in-themselves which might be the *source* of such representations. He established universal criteria for human knowledge—space, time, and catego-

---

5. See Hacking 1975, 26–33, for a discussion of the central role played by *ideas* in early modern philosophy. Kant distinguishes *Vorstellungen* or representations from "ideas" at CPR A320/B377 because he wants to save the term for the special objects of pure reason, but what he means by *Vorstellungen* is what his predecessors mean by "ideas." Representations/ideas may involve the thinker in existential commitment to an object (*Objekt*) recognized as their ground, or may be joined to form an object (*Gegenstand*) in thought, with or without existential commitment.

ries—with both a logical and a transcendental application to objects of intuition. While he believed that the idea of God was an important support for moral action, he henceforth divorced questions concerning divine nature from questions of knowledge—and, to a great extent, politics. In this way, he lent invaluable metaphysical support to scholars and statesmen who wanted to reorient governmental practice around the health of populations and the rights of those being governed rather than royal sovereignty. While he accepted the principle of monarchy in existing absolutist states, Kant believed that even the legislative sovereign is deprived of insight into things-in-themselves and must be governed by natural and moral laws, just like his subjects.

Powers of the mind such as sensibility, understanding, and imagination are for Kant peculiarly *human* sources of *human* representations. (Reason is shared with other moral beings, according to the *Groundwork*; see *GR* 4:408.) This means, as I understand it, that Kant begins his work in an ordinary experience of reflection in an ordinary world of phenomenal objects. He is not explaining how that world is constructed from "scratch" using concepts, intuitions, and faculties, for his reflection, like his knowledge, is bounded by the conditions of possible experience.[6] Nor is his ultimate goal to explain why statements concerning observable experience should count as knowledge, as many twentieth-century readers presume. Rather, he is trying to describe the logical and perceptual structures that must be assumed to operate if his thought and action are to achieve as much genuine, communicable and cumulative scientific knowledge and as much practical self-determination as possible. In other words, his analysis is driven by his goals: reliable knowledge and autonomy.[7] Excluding a "problematic object" from the field of philosophical consideration is the first boundary-setting gesture of Kantian reason, enabling it to carry out the "natural" pursuit of unity among the domains into which elements of experience are eventually grouped.

One is tempted to ask how the world of representations described by

---

6. However, this is the way Foucault seems to interpret Kant's *Critique of Pure Reason*, restricting the more phenomenological analysis of experience and knowledge I describe here to the "anthropological" project (1961, 59).

7. The reading of Kant I employ focuses on his image of thought as *critique*, rather than on epistemological or ontological issues. This is a strategy employed by readers such as Fenves (1991), Goetschel (1994), Kerszberg (1997), and Shell (1996), who prioritize Kant's work as a whole and his long-standing interest in philosophical anthropology (including the justification of aesthetic judgment found in the *Critique of Judgment*) over the specific contributions to modern theories of knowledge offered by the *Critique of Pure Reason*.

early modern philosophy relates to or resembles the approach to reality as representation in "postmodern" experience. Postmodern experience is variously described as the awareness or active celebration of contingency and contextuality produced by changes in the speed of production and communication or the decline of grand narratives (Harvey 1990, 284–307; Lyotard 1989, xxiv). These sociological conditions enable physical reality to be quickly and easily altered on the basis of representations such as economic value. According to Jean-François Lyotard, postmodernity is implicit in the notion of modernism as self-consciousness about one's situation in time and space, revealed by changes in the arts and mode of knowledge production (1989, 79). Perhaps one could summarize by saying that people in all parts of the world, including the older industrialized democracies, have become aware that they inhabit many narratives at once, or have lives and desires that bear different meanings in different systems of representation, which they must decide whether or not to reconcile. The ability to reconcile these different representations of their own desire and others' desires is a measure of power and has very concrete effects on someone's bodily and emotional well-being, on his or her ability to enjoy community as well as solitude. Self-consciousness is the symptom or effect, not the cause, of modern subjects' obligations to identify with a plurality of spaces at once.

When Kant thought about the plurality of possible worlds/monads, by contrast, he assumed the representations each had a single "style" of being, and he justified belief in their homogeneity by subordinating them to the understanding, pure forms of intuition (including time), and unity of apperception. Most philosophers have assumed that normal experience is unified, and Kant's philosophy has been used to *make* unity the norm. But more and more of the world's citizens, even in the industrial democracies, have become self-conscious about the disunity and collective nature of imagination. Capitalism, moreover, encourages us to interpret discontinuities between narratives and practices as motives for work and consumption. In this situation, it becomes possible for us to see the fractures in Kant's thought. These fractures do not necessarily haunt the specific "experience" whose conditions of possibility are described in the *Critique of Pure Reason*—an experience that best applies to experimental scientific practice—so much as they haunt the spaces his text generates and brings into relation. Kant did not believe imagination and experience were heterotopic, but his thought moves "heterotopically," and this is the aspect of his thought from which we can learn most today.

* * *

What were some of the specific intellectual and political issues at stake in 1770, when Kant was first proposing the distinction between phenomena and things-in-themselves in his "Inaugural Dissertation"? A few pieces of historical information help us understand the tensions affecting Kant as a thinker. Frederick II, the reigning monarch of Prussia, had used a policy of principled religious tolerance to build national unity from fragments of the former Holy Roman Empire brought together by his father (Brunschwig 1974, 9–21). Frederick hoped that such tolerance would enable teachers, merchants, and state officials to improve Prussian education and technical competence. Unlike France, moreover, he was forced to rely on universities for administrative expertise rather than on an established state apparatus (*STP* 324–26). Thus the educated segment of Prussian civil society had good reason to hope for a gradual liberalization and modernization of the realm, especially in matters of conscience. Kant's writings were circulated by the "literate bourgeois public sphere" described by Jürgen Habermas and others as contemporaneous with modern nation building (Habermas 1999; Calhoun 1992).

But Frederick was not unopposed in these policies. During the 1760s and 1770s, German religious authorities opposed the Enlightenment on the grounds that rationalism—as represented by Leibniz and Wolff on the one hand, and Lessing on the other—promoted atheism and/or pantheism by refusing final authority to scriptural revelation and grounding all knowledge in reason alone (Zammito 1992, 18). Christian Wolff, the chief proponent of Leibnizian rationalism during the first half of the eighteenth century, had been banned from teaching in Prussia before Frederick II, desirous of creating an intellectual environment comparable to the French, invited him to resume a chair at the University of Halle. Many *Aufklärer* were also accused of being secret admirers of Spinoza, whose pantheism and denial of free will offended basic Christian beliefs.

Although Kant was critical of certain rationalist philosophers such as Leibniz, Spinoza, and Wolff because they equated human knowledge with confused and limited access to things-in-themselves, he was equally concerned to defend the Enlightenment against Hume and against younger protoromantic German thinkers like Hamann and Herder. Hume's skepticism regarding the existence of causal regularity in nature threatened the very possibility of reliable human knowledge concerning the physical world, as well as the existence of God. Ironically, his provocative claim that physics was nothing more than habitual "faith" in the persistence of established natural phenomena appealed to religious thinkers who wanted to render science subordinate to belief. Hamann and Herder, whose ideas would later be associated with the birth of modern history and anthropology, argued that the *language* of thought and its natural, bodily, or historical origins irrevocably shaped the content of reason and rendered it "impure." Herder conceptualized reason itself as the flower of historical human evolution; Hamann, who was influenced by Hume, regarded reason as subordinate to the emotional force of revealed religion. Although he was attracted by the values of self-rule and moral universalism found in another naturalist philosopher, Jean-Jacques Rousseau, Kant refused to ground these values in a theory of human nature. In his view, neither religion, metaphysics, nor anthropology should have methodological priority over a self-certifying, self-critical reason.

According to Kant, these positions shared the defect of claiming to have insight into the nature of *things-in-themselves* rather than simply into the structure of *appearances* confronting human knowledge. The Leibnizian, for example, judges all human knowledge on the standard of God's intellectual intuition regarding the essences of created things, regarding it as "accurate" but confused. If the standard is humanly inaccessible, however, Kant would be loath to consider this real "knowledge" in any sense. Religious enthusiasts—whether Protestant clergymen or Platonic revivalists—believe that they have an incommunicable and therefore unchallengeable insight into God's nature and plan for humanity. Empiricists and skeptics make claims based on the authority of experience, without inquiring into the conditions that make experience and knowledge possible in the first place. They behave, in short, as if the conditions for experience were "given" in a direct and unquestionable way along with the phenomena. Each in its own way, these positions prevent thinkers from claiming the knowledge necessary for individual and collective self-governance or, more insidiously, give them a right to rule others on the basis of insights whose validity cannot be tested. Kant denounces them as *Schwärmer*, "enthusiasts" or "fanatics."[8] Only critique can effectively

---

8. Kant describes enthusiasm as an affect accompanying "unbridled" imagination, where fanaticism involves "deep-seated and brooding passion" without a rule. Like madness (*Wahnsinn*), enthusiasm is a temporary state, whereas fanaticism is a long-lasting and destructive illness (*CJ* 5:275). For the religious and philosophical context in which *Enthusiasmus* and *Schwärmerei* were distinguished during Kant's lifetime, especially in relation to Neoplatonism, see Losonsky 2001, 105–31, and Fenves 1991, 241 n. 42. Monique David-Ménard (1990) argues that Kant was driven to identify transcendental conditions for the possibility of experiencing *objects*, at least in part, by the desire to distinguish himself from philosophers like Leibniz who bore an uncanny resemblance to Swedenborg insofar as they confused intuition,

counteract *"materialism, fatalism, atheism* [and] freethinking *lack of faith,"* while checking these enthusiastic tendencies, which render the general public superstitious and turn intellectuals into skeptics (*CPR* Bxxxiv). Critical philosophy, therefore, has implications for the constitution of a public sphere and the identification of legitimate members of this public.

Drawing the Boundaries of Pure Reason

Early in his career, Kant hoped that the body's boundaries could serve as a phenomenological and metaphorical defense against the temptation to claim insight into things-in-themselves. For one thing, he was unconvinced by the Leibnizian-Wolffian claim that human bodies and sensible experiences were merely confused, imaginatively mediated impressions of things-in-themselves. Leibniz believed that the cosmos was composed of simple substances or "monads" (some souls, some merely physical), which did not really interact, but only appeared to do so from the limited perspectives of the souls. Kant's first essays were attempts to reconceptualize the interaction of mind and body in such a way that material bodies could conflict with and change one another, producing psychic as well as physical effects. They also tried to account for the *reality* of moral and psychological conflict, which Kant analogized to opposing and canceling physical forces or intensive magnitudes.[9] For another thing, Kant suggested that the human body could provide a "focus imaginarius" enabling philosophers to distinguish between the material world known by physical scientists and the metaphysical or purely imaginary world of "souls." This is the function it served in another early satirical piece, "Dreams of a Spirit-Seer Elucidated by Dreams of Metaphysics."[10] Banishing souls

---

imagination, and reason. Having an object of thought is as much a *defense* against certain forms of passivity as it may be a demonstration of *agency*.

9. See "Attempt to Introduce the Concept of Negative Magnitudes into Philosophy" (1763, in Kant 1992c) and "True Estimation of Living Forces" (1747, in Carpenter 1998). Kant's argument for interaction assumes a critical form in "On the Form and Principles of the Sensible and the Intelligible World" (1770, in Kant 1992c, 2:407) and is explained in the *Critique of Pure Reason* at A273/B329, where he protests that Leibnizians and Wolffians recognize "no conflict [*Widerstreit*] other than that of contradiction [*Widerspruchs*](where by the concept of a thing is itself annulled [*aufgehoben*]) such as "the conflict of reciprocal impairment, where one real basis annuls the effect of another."

10. This image recurs in the *CPR* at A644/B672. Swedenborg claims to participate consciously in a community of immaterial spirits in addition to his obvious participation in the material life of the human community. Kant does not deny the rational possibility of Swedenborg's claim. But just as Kant's solution to the Third Antinomy will make room for the

from the realm of possible knowledge caused Kant some chagrin, for he admired Rousseau's image of a polity formed by communicating souls.[11] His later ethical philosophy tried to provide such a moral vision with grounds that were neither naturalist nor supernaturalist.

Finally, Kant's early writings suggest that the human body has an inner capacity to differentiate right from left, proving that space is not merely a confused, imaginary impression of things-in-themselves, but has (given his pre-critical perspective) as much reality as the objects of experience. The orientation of right and left hands also provided Kant with a metaphor for orientating oneself with respect to conceptual and social phenomena.[12] A later essay, "What Is Orientation in Thinking?" criticizes fellow *Aufklärer* Moses Mendelssohn for assuming that the ability to orient oneself in thought is a straightforward natural capacity (*KPW* 237–49). His goal in this essay was to defend Mendelssohn against accusations of atheism (by association with Spinoza), but also to affirm that the balance between reason and faith was best maintained by critical philosophy rather than by philosophy plus "common sense." Mendelssohn, he argued, relied on "reason *alone* . . . as a necessary means of orientation," rather than on intuition or faith (*KPW* 238). He compares the idea of an "invisible" distinction between east and west enabling travelers to use the stars as a guide, the subjective distinction putting us in touch with an otherwise reversible geometrical space, and finally a *felt* need or sense *internal to reason* itself, which encourages it to use the boundaries of possible experience as map for separating knowledge and error (238–40). In these essays, orientation is the touchstone of a "responsible" rational-

---

possibility of freedom while leaving it questionable how we might recognize the work of freedom in human action, "Dreams" makes room for the possibility of contact with spirits while leaving it questionable how we might recognize that a representation has a supernatural origin rather than simply being a product of imagination.

11. In this essay, therefore, one can see the germ of ideas with which the *Critique of Judgment* will be preoccupied: the relation between a community of feeling underlying reason's capacity to make distinctions and the community formed by rationally considering one's own opinions from the standpoint of others. Although the tendency to demand others' approval is generally selfish, it is also possible that one might sincerely and selflessly "compare that which one knows for oneself to be *good* or *true* with the judgement of others, with a view of bringing such opinions into harmony"; thus "we sense our dependency on the *universal human understanding*, this phenomenon being a means of conferring a kind of unity of reason on the totality of thinking beings" (1992c, 2:334).

12. On the theoretical import of this inner difference, see "Concerning the Ultimate Ground of the Differentiation of Directions in Space" (1768, in Kant 1992c) and in *Prolegomena* (PR 4:285–86); for its political import, see "What Is Orientation in Thinking?" (1786, in *KPW*).

ism that can resist dogmatism, in either its spiritualist or naturalist forms. Kant does associate responsible rationalism with the body, but not a "natural" empiricist body. Rather, he associates it with a body situated at the crossroads of multiple spaces, including but not limited to the form of outer intuition.[13]

With the *Critique of Pure Reason*, Kant leaves mind-body interaction aside, both as a topic of metaphysical speculation and as a metaphor for responsible rationalism.[14] Planetary motion, rather than proprioception, provides Kant's most powerful metaphor for creative, well-delimited displacement in thought. "Having found it difficult to make progress [in astronomy] when he assumed that the entire host of stars revolved around the spectator, [Copernicus] tried to find out by experiment whether he might not be more successful if he had the spectator revolve and the stars remain at rest" (Bxvi). The idea of experimentation suggests that Kant regards his situation as provisional, although his goal is to discover a stable landscape. It brings out the active sense of "orienting" oneself in thought. "Now," Kant continues,

> we can try a similar experiment in metaphysics, with regard to our *intuition* of objects. If our intuition had to conform to the character of its objects, then I do not see how we could know anything a priori about that character. But I can quite readily conceive of this possibility if the object (as object of the senses) conforms to the character of our power of intuition. However, if

---

13. On the basis of these passages, Lyotard argues that thought as well as the body require "feeling" for orientation, and identifies this feeling with "state of mind" with which critical reason attributes a representation to sensibility or understanding as powers of thought in the act of "transcendental deliberation" (*CPR* A261/B317; Lyotard 1994, 7, 37–38). Drawing on the idea that the proper domain of judgment is the *feeling of pure pleasure and pain* (from the *Critique of Judgment*), Lyotard argues that we see both judgment and feeling at work in transcendental deliberation. Indeed, Kant also makes reference to *feeling (Gefühl)* in the *Critique of Practical Reason*, where *Achtung* or respect is one of the few indicators that an ostensibly moral act has indeed been performed by reason for the sake of reason, rather than for the sake of inclination.

14. The only mention is in "On the Paralogisms of Pure Reason" (A386–96, B427–28). His discussion of interaction and the problem posed by the original "source" of intuitions is more extensive in the A edition; in the B edition he sums up the problem quickly by saying its answer will hang on whether "real communion" of substances is possible, a topic handled in the Analogies of Experience (especially the Third Analogy, A211–18 / B257–65). In "Refutation of Idealism" in the B edition, Kant explains that the determination of any inner (temporal) appearance has relation to some external information that connotes comparative stability or change and thereby reveals the thinker as active or passive in relation to a potential manifold of intuition (B277).

> these intuitions are to become cognitions, I cannot remain with them but must refer them, as presentations, to something or other as their object, and must determine this object by means of them. (Bxvii)

In other words, it is natural to assume that our concepts must conform to the objects of experience, but it is nonetheless much easier to explain the regularity of experience (the fixedness of stars, including the sun) if we assume that the most fundamental rules governing the object of our intuitions lie within our human understanding, not the world in itself. If separating objects of intuition from objects of pure cognition, only some of which are suited for explaining the unity of intuited experience, solves long-standing philosophical problems, then this success will be worth the loss of access to things-in-themselves. This is Kant's "Copernican revolution," which neutralizes conflicts among rival doctrines by referring them to the *limits* of pure reason rather than to pure reason "in itself." It is not a revolution that destroys existing practices and institutions, any more than Copernicus's insight destroyed or moved the sun and stars. Rather, it alters what can be *done with* those practices and institutions.

Kant distinguishes between useful and unproductive cognitive limits by comparing his own method to Hume's. "Consciousness of my ignorance," Kant writes in the *Critique of Pure Reason*, "instead of ending my inquiries, is rather the proper cause to arouse them." *Contingent* ignorance can either be satisfied through dogmatic (empirical) investigation, or by investigation which is tempered by critical scrutiny of the cognitive powers involved (*CPR* A758/B786). These are the tasks for natural science and philosophy, respectively. However, there is another kind of limitation on knowledge that can only be conceived of *once we have learned more:* this is the limitation Paul Veyne analogizes to the "automobile driver who *does not see that he does not see*" and "does not know that he is driving too fast for an unknown stopping distance" (Davidson 1997, 158). Inevitably, even the best philosophy and natural science in the world will still have more to learn. One can think of this ignorance in a concrete, positive sense as referring to problems we have been unable to solve in the past, or in a more abstract, negative sense as referring to fields of knowledge whose existence we have not yet conceived but may conceive in the future. The finitude of human knowledge and its infinite capacity for improvement or expansion are twin symptoms of Kantian thought—a *discursive* image of thought.

But there is, Kant continues, a further sort of ignorance, which delimits the realm of infinite exploration as such and thereby serves an orienting function for philosophers and natural researchers:

> Experience does teach me that wherever I may go, I always see a space around me in which I could proceed farther. Hence I cognize the limits of what is in each case my actual geography, but I do not cognize the bounds of all possible geography. But if I have indeed got as far as to know that the earth is a sphere and its surface spherical, then I can also from a small part of it—e.g., the magnitude of a degree—cognize determinately and according to a priori principles the diameter, and through it the complete boundary of the earth, i.e., its surface area. (*CPR* A759/B787)

Although all empirical attempts to reach the boundary of our possible knowledge and grasp the totality of knowable entities fail, nonetheless, "all questions of our pure reason still aim at what may lie outside this horizon, or—for that matter—at least on its boundary line" (A760/B788).

Hume discovered that reason was limited—that portions of the earth, to follow Kant's earlier analogy, were unexplored—but did not go far enough to ground his inquiry in the certainty that might have been provided by knowledge that his *contingent* ignorance concerned a spherical world. Hume did not, in other words, find *informative* limits on reason, and for this reason Kant considered his skepticism unsatisfying. The distinction between skepticism and critique lies in this, therefore: that skepticism refuses to see the *positive import* of those limits, conceived as transcendental *bounds* directing reason away from certain speculative ventures and toward others which are more legitimate. Thus skeptics throw up their hands in the face of limits that they can only conceive as empirical. By contrast to limits (*Schranken*), which are provisional, bounds (*Grenze*) are fixed and reliable, enabling Kant to prove that "all possible questions of a certain kind" will forever remain beyond reason's abilities (A761/B789). Critique turns *limits* on thought into positive guidelines or *bounds*. It gathers together the elements of "experience" and opposes them to the intellectual objects of pure understanding or reason, but it also enables experience to be scientifically explored and enriched.

* * *

The most important tool in Kant's critical kit is the wholly *problematic* concept of the noumenon:

> I call a concept problematic [*problematisch*] if, although containing no contradiction and also cohering with other cognitions as a boundary of given concepts involved in them, its objective reality cannot be cognized in any way. The concept of a *noumenon*, i.e., of a thing that is not to be thought at all as an object of the senses but is to be thought (solely through a pure understanding) as a thing in itself, is not at all contradictory, for we cannot, after all, assert of sensibility that it is the only possible kind of intuition. ... Yet, in the end, we can have no insight at all into the possibility of such noumena, and the range outside the sphere of appearances is (for us) empty. (A254/B310, see also A286/B342)

The "object" of this concept is not really an "object" in the sense that Kant hopes to reserve for objects of experience. The concept of the noumenon is only a *"boundary concept* serving to limit the pretension of sensibility" enabling the sphere of sensible concepts (as opposed to merely intellectual ones) to be correlated with the sphere of experience itself (A255/B311). Thus it is more a kind of imaginative placeholder than an actual terrain or entity. Since "problematic" is one of the modalities in which a judgment may be posed, when Kant says this concept is "problematic," he means that it is an element in a judgment "where the affirmation or negation is taken as merely *possible* (optional)" (A74/B100). Here, his suggestion that the "world" or sphere of all conceivable metaphysical entities be divided between "phenomena" and "noumena" (A249, note 169)—even though the latter refers to nothing "real"—makes this world the object of a disjunctive judgment, both of whose elements are merely "problematic." Renewing the metaphor of orientation, this hypothesis "serves us in finding the true proposition (just as indicating the wrong road serves us in finding the right one among the number of all the roads that one can take" (A75/B101). According to Kant, then, the first step in understanding may be a problematic judgment, for "at first we judge something problematically; then perhaps we also accept it assertorically as true; and finally we maintain it as linked inseparably with the understanding, i.e., as necessary and apodeictic" (A76/B101).

Proposing that the world is divided exclusively between the (empty) extension of the noumenon and the (indefinite) extension of the phenomenon allows Kant to create a "community of cognitions" (A74/B99). It also lets him make an infinite judgment concerning the objects of these cognitions; that is, a judgment whose quantity is "infinite." An infinite judgment asserts that some object lacks or stands outside the range of a given, known property. Thus it leaves the field of experience as indefinitely open as ever, while qualifying it as merely "phenomenal" and permitting it to be brought in relation to transcendental conditions of possibility.[15] It is an apparently affirmative judgment that gives its object only a negative determination. For example, if I say

> The soul is nonmortal [*nicht sterblich*], then I have indeed, in terms of logical form, actually affirmed something, for I have posited the soul in the unlimited range of nonmortal beings [*nichtsterbenden Wesen*]. Now what is mortal comprises one part of the whole range of possible beings, and what is nonmortal [*das Nichtsterbende*] comprises the other. . . . But to say that is only to limit the infinite sphere of all that is possible, viz., to limit it to the extent that what is mortal is separated from it and the soul is posited in the remaining space of the sphere's range. But despite this exclusion [of what is mortal from it], this space still remains infinite. (A72/B97)

Kant does not deny that some other kind of (nonhuman) being might have actual objects for a concept that is employed merely *problematically* by humans (A253/B307). Admittedly, an understanding capable of grasping intelligible objects, "not discursively through categories, but intuitively in a nonsensible intuition," "is itself a problem" for Kant (A256/B312; A287 / B343–44). The appropriate infinite judgment for Kant's crit-

---

15. David-Ménard 1990, 33, 148; Žižek 1993, 113–14; Cutrofello 1997, 55–56. In the Jäsche *Logic*, Kant explains, "Everything possible is either A or *non* A. If I say, then, something is *non* A, e.g., the human soul is *non-mortal*, some men are non-learned, etc., then this is an infinite judgment. For it is not thereby determined, concerning the finite sphere *A*, under which *concept* the object belongs, but merely that it belongs in the sphere outside *A*, which is really no sphere at all but only *a sphere's sharing of a limit* [*Angrenzung*] *with the infinite*, or the *limiting itself* [*die Begrenzung selbst*]" (Kant 1992b, 9:104). Because an indefinite judgment declares that X is non-A without saying whether this means X is B, Kant refers to it as a *conflict* at the level of general logic which must be referred to transcendental logic to ascertain whether X can only be A or B. Thus an indefinite judgment *structures a logical field* within which further determinations (as to existence or nonexistence of particular objects, for instance) can then be made: logically speaking, the world is divided into A and non-A, *even though A might only be a logical operator and not refer to an actual object.*

ical turn is not "noumena are objects of a nonsensible intuition" (A255/B311) but "phenomena are objects of a non-intellectual intuition," that is, one mediated by sensibility and understanding. What this means is that no matter what else can be said of phenomena, they are all products of imagination, a discursive understanding, and forms of intuition; they will never be completely determinable in themselves.

What Kant *knows* is that Leibnizians, Wolffians, and other dogmatic philosophers believe that such an "intuitive understanding" *does* have an object for this concept in the same sense as we have objects of experience. They exempt themselves from participation in the community of scholarly and scientific discourse whenever they claim to have more or less insight into such "objects." In a way, therefore, Kant can grapple with philosophers he regards as *Schwärmer* by acknowledging the concept of the noumenal but rejecting its object as a "non-object" or "pseudo-object." At the same time, he can define the boundaries of his *own* possible experience—the kind of space he thinks we all actually share, in fact. Thus Monique David-Ménard (1990), for example, argues that positing this concept enables Kant to distance himself from any personal tendencies to *Schwärmerei* and to substitute a positive object of genuine intuition and understanding for any connection to that troubling "other," pre-critical way of thinking. As we will see, many kinds of thought and many troubling resemblances can be mastered or at least placed at a distance by analogizing them to the object of this "problematic concept"; and in the Paralogisms (*CPR* B409–10) Kant himself warns readers against settling the "soul" on its empty territory. But the negative work of this concept is not as important as the positive work it does. The noumenon occupies a crucial spot in Kant's system insofar as it "closes" the sphere of conceivable metaphysical entities to which concepts, principles, and regulative Ideas can apply, i.e., the sphere of experience.

The object of this problematic concept should not be mistaken for the "transcendental object = x" (A109; A249). Nor should the noumenon be identified with the *source* of intuitions (A387; *PR* 4:289). The noumenon simply limits the experiential "signified" insofar as it might be an object of divine intuition, while leaving it infinite for human knowledge. The "transcendental object = x," however, limits the domain of representations (comparable to the "signifier" in Saussurean or Lévi-Straussian structuralism) and gathers them in a single manifold: it serves only "as a correlate of the unity of apperception" (*CPR* A249).[16] The judgment that

16. In his reading of Kant, Slavoj Žižek identifies the "transcendental object = x" from

"all objects of (any) intuition either fall under the concept of the noumenon or of the phenomenon" together with "phenomenal objects are objects of a non-intellectual intuition" leaves us with a realm of phenomenal objects from whose features the capacities of a fundamentally mediated, sensible, finite intuition can be identified, without requiring Kant to say anything at all about God's knowledge.

The noumenon is not the only concept Kant deployed problematically. God, the world as a whole, and the soul (as seat of all thought, experience, and desire) are Ideas (also problematic objects) of a self-limiting *reason*. Theoretical reason employs such Ideas in an effort to think the unconditioned ground, totality, or origin of phenomena as objects in their own right, as well as to actualize desire. For example, "the absolute whole of appearances *is only an idea;* for since we can never outline such a whole in an image, it remains a *problem* without any solution" (an Idea of practical reason, however, is more than "only" an idea because it may inspire us to produce its object) (*CPR* A328/B385; A417/B445). The Ideas are products of reason's tendency to go beyond the realm of experience in search of unity and to limit the cognitive impact of such excesses—in other words, to act on its own actions (*PR* 4:333; *CPR* A339/B397). But these problematic concepts only perform their crucial task of guiding the understanding when they have definitively been detached from their meanings in the Leibnizian intellectual framework. Simultaneously proposing and canceling the prospect of "noumenal" intuition delimits a human "world" that is both infinite and whole, one in which Ideas of reason can mediate between the signifier (concepts and intuitions) and whatever lies at their root (the objects of a nonintellectual intuition).

Ideas orient Kantian reason toward unity in the series of judgments made by the understanding, even when employed only regulatively or critically. This has enabled some scholars, such as Rudolf Makkreel (1990) and Dieter Henrich (1994), to read reason as essentially reflective in the *Critique of Pure Reason* as well as the *Critique of Judgment*. In other words, reason compares presentations in order to *generate* a concept, rather than *applying* an existing concept to appropriate representations

---

the A edition of the *Critique of Pure Reason* with that "form" of the understanding which prepares thought to receive empirical content and which expresses the finite understanding's *reliance* on experience for content. Žižek describes the transcendental object as "the semblance of an object . . . [giving] a body to the gap which forever separates the universal formal-transcendental frame of 'empty' categories from the finite scope of our actual experience" (1993, 18, 37).

(*CJ* 20:211'). I agree that reflection has priority over determining judgment in both texts. But I want to focus on the act of reflection as dividing, sorting, and ultimately constructing something that was absent in the pre-critical world rather than "reading" nature like a text from part to whole and back again.

As a concept of critical philosophy, the noumenon allows reason to define its own critical employment and domains, to become the "agent" rather than the mere "patient" of its own operations, but without denying its reliance on "affection" or intuition (*CPR* B166–67, B275–76). Within the realm of phenomenal objects, but only within that realm, can we employ "God" as a regulative Idea guiding our appreciation of physical laws and organization in nature. Reason's ideal of unification is initially expressed in the act of exclusion (the first among many dividing practices) enabling a whole to be projected in the first place. This foundational opposition between the phenomenal world and its "other" (even if technically there *is* no other, the concept being empty) reflects Kant's desire to acknowledge the reality of conflict and interaction while avoiding the excesses of spiritualism that troubled him in the pre-critical essays. Although he may hope for completeness in the sciences and be guided in this pursuit by harmony among the faculties, this harmony is not to be found at the origin, in the conditions for unified experience.

Transcendental and Other Topographies

"Where" is Kant, in everyday terms, when he is struggling to orient himself? "Dreams of a Spirit-Seer" and other pre-critical works give significant attention to the location of the soul, which for many years he considered to occupy the space of the body. But critical reason is both dis-embodied and de-psychologized, which puzzles contemporary readers. Given that Kant employs so many metaphors involving landscapes (for example, A3–5 / B6–9, A235–36 / B294–95, and A395–96), what does his landscape look like? The answer: he is situated amid the representations of phenomenal experience, some lawfully ordered and others seemingly contingent, some sensual and others purely conceptual, some productive of action and *all* linked by judgments. He is not "apart from" space, described in the Transcendental Aesthetic as a uniform and qualitatively indifferent structure relating the subject and the elements of his or her world, but he is a *subject* of spatial representations. "Space is a necessary a priori presentation that underlies all outer intuitions" and to which all phenomenal appearances must be related; it is given as "one and the same unique space" and as an "infinite *given* magnitude, out of which other spaces may be marked off or upon which various measures and divisions may be imposed (A24–25 / B39–40). In some mundane sense, Kant has a body "in space," although he only knows what this means and can correct misperceptions about his body (such as those suffered by anorexics, psychotics, hypochondriacs, and religious fanatics) if he can first identify these misperceptions in relation to faulty *concepts* (representations of the understanding), *intuitions* (representations of sensibility and geometry), and products of *imagination* or *memory*.[17] But Kant's mind *and* body also have an unknown dimension, because introspection and medicine never have unmediated access to either as things-in-themselves. Indeed, he expresses concern that "mind" and "body" are abstractions whose opposition should not be taken too rigidly, differing "from each other intrinsically but only insofar as one extrinsically *appears* to the other (B427–28). They also have a social meaning that can only be identified through reference to the perception and judgment of other persons, although the *Critique of Pure Reason* does not discuss this aspect (we will consider it later).

In the *Critique of Pure Reason*, Kant refers to the "place" of reflection as a *transcendental topography* (or *topic*), and the "act" of reflection as a transcendental *Überlegung* (A268/B324). Note that *Überlegung*, which is translated as "deliberation," also has connotations of "overlapping." The role of transcendental deliberation in the first *Critique* is to differentiate "types" of judgment according to the kind of representations that are brought together by the judgment, and to determine whether the judgments are acts of general or transcendental logic (B317). In other words, it teases apart overlapping representations. In "On the Amphiboly of Concepts of Reflection," Kant describes *Überlegung* as our "state of mind [*Zustand des Gemüts*] when we first set about to discover the subjective conditions under which [alone] we can arrive at concepts" (*CPR* A260/B316). "All judgments—indeed, all comparisons—require a *deliberation*, i.e., a distinction of the cognitive power to which the given concepts belong," and Kant therefore defines "transcendental deliberation" as the "act whereby I hold the comparison of presentations as such up to the

---

17. Kant appears to have fought the temptation to hypochondria throughout much of his life. Key texts include "Investigation Concerning Diseases of the Head" (1764), the *Anthropology* (*AN* 7:212–13), and the "Conflict of the Faculties" (1992a), especially chapter 3. See also Cutrofello 1994, 45–56; David-Ménard 1990, 113–20; and Shell 1996, chap. 10.

cognitive power in which this comparison is made, and whereby I distinguish whether the presentations are being compared with one another as belonging to pure understanding or to sensible intuition" (A261/B317). The division of representations, chiefly those referring to understanding alone or to understanding in conjunction with sensibility, exhibits a "transcendental topic" with distinct logical locations (A281/B324).

As Lyotard (1994), Makkreel (1990), and others have observed, the function of reflective judgment in the *Critique of Judgment* is similar to the role played by transcendental deliberation in the *Critique of Pure Reason*.[18] Judgments of reflection are only discussed at the end of "On the Amphiboly of Concepts of Reflection," because Kant's goal is to situate judgments with respect to the faculties rather than to understand judgment itself. But reflective judgment is central to Kant's argument in the third *Critique*. Unlike "determinative judgment," which applies existing concepts to known particulars, reflective judgment identifies concepts and meaningful forms from the welter of presentations available to consciousness. In the *Critique of Pure Reason*, Kant described deliberation as "our consciousness of the relation of given presentations to our various sources of cognition—the consciousness through which alone the relation of these presentations to one another can be determined correctly" (A260/B316). In the first introduction to the *Critique of Judgment*, however, Kant distinguishes between judgment "as an ability to *reflect*, in terms of a certain principle, on a given presentation so as to [make] a concept possible," and judgment as an ability to "*determine* an underlying concept by means of a given empirical presentation. . . . To *reflect* (or consider [*überlegen*]) is to hold given presentations up to, and compare them with, either other presentations or one's cognitive power [itself], in reference to a concept that this [comparison] makes possible. The reflective power of judgment [*Urteil*] is the one we also call the power of judging [*Beurteilung*] (*facultas diiudicandi*)" (*CJ* 20:211').

Both transcendental deliberation and reflective judgment enable Kant to define the mental powers exercised in classifying representations.[19] In

18. In *Lessons on the Analytic of the Sublime*, Lyotard argues that Kant distinguishes between types of representation in the act of transcendental deliberation by determining these representations *aesthetically* rather than *categorically*: "Critical thinking has at its disposal in its reflection, in the state in which a certain synthesis not yet assigned places it, a kind of transcendental pre-logic. The latter is in reality an aesthetic, for it is only the sensation that affects all actual thought insofar as it is merely thought, thought feeling itself *thinking* and feeling itself *thought*," a determination he calls "tautegorical" (1994, 32).

19. An objection may be raised that transcendental deliberation involves only the discrimination of representations relating to understanding and sensibility, not all the powers of

sorting those representations which pertain only to the understanding from those which form part of sensible experience, transcendental deliberation shapes our images of understanding and sensibility themselves. So too, in analyzing the form which provokes an aesthetic or teleological judgment in terms of a potentially communicable harmony or attunement among the mind's faculties, Kant further develops his account of imagination, understanding, and sensibility. Transcendental deliberation does not assume that these spaces of imagination, perception, and understanding (jumbled in the Classical[20] worldview) are necessarily unified in a single manifold. Nor, as we will see, did Kant seem to want to say whether they originate in one source or in many. But by the time Kant wrote the *Critique of Judgment*, such unity was his primary goal.

Transcendental topography is one of the "quasi-metaphorical" spaces of Kantian thought, along with others mentioned in the introduction to the *Critique of Judgment*, such as *domain (Gebiet)*, *territory (Boden)*, and *World (Welt)*. The task of critique is to tease out from an apparently unified experience, still riddled with ignorance and error, the powers of the mind capable of being genuinely ordered by critical reason. The transcendental analysis of thought into its various powers or faculties should not be understood as a "psychological" act of introspection. It is a properly "impersonal" exercise, not because it *abstracts* from individual contents of experience (as in Husserl's transcendental reflection), but because it ultimately uncovers the structures that enable us to know ourselves as individual beings, distinct from other thinkers in phenomenal experience. Kant did not focus on this process in the first *Critique* because he was more interested in justifying claims about the natural-scientific world than in psychological or bodily individuality, although his lectures on anthropology and the last section of *Conflict of the Faculties* touch upon it. To define styles of thought such as the power of judgment (the understanding), the urge to systematicity (theoretical reason), desire (practical

---

thought. Here I follow Lyotard, who believes that the account of transcendental deliberation gives insight into Kant's general style (or *manner*) of philosophical work (see previous note). Lyotard suggests that if intuition is necessary for all human knowledge, according to Kant, some form of intuition may be just as decisive for knowing the *transcendental conditions* of experience as for knowing ordinary states of affairs (1994, 9). However, Lyotard's account in no way implies that transcendental deliberation is *constitutive* of experience, or that every element of the manifold must pass through an explicit process of reflection.

20. In this text and in Foucault's usage, "Classical" refers to French politics and culture of the seventeenth and eighteenth centuries, and by extension to European politics and culture at that time, rather than to ancient Greece and Rome.

reason), imagination, and sensibility in terms of their respective contributions to experience is very different from analyzing that same experience in terms of actual things said, memories, moods, reactions, and interpretations of those reactions. The latter exercise, which one might want to call empirical reflection as distinct from transcendental or logical deliberation, takes place in time and involves an actual synthesis of the manifold of inner sense as psychological object. Although commentators disagree about the mechanics of this process, Kant indicates that the synthesis of inner sense (that is, all representations involving time) continually changes the time relations of the synthesized contents. Reviewing the contents of experience produces new mood changes and reactions, and thus Kant refers it as "self-affection" (*CPR* B152–53). (The results of transcendental and empirical reflection, obviously, should not be confused with physiology, which concerns the motion of brain matter in phenomenal space *according* to the knowledge made possible by transcendental faculties of the mind—the mind, for instance, of a neurophysiologist).

The Classical era (to which Leibniz, Locke, and Hume belonged) assumed representations to be the property of an "I." Although philosophers debated the relationship between linguistic or visual representations and those of pure thought, they could not imagine language and art as composed of anything other than representations, or of thought as anything but a reorganization and analysis of representations (*OT* 78–87; see also Hacking 1975, 13–53, and Ricken 1994). Kant accepted Hume's skepticism regarding the illusory nature of psychological unity. It is true that humans demand consistency in their psychological experience (for example, in the sequence of memories or similarity in the emotional expressions of a single "personality"). But as Hume rightly observed, this demand reflects the *aspirations* of thought rather than empirical experience or a priori knowledge of psychological coherence. We are compelled to "connect all appearances, actions and receptivity of our mind *as if* this mind were a simple substance that (at least in life) exists permanently and with personal identity," and to regard all its states, bodily or mental, as contingent and logically separable from the substance itself (A672/B700). However, this ideal of self-knowledge will never be satisfied.

This means, among other things, that "empirical reflection" or introspection will never arrive at the simple taxonomy of faculties defined through transcendental reflection, much less a reduction of complex inner experiences to a single power of thought. Introspection perpetually discovers new complexities and obscurities in the thought material that it aspires to unify, but never achieves "a systematic unity of all appearances of inner sense." Reason treats the empirical unity and coherence of psychic states as expressions of a deeper unity, "unconditioned and original," ultimately "the rational concept (idea) of a simple substance that, in itself, immutable (personally identical), stands in community with other actual things outside it" (CPR A682–83 / B710–11). Reason's principles tell us "to regard all determinations as [united] in a single subject; to regard all powers as much as possible as derived from a single basic power, to regard all variation as belonging to the states of one and the same permanent being; and to present all *appearances* in space as entirely different from the actions of *thought*." Kant thereby concludes that the psychological Idea, or idea of the mind as a coherent whole, "cannot signify anything but the schema of a regulative concept (CPR A684/B712; see also B422).

If representations seem to cohere, this is not because an *ego cogitans* exists as their substrate, but because these representations are all potential correlates of an "I think" whose fields of operation (intuition, judgment, speculation) were defined through the exclusion of a problematic object. The faculties or powers of thought are like impersonal "folders" into which representations can be filed by anyone familiar with an office's activities. But they also limit and anticipate the concepts that can be used communicably by that thought (or those office workers) and the kinds of sensory particulars to which those concepts might be applied. Representations that do not fit the purpose of the files are either piled on individual workers' desks or put outside in the hallway as trash; the order of the filing system depends on their exclusion.

For an analogy to the act of transcendental deliberation, let us consider Foucault's description of the Velázquez painting *Las Meninas*, which is in the Prado (OT 3–16). The painting depicts the Infanta of Spain, surrounded by maids and dwarfs, gazing at the viewer alongside a court painter. The painter is intent on some subject in the foreground (where the actual viewer stands) and works on a large canvas whose back is to the viewer. Behind the Infanta, attendants, and artist stretches a large hall whose far wall is hung like a gallery with paintings and a single mirror, in which the subjects of the artist's and child's attention are presumably reflected.

The Prado is a space of representational objects. Foucault points out that much of the interest in this painting is generated by its superimposition of three seemingly distinct spaces of representation: the Prado gallery itself (in which the contemporary viewer stands), the space of Velázquez's studio (including the mirror in which he must have observed his own image, some five centuries ago), and finally, the space of the hall at court *depicted* in *Las Meninas* and fictively "inhabited" by the sovereigns and their royal daughter. Just as Kant attempts to organize his world of representations into discrete faculties and domains, Foucault tries to break down the constituent spaces of his experience in the painting *Las Meninas*. He cautions his reader not to imagine that the significance of the painting resides primarily in the persons it represents (*OT* 9). "It is in vain that we say what we see," because acts of determinate judgment cannot capture the *state of reflection* allowing representations and spaces to be teased apart.

In the central point of the mirror in the background of *Las Meninas*, Foucault notes,

> occurs an exact superimposition of the model's gaze as it is being painted, of the spectator's as he contemplates the painting, and of the painter's as he is composing his picture (not the one represented, but the one in front of us which we are discussing). These three 'observing' functions come together in a point exterior to the picture; that is, an ideal point in relation to what is represented, but a perfectly real one too, since it is also the starting point that makes the representation possible. (*OT* 14–15)[21]

After all, *Las Meninas* is not painted from the standpoint of the painter we see at his easel, but that of the royal couple who are only depicted as reflections of the *painter's* (and contemporary viewer's) point of view. By superimposing the actual painter's standpoint on that of the king and queen whose portraits are in the process of being painted, *Las Meninas* sutures together the space "in" the painting (the space depicted), the space in which it is painted (a space from which the painter is excluded by his own point of view, unless he, too, observes himself in a mirror), and the space from which it is viewed today. No single person could possibly stand in all those scenes and times at once. Foucault argues that the painting

---

21. This ideal point, from which all times and spaces involved with the picture could be surveyed equally, is an Idea or purely problematic object of reason. Lacan would identify this point with the *gaze*, which is an *object* of the viewer's drive, whereas the viewer's unreflective *involvement* with the depicted scene, including identification with the position once occupied by the two sovereigns, would describe the space of the *imaginary*. See Copjec 2000 for a discussion of Kantian reason and imagination in art history, including an analysis of *Las Meninas*.

can only function effectively to link these spaces because it does not represent one person for whom these spaces come together (*OT* 16).

Like the *cogito* in Kant's "On the Paralogisms of Pure Reason," the mirror on the far wall of the painted gallery is a lure—positioning the viewer as royal subject (and object) of the space of representation. This lure, which excludes as much as it includes, seduces the viewer into an experience of the painting's coherence that blends seamlessly into his or her experience of the museum as a whole. For by rights, the mirror should reflect the inhabitants of these other spaces too: the artist and the viewer. Does this mean that the sovereigns are the problematic object whose exclusion enables this painting to bring the viewer into the Infanta's presence? No, for they are clearly phenomenal entities (representations) just like the missing artist and viewer. The problematic object is not a representation, although its existence conditions the resulting system of representations. The mirror is a token or phenomenal substitute for this excluded object, within the system of representation it makes possible (the mirror is also the image privileged by Kant's method of transcendental reflection). The hidden side of the canvas whose back occupies the left-hand portion of the painting would be another, less reassuring token of what must be excluded for these spaces to cohere. We can only speculate on the contents of this canvas, which might include sovereigns, viewer, and artist, something completely unknown, or "no-thing," the "formless."

For further applications of the problem of sorting or "carving" spaces, let us consider the work of Henri Lefebvre and Arjun Appadurai. In *The Production of Space* (1991), Lefebvre describes a similar kind of "overlapping" between domains. Lefebvre describes "social" and "mental" spaces as forms of perception and movement that link individuals but also leave certain aspects of experience in the privacy of psychological interiority or the abstraction of shared discourse with no clear referent. Lefebvre regards *social* space as a historically produced "concrete abstraction" in which affective, motor, and cognitive relationships are condensed and reproduced. Appadurai, likewise, speaks of "locality" as a way of relating to some topic (conceptual or material) through repeated attention and acts of complication, as when a bird creates a nest by turning repeatedly in some hollow and lining it with grass, or a person makes a text, idea, or plot of land "their own" by writing about it or using it in rituals (Appadurai 1996, 18, 178). Lefebvre's distinction between "representations of space" (geometric, geographic, and financial), "spatial practice" (the production of space through repeated acts of inhabiting), and "representational space" (symbolically organized space and theoretical reflections on spatial practice) helps us understand why the "form of outer intuition" is not the only space that matters in Kant's work (Lefebvre 1991, 33). The form of outer intuition is only one *way* in which "locality" is produced. It also helps us understand how a conflict can arise between the presumed unity of the form of outer intuition (a representation of space), the presumed unity of the anthropological/historical "world" (the site of spatial practice), and the unity Kant hopes to find in reason itself, the *representational space* that constitutes the *lived milieu of reflection* for Kant and his readers.

Lefebvre would consider the space of pure intuition a "representation of space" rather than an epistemologically primary form of intuition. Also called "conceptualized space," such a representation is the primary orientation for those (like Leibniz, say) who "identify what is lived and what is perceived with what is conceived" (Lefebvre 1991, 38). According to Kant, people engage with space as a "pure" form of intuition only when contemplating geometry; the spatiality of lived experience is always part of a complex manifold deployed through time. Lefebvre does not regard geometry as the only possible representation of space; money, for example, also creates equivalences and inequalities among entities and places (49). Representations of space would also include maps illustrating demographic patterns and other phenomena that give viewers power over those who inhabit a certain (physical) space (Scott 1998, chaps. 1 and 2).

"Spatial practice," Lefebvre's second category, "embraces production and reproduction, and the particular locations and spatial sets characteristic of each social formation" (1991, 33). It involves differentiating interpersonally significant locations through required "competences" and possible "performances," only some of which are discursively structured, and renders space cohesive but not coherent "in the sense of intellectually worked out or logically conceived." At first, Kant would seem unlikely to characterize the activities through which members of a community share a more or less common division and perception of spatial places and meanings as "production."[22] But Enlightenment, like any ritual of initiation

---

22. See Heidegger 1990 and Melnick 1989 for two very different approaches to intuition and imagination that nevertheless stress their performative, indeed *productive* character. When critical thought, according to Heidegger, "inquires into nature after a plan of its own," this thought is fundamentally imaginative and shapes nature in the very act of apprehending. Melnick, far from considering intuition and understanding as dual aspects of imaginative

into maturity, is a way of giving individuals "who are, paradoxically, already within it," access to "a space in which they must either recognize themselves or lose themselves, a space which they may both enjoy and modify" (Lefebvre 1991, 35). Unlike Kant, Lefebvre thinks of spatial practice as taking place at many levels and scales, which he refers to as "superimposed" on one another and which require time or effort to translate into one another (93–95). Insofar as Enlightenment and a cosmopolitan world of international right have yet to be achieved, however, Kant does acknowledge that social space is both historically given and produced in a manner very different from space as "pure form of outer intuition." Enlightenment remains one of the most important contemporary "ideoscapes" being worked over and "localized" or "made local" in different parts of the world (Appadurai 1996, 36).

"Representational spaces," Lefebvre's third classification, involve

> space as directly *lived* through its associated images and symbols, and hence the space of "inhabitants" and "users," but also of some artists and perhaps of those, such as a few writers and philosophers, who *describe* and aspire to do no more than describe. This is the dominated—and hence passively experienced—space which the imagination seeks to change and appropriate. It overlays physical space, making symbolic use of its objects. Thus representational spaces may be said, though again with certain exceptions, to tend towards more or less coherent systems of non-verbal symbols and signs. (Lefebvre 1991, 39)

For Lefebvre, representational space is the affective and instinctive experience of space that is transformed by knowledge (representations of space) as a child or society grows and changes. But the authors of symbolic works also draw upon this residue to alter spatial practice and representations of space. For example, Kant's numerous metaphors involving landscape in the *Critique of Pure Reason* could be read as "representational space" from which his other more rigorous constructions borrow (see also Sallis 1987, 67–74).

---

thinking, treats space and time as topological surfaces described by patterns of exteriorizing and interiorizing movement and concepts as rules delimiting "legitimate" moves which will result in "movers" at many points (or points of view) achieving sensation on those surfaces. Space is the "form of getting affected or making contact" (1989, 9). The fundamentally spatial nature of both space and time in Melnick's reading is an interesting contrast to the fundamentally temporal nature of the forms of intuition in Heidegger's.

The *transcendental topography* or *topic* of deliberation functions for Kant as a "representational space" out of which the various elements of experience are drawn, including the nature of mental faculties, abstract forms of intuition, and conditions for the possibility of shared social practice. The representational space of self-limiting reason in the first *Critique* is not social, although any finite rational being should be able to enter into its structure and recognize, there, the architecture of his or her experience. Lefebvre attempts to explain how these three forms, despite Kant's egalitarian ambitions, shape one another in ways that essentially permit some agents to limit the imagination of others through spatial practices in the service of financial, sexual, or ethnographic "representations of space." Appadurai adds "ethnoscapes," "mediascapes," "technoscapes," "financescapes" and "ideoscapes" to the range of representational spaces described by Lefebvre:

> These terms with the common suffix—*scape* . . . indicate that these are not objectively given relations that look the same from every angle of vision but, rather, that they are deeply perspectival constructs, inflected by the historical, linguistic, and political situatedness of different sorts of actors: nation-states, multinationals, diasporic communities, as well as subnational groupings and movements. . . . Indeed, the individual actor is the last locus of this perspectival set of landscapes, for these landscapes are eventually navigated by agents who both experience and constitute larger formations, in part from their own sense of what these landscapes offer. (1996, 33)[23]

Emotional investment and even physical individuality are products of "spatial practice" that position people with respect to representational spaces, of which "conceptual space" tends to be the most impoverished ("localized" in the specific work of mathematicians and engineers). The *Critique of Judgment*, however, addresses the possibility of conflict and the conditions of concord or congruity between the representational

---

23. Lefebvre would contend that distinguishing kinds of spaces in this way contributes to the ideological fragmentation of experience through which some financial and conceptual abstractions deprive ordinary citizens of "orientation" in action as well as thought (1991, 91). Without denying the problems of power involved in fragmentation and abstraction, I question the necessity of presuming that all spaces are ultimately unifiable and homologous—which is one reason I have focused on *imagination* rather than space.

spaces of diverse persons. It makes explicit the role played by "spatial practice" in the ordering of representational space and vice versa.

In summary, Kant's transcendental reflection takes place within phenomenal experience but reveals that this experience is composed of elements and styles of thought that are not entirely continuous with one another, such as introspection, moral action, and scientific knowledge. The complexity of "space" in Kant's thought can be compared to the diverse spatialities unifying a historical or a practical, economic manifold in the studies of Foucault and Lefebvre.

The Quest for Unity

The problematic noumenon divides the realm of appearances to be analyzed by transcendental deliberation (mostly concepts and intuitions) from the (purely hypothetical) realm of things-in-themselves. Asserting the finitude of human understanding, that is to say, limiting its application to the realm of appearances (or nature), leaves room for freedom to operate somewhere outside or alongside nature, which Kant conceived mechanistically in the *Critique of Pure Reason*. Such freedom would be the object of another *type* of reason—self-legislating practical reason, which is incapable of altering nature's laws, but nevertheless brings about new states of affairs in the phenomenal world. But giving reason a second domain does not allow Kant to achieve the comprehensive systematization of philosophical problems he had initially hoped for.

Kant begins the first edition to the *Critique of Pure Reason* by boasting of the unity that his critical method will grant to reason. By beginning from the standpoint of human finitude, he can make a "complete specification" of "reason's questions" rather than having to plead the "incapacity of human reason": "Pure reason is so perfect a unity that, if its principle were insufficient for the solution of even a single one of all the questions assigned to reason by its own nature, then we might just as well throw the principle away; for then we could not fully rely on its being adequate to any of the remaining questions either" (Axii–xiii).[24] "The

---

24. Kant makes a similarly extreme statement in the *Prolegomena*: "In the domain of this faculty one must determine and settle either *all* or *nothing*" (PR 4:263). The meaning of reason's unity varies depending on the context of discussion; the disunity posed by the existence of antinomies differs from the disunity posed by reason's division into practical and theoretical employments; in the section just quoted from the *Prolegomena*, for example,

Architectonic of Pure Reason" contrasts a system of knowledge that develops piecemeal from one that originates in a whole, such as critical reason: "We start only from the point where the general root of our cognitive power divides and thrusts forth two stems, one of which is *reason*" (the other being sensibility) (A835/B863). Nonetheless, Kant was happy to have found room for practical reason within the world described by science (see the solution to the Third Antinomy), and this unity was immediately threatened by the division of reason into a theoretical and a practical employment.

Many readers share Kant's hope that these disparate spaces and points of view can be attributed to a common origin as well as unified in a common world. Heidegger, for example, suggested that the apparent dualisms of Kantian philosophy resulted from Kant's decision to subordinate imagination, originally the "hidden root" of thought's unified and unifying grasp on the objects of experience, to an understanding conceived more along the lines of a rational *moral agent* than a *sensitive recipient* of the phenomenological given. In *Kant and the Problem of Metaphysics*, Heidegger portrays Kant's difficulties as originating in his unwillingness to address the *ontological* difference between the being of objects received by intuition and the being of those who intuit and represent them; namely, human beings. Basing his argument on comments in "The Architectonic of Pure Reason" rather than the introductions to the first *Critique*, Heidegger contends that Kant's overall goal was not to identify transcendental presuppositions of legitimate judgments concerning objects but to "lay the ground for metaphysics"—that is, explore the limits within which human beings are capable, a priori, of anticipating the being of entities other than themselves. What is important about the Kantian subject is less the legitimacy of its judgments than a certain ontological knowledge or "preliminary understanding" of how its thought must conform to entities which appear to it incompletely and reluctantly.

A finite knower, Heidegger explains, receives and determines such entities through a synthesis that is neither simply aesthetic nor merely logical, but takes place over time. In the first edition of the *Critique of Pure Reason*, Kant assigned this synthesis to a faculty of *transcendental imagination* (88, 91). But in the second edition, in Heidegger's opinion,

---

Kant's goal is to isolate reason as an "organized body" whose individual functions must be reciprocally determined in relation to the whole. The constant among these variations is that reason is the style of human thought which strives for maximal unity.

he "shrank back" from this insight and defined imagination as only one among several components of an essentially *conceptual* synthesis of experience carried out by understanding. As a result, Kant identified thinking with judgment and gave logic an "incomparable priority over the aesthetic" (Heidegger 1990, 44, 115).

The consequences of his decision were twofold. On the one hand, according to Heidegger, Kant's pure reason was embarrassed by an irreducible dualism between two kinds of mental powers: (1) the (active) powers of understanding and reason (with imagination as their handmaid); and (2) the (receptive) powers of intuition and feeling. On the other hand, Kant was unable to account for the central role which the peculiar being of humanity must play in any understanding of metaphysical thinking. He was forced to regard the person's finite human being-in-the-world as an "anthropological" accident befalling a creature of imperfect theoretical reason, but pure moral reason (1990, 115–16). Paradoxically, this reorientation toward morality pushed Kant to conceive of the faculties, including imagination, in a *more* psychological, *more* anthropological way than he might otherwise have done.[25]

Other scholars have argued that far from "erring" when he refused to make imagination (or any other faculty) the "common root" of understanding and intuition, the asymmetry among Kant's faculties results necessarily from his insistence on the finitude of human perception and knowledge. Certainly the priority of logic in Kantian metaphysics and ethics is open to criticism. But Kant's reason is subject to interminable bifurcation because it is *finite, always* condemned to a partial view. It only generates positive knowledge and determinate experience to the extent that it *excludes* things-in-themselves behind the veil of the noumenon. The multiplicity of Kantian reason is, paradoxically, the *condition* for its unity, both real and anticipated.

According to Henrich, if the plurality of faculties is an irreducible aspect of our empirical self-understanding, then to seek the unifying root of these faculties would be to try and grasp the condition for the possibility of self-consciousness as an *object* of consciousness, which is impossible

---

25. Foucault and Heidegger share a commitment to antihumanism, but differ over the meaning of this commitment. Heidegger and Cassirer disputed the importance of ontology and imagination in Kant's work at Davos in 1929 (see Heidegger 1990, 171–85). In his review of Cassirer's *Philosophy of the Enlightenment*, Foucault situates the dispute in the context of German politics and sides with the latter ("Une histoire restée muette," in *DE* 1:545–49).

for a finite understanding (1994, 19–30).²⁶ Consequently, the unity of the faculties can only be a *projection* of the unity of transcendental apperception, a logical rather than metaphysical basis for the unity of self-consciousness. John Sallis argues that critique begins always "to one side" of any common root, emerging from within a division, because its goal is to elaborate the powers of reason beginning with a fundamental self-limitation (1987, 78). Pierre Kerszberg, finally, regards Kant's *exclusion* or *suspension* of things-in-themselves as a perpetually renewed commitment to finitude, an ethical act more foundational than his analysis of pure practical reason (1997, 18). For Kerszberg, the reflective delimitation of faculties and their domains must rest on the gamble of a "first reflection," a provisional "construction" in which critical reason proposes to analyze its own limits *as if* it were an organism or transparently intuitive understanding (1997, 32). I have argued that this "first reflection" is oriented by the problematic object hypothetically *posited* and simultaneously *excluded* by critical reason. Thus imagination is already implicated in reason's effort to reflectively distinguish among its own powers and representations—including, and perhaps especially, those of the imagination.²⁷

The *Critique of Judgment* suggests that the theoretical and practical domains are unified in the seeming purposiveness of natural forms (including humanity as natural form) and in the state of mind in which both styles of reason can set themselves goals. If "the concept of freedom is to actualize in the world of sense the purpose enjoined by its laws," as Kant believed, "there must after all be a basis *uniting* the supersensible that underlies nature and the supersensible that the concept of freedom contains practically, even though the concept of this basis does not reach cognition of it either theoretically or practically, and hence does not have a domain of its own" (*CJ* 5:176). Kant hopes that locating the peculiar object of a capacity to judge whose employment is neither purely cognitive nor

---

26. See Zammito 1992 for further discussion of pre-Kantian arguments in favor of a single cognitive faculty.

27. Gibbons (1994) concurs with Kerszberg in viewing Kant's transcendental philosophy as a "construction" through which reason imaginatively investigates its own capacities just as scientific inquiry investigates nature by presenting it with questions of our own design. Gibbons regards the critical project as one that seeks to build a "house of reason" in which each individual may recognize the operation of his or her powers of thought, and thus as a text oriented toward sustaining the communicability it helps to explain (1994, 5). This resulting image of Kantian philosophy as an *imaginative scaffolding* for the direction of human history (12–13) resembles the Hegelian/Marxist account of imagination presented with great eloquence by Elaine Scarry (1985).

purely moral will enable him to complete the work of transcendental deliberation by pointing the way to such a "supersensible."

Since this concept should bring together domains of theoretical reason, moral agency, and anthropological habit, it must be an *indeterminate* concept, a little like the hidden side of the canvas depicted in *Las Meninas*. In the *Critique of Pure Reason*, Kant used the distinction between phenomenal and noumenal realms to expose the error of treating the Transcendental Ideal (i.e., God, one of the Ideas of pure reason) as the ontological totality underlying determinate appearances, rather than as a logical correlate of our intuitive and conceptual grasp on appearances. In the "Critique of Teleological Judgment," Kant further contends that our experience of *actuality* will forever be severed from what we can *think* as mere *possibility*, because we can only access the world through the mediation of concepts (*CJ* 5:402–4). The object of the indeterminate concept governing reflective judgment—"the supersensible substrate of appearances"—is even more shadowy than the noumenon or Ideas of reason. If an object were to exist for this concept, our apprehension of nature could be reconciled with the possibility of effective moral action on nature's own terms. But such an object is no more than the corollary of a problematic concept.

Why then should we think any object corresponds to this concept? Kant believes reference to the supersensible is justified because it helps explain why critics do (and should) expect correct judgments of taste to be universally valid. A judgment of taste is based on a concept—the concept of a general basis of nature's subjective purposiveness for our power of judgment. But this concept is indeterminate and inadequate for cognition, leaving us unable to cognize and prove anything concerning the art object or beautiful form (still less the sublime). How, then, can it validate the critic or observer who demands universal assent for his or her judgment of beauty, just as if he were pronouncing on the species to which the beautiful form belongs? More perplexing, no *empirical* concept has been found for the particular to which this indeterminate concept supposedly applies—because each person's judgment is singular and directly accompanies his or her intuition. However, Kant argues that the fact we expect and can sometimes train people to agree on the beauty or ugliness of particulars they have never before encountered proves that something like a "supersensible substrate" must exist as condition for the possibility of this agreement (*CJ* 5:340). The pleasurable feeling associated with beauty or sublimity, and situating the viewer with respect to other people's feeling for what is most communicable and worthy of being communicated in natural and cultural experience, is a symptom of *some* object corresponding to the "supersensible."

What Kant is getting at here is the important insight that "reality testing" involves an appeal to other perceivers' sense of reality. Collective agreement on what is worth noticing, which precedes the assignment of a concept, may ultimately be more epistemologically significant than any individual person's judgment that this or that representation is "real."[28] Whether or not he needs the supersensible substrate to solve it, the problem is one that has haunted Kant since "Dreams of a Spirit-Seer"—how psychologically and physically individuated reasoners can inhabit a common world and recognize their fellows as moral and epistemological subjects. Kant presumes the universal applicability of the categories in the *Critique of Pure Reason* and explicitly mentions them in the *Prolegomena to Any Future Metaphysics* (PR 4:298), but never explicitly justifies them until the *Critique of Judgment*.

Now, the fact that aesthetic judgment involves an *indeterminate* concept permits Kant to acknowledge and resolve the differences of taste that inevitably do break out among viewers of an aesthetic phenomenon. It also enables him to acknowledge, albeit obliquely, the epistemological difficulties such disagreements might create. For we could hardly dispute over the necessary features of a beautiful form if every critic of beauty referred to a determinate concept with a clearly delimited range of application ("This painting is *Las Meninas*"). Nor could aesthetic judgment be considered judgment *at all* if every judgment involved the application of determinate concepts to particulars. Because aesthetic judgment refers to an indeterminate concept, however, the requirement that judgment refer to some concept can be reconciled with the *sociological fact* of quarrels over the feeling associated with aesthetic forms (CJ 5:338). In a move that may be familiar from structural anthropology, Kant's solution consists in

---

28. In the *Prolegomena*, Kant states that "all judgments of experience are empirical," involving the senses, but he adds that some *empirical* judgments are not yet judgments of experience, inasmuch as the object has not been schematized through categories that would allow the judgment to "be valid at all times for us and for everyone else" (PR 4:297–98). A judgment of *perception* is valid only for the individual subject/thinker. Kant implies that the categories of understanding dictate which representations will be involved in judgments of experience and judgments of perception. This seems sufficient to rule out phantasms but insufficient to guide judgment in ambiguous situations requiring interpretation and contextualization.

finding a parallel between the *gaps* in two ways of posing the same problem.[29] Thus the indeterminate concept enables Kant to resolve differences within the social and political domain as well as purely "experiential" disputes over the classification of representations with which nature confronts us.

Let us consider three peculiarities of an aesthetic judgment. First, the universality of an aesthetic judgment cannot be predicated of the form under consideration, for this form has not been recognized as an "example" of a class using a determinate concept. Rather, the judgment extends the predicate of beauty or sublimity "over the entire sphere of *judging persons*" (*CJ* 5:215). The harmony among faculties that Kant himself reflectively "discerns" within the upsurge of pure aesthetic pleasure eludes particular idiosyncrasies of individual preference and need:

> For if someone likes something and is conscious that he himself does so without any interest, then he cannot help judging that it must contain a basis for being liked [that holds] for everyone. He must believe that he is justified in requiring a similar liking from everyone because he cannot discover, underlying this liking, any private conditions, on which only he might be dependent, so that he must regard it as based on what he can presuppose in everyone else as well. (*CJ* 5:211)

Second, the universal communicability of aesthetic judgment cannot be grounded in a concept but allows us to anticipate the generation of such concepts:

> A presentation that, though singular and not compared with others, yet harmonizes with the conditions of the universality that is the business of the understanding in general, brings the cognitive powers into that proportioned attunement which we require for all cognition and which, therefore, we also consider valid for everyone who is so constituted as to judge by means of under-

---

29. In "The Structural Study of Myth," Lévi-Strauss argues that although myths may differ from one version to another and one culture to another, many contain the same self-contradictory elements or problems and can therefore be regarded as versions of the same myth: "Contradictory relationships are identical inasmuch as they are both self-contradictory in a similar way" (1963, 216–17).

standing and the senses in combination (in other words, for all human beings). (CJ 5:219)[30]

Finally, this pleasure, Kant asserts, is a pleasure in *reflection* rather than perception and

> must of necessity rest on the same conditions in everyone, because they are subjective conditions for the possibility of cognition as such, and because the proportion between these cognitive powers that is required for taste is also required for the sound and common understanding that we may presuppose in everyone. That is precisely why someone who judges with taste . . . is entitled to assume that his feeling is universally communicable, and this without any mediation by concepts. (CJ 5:292–93)

Concepts and judgments involved in everyday experience, as well as the more arcane judgments involved in specific fields of scientific knowledge, are communicable because they somehow reflect the harmonious relation among faculties manifest in a judgment of taste. The fact that we manage to communicate suggests to Kant that a "supersensible" does, indeed, unify the multiple spaces of social perception, scientific knowledge, moral action, and philosophical reflection, just as the excluded "obverse" to the canvas in *Las Meninas* brings incompatible spaces together into a single work of representation. Kant is perfectly aware that the universal assent a judgment of taste demands from others is "rejected often enough" (CJ 5:214). But even critics who disagree in their judgments do not believe such claims are impossible, merely that their opponents judge poorly. Since we are in fact capable of generating and communicating such concepts, he concludes that we have a transcendental reason for assuming the existence of a "common sense," and *"for assuming it without relying on psychological observations,* but as the necessary condition of the universal communicability of our cognition, which must be presupposed in any

---

30. Again, in section 21, Kant asserts that "cognitions and judgments, along with the conviction that accompanies them, must be universally communicable. For otherwise we could not attribute to them a harmony with the object, but they would one and all be a merely subjective play of the presentational powers, just as skepticism would have it. But if cognitions are to be communicated, then the mental state, i.e., the attunement of the cognitive powers that is required for cognition in general . . . must also be universally communicable" (CJ 5:238).

logic and any principle of cognitions that is not skeptical" (239, my emphasis).

But Kant's account of the *sensus communis* involves some important ambiguities. Specifically, it is not clear (1) whether communicability is only implicit in the individual consciousness but actualized in (presumably cognitive) communicative acts; (2) whether communicability is actualized for the individual consciousness as a result of a play of faculties occasioned by the beautiful presentation; (3) whether communicability is actualized following a judgment concerning the likely responses of other individuals to the same representation, or (4) whether communicability is actualized following a disidentification with one's own interests in the presence of the representation. In sections 9 and 20, Kant claims that the communicable state of mind is provoked by the beautiful representation but can only be communicated in some other (cognitive and presumably linguistic) medium (*CJ* 5:217). Thus it may inspire communication while differing from critic to critic. But in section 22 he proposes that the "objective necessity that everyone's feeling flow along with the particular feeling of each person" may only be a regulative demand of reason, suggesting that he previously *did* expect the qualitative mental state underlying a communicating concept or image (that is, the feeling) to be identical in each individual. This would have been a stronger and more surprising claim.

In the deduction of pure aesthetic judgments, Kant repeats that a judgment of beauty is grounded in a proportion of the mental powers (*CJ* 5:287) but adds that the universal rule for the power of judgment is not the pleasure but the "*universal validity of this pleasure,* perceived as connected in the mind with our mere judging of an object" which legitimates the universal claim on others' agreement (289). The *validity* of a pleasure (which may never give rise to actual cognition or communication) is not a pleasure, even a pleasure that results from a proportion between imagination and understanding—much less a pleasure in the anticipation of communication. While validity might indeed be one and the same for all critics, the forms of pleasure might vary from individual to individual, even if all empirical influence were excluded, much as the intuitions associated with an object of empirical perception may differ for viewers even when they are brought under one and the same concept.[31]

---

31. See Guyer 1997, chaps. 7–9, for a detailed account of inconsistencies in Kant's deduction.

To make matters more complex, Kant states that this pleasure follows upon a judgment comparable "not so much with the actual as . . . the merely possible judgments of others." The critic puts him- or herself "in the position of everyone else, merely by abstracting from the limitations that [may] happen to attach to our own judging" (*CJ* 5:294). Here, it seems to be pleasure not in the representation, but in one's ability to identify with the likely mental state of others or to disidentify with the interested aspects of one's own mental state. Certainly, it is pleasurable to be part of a shared way of perceiving or to feel sufficiently sure of one's views that one can experiment with setting them aside and adopting someone else's stance. When Kant refers to taste as an ability to "judge a priori the communicability of the feelings that . . . are connected with a given presentation," one also wonders what these feelings would be, since the *pleasure*, at any rate, followed *from* communicability or validity rather than preceding it (296). This is not even to mention the problem of whether and in what medium they might be communicated.

Artists and critics are unlikely to agree that pleasure in one's ability to identify with other spectators or to disidentify with our own standpoint has much to do with the basic content of aesthetic feeling. But Kant is not so much trying to understand a successful "work" as to identify something missing from his previous epistemology and necessary to its functioning, an interpersonal "sensus communis" rarely isolated from other elements of experience *except* in human responses to beauty and sublimity. Kant denies that feeling for the sublime is "initially produced by culture and then introduced to society by way of (say) mere convention. Rather, it has its foundation in human nature," specifically in the predisposition making moral action possible (*CJ* 5:265). But he also considers the possibility that the common sense may only be a regulative ideal, rather than a real condition for objective (interpersonally valid) judgments of experience, especially since empirical observation seems to reveal as much disagreement as agreement regarding the sort of forms that provoke disinterested pleasure or evoke a state of free, lawful harmony among understanding and imagination. And finally, he asks whether taste is truly "an original and natural ability," or whether it may be "only the idea of an ability yet to be acquired and [therefore] artificial, so that a judgment of taste with its requirement for universal assent is in fact only a demand of reason to produce such agreement in the way we sense?" (240). In other words, rather than producing a concept for a shadowy object, he may end by having to commission the *creation* of an object to go along with the concept required by his philosophy.

Let us therefore consider the problem from the anthropological rather than transcendental angle. On the one hand, the *sensus communis* refers to the "raw material" of the faculties whose play incites a communicable pleasure in us. This raw material justifies vehemence in quarrels over taste. In the case of the sublime, this raw material gives pure reason a dramatically *creative* power. On the other hand, the *sensus communis* refers to the "common sense" for what is universally pleasing, brought about through refinements of culture and necessary in a direct way for humanity's historical progress toward its moral purpose, even if it is only indirectly required for the moral development of individual humans.[32] "We must," he writes,

> take *sensus communis* to mean the idea of a sense *shared* [by all of us]; i.e., a power to judge that in reflecting takes account (a priori), in our thought, of everyone else's way of presenting [something], in order *as it were* to compare our own judgment with human reason in general and thus escape the illusion that arises from the ease of mistaking subjective and private conditions for objective ones, an illusion that would have a prejudicial influence on the judgment. (*CJ* 5:293)

In other words, the *sensus communis* is a *positive* way of reifying or thinking about detachment from our own empirical standpoint. Psychoanalysts might call this a release from narcissistic self-conceptions, which does not mean a loss of individuality. In this passage, however, the *sensus communis* is not *in fact* brought about by such a comparison, since the concepts required for a deliberate comparison and abstraction from one's personal situation and interests are lacking in aesthetic judgment (*CJ* 5:293, 295).[33] It only arises *as it were* through such comparison because

---

32. Kant distinguishes between an intellectual interest in natural beauty, which gives rise to "voluptuousness for the mind in a train of thought that he can never fully unravel," an intellectual interest in beauty as evidence for the possible harmony between our moral vocation and the world's structure, and an "empirical" interest in taste as a method of socialization (*CJ* 5:298–302; see also 356 and 430–34 for other discussions of the development of culture and the conditions for moral action). The disinterested harmony between form and human faculties embodied in a beautiful object, moreover, is analogous in many ways to the relationship between the faculties and the maxim determining a moral action (351–56).

33. This process, one could argue, is the political equivalent of the sort of comparison undertaken in the act of transcendental deliberation described in "The Amphiboly of Concepts of Reflection" in the *Critique of Pure Reason*. It forms the basis for Arendt's understanding

the possibility of such comparison and abstraction *presupposes* a common world of intersecting situations. Even if the formation of a broadened or common human understanding does not require us to actually investigate the opinions of others, the practice implies faith in our "community" with others from whom we may later be differentiated through acts of empirical observation. Thus "taste can be called a *sensus communis* more legitimately than can sound understanding, and . . . the aesthetic power of judgment deserves to be called a shared sense more than does the intellectual one. . . . Hence taste is our ability to judge a priori the communicability of the feelings that (without mediation by a concept) are connected with a given presentation" (*CJ* 5:295). It is because the communicability of a feeling of pleasure in the presence of the beautiful or sublime makes no reference to concepts, moreover, that we are prone to quarrel over whether a judgment of taste has successfully abstracted from the critic's personal, empirical preferences. No dialectic is possible concerning judgments of taste, for they cannot be contradictory with respect to a determinately understood *sensus communis*, but only with respect to the method by which we *criticize* others' judgments as unacceptably partial.

But taste can also be cultivated, and Kant argues that training individuals to recognize the forms that are likely to give universal pleasure contributes to the development of human *sociability* in an important way. Learned appreciation of beauty—refinement—*teaches* us to communicate; even if we communicate *about* what is communicable, we do so by virtue of a trained sense *for* the communicable in our very phrasing: "For we judge someone refined if he has the inclination and the skill to communicate his pleasure to others, and if he is not satisfied with an object unless he can feel his liking for it in community with others. Moreover, a concern for universal communicability is something that everyone expects and demands from everyone else on the basis, as it were, of an original contract dictated by our very humanity" (*CJ* 5:297). In "On Methodology Concerning Taste," Kant elaborates upon this social contractarian theme by suggesting that a people attempting to form a "body politic" (my phrase) "had to begin by discovering the art of reciprocal communication of ideas between its most educated and its cruder segments, . . . finding in this way that mean between higher culture and an

---

of judgment as emerging from a deliberate effort to enter imaginatively into the perspective or standpoint of others. But in fact, Kant's desire is as much to explain how we become the sort of subjects capable of such comparison as to recommend that we cultivate this kind of "enlarged mentality."

undemanding nature constituting the *right standard, unstatable in any universal rules, even for taste, which is the universal human sense"* (*CJ* 5:355–56, my emphasis).[34] Abstracting from everything particular but the *feeling* they share or imagine themselves to share with others grounds the social contract in something like a "web of ignorance" rather than the "veil of ignorance" dividing the participants in Rawls's original position. This web of feeling must be created if it does not already exist, and in Kant's critical philosophy coexists with the spaces of moral action, scientific knowledge, and introspection or private fantasy.

Post-Kantian thinkers, therefore, have tried to resolve the dualisms plaguing Kant's desire that reason form a perfect unity by appealing to imagination as a "common root" or by appealing to a unifying ethical strategy such as critique. Kant appears to resolve the problem by appealing to an "indeterminate concept" of supersensible humanity whose effects are felt in aesthetic judgment and its communicable state of mind, though he also suggests that the object of this indeterminate concept must be *constructed* historically through artistic culture.

Discursivity and Materiality

The previous section described the multiplicity of domains resulting from Kant's critical image of reason. This section will look at the same problem from a different standpoint: Kant's commitment to the *discursivity* of the understanding.

Kant's understanding is imaginative; that is, imagination is essential if humans are going to "understand" their environment even in the minimal sense of recognizing objects, much less communicating their recognition with others. The same finitude that makes the understanding

---

34. Kant argues that the human ability to set purposes (and to produce culture) must be endowed by nature, even though natural human predispositions to warfare tend to obstruct these purposes. The first aspect of culture is the *skill* to achieve purposes chosen by the will; the other is *discipline*, which enables the human creature to abstract from natural desires. Unfortunately, the technical division of labor required for a high level of skill in a society leads to inequalities and war, so that the promotion of culture also demands the implementation of progressively more international civil constitutions. Over time, however, Kant believes that inequalities lead to the improvement of conditions for the worst off, and suggests that even warfare may serve the purpose of increasing a society's level of skill. As for discipline, the fine arts play an important role in training someone to act on the basis of communicable rather than sensory, private pleasures, "and so prepare him for a sovereignty in which reason alone is to dominate" (*CJ* 5:429–34).

imaginative renders it "discursive"—unable to represent particulars except through reference to a universal, or to consider several aspects of a particular without successively subsuming it under different universals (*CJ* 5:404; Heidegger 1990, 20). It also ensures that objects synthesized by the understanding or representations linked in judgments will be "material"—that is, unable to cohere unless other elements of the conceptual, perceptual, and imaginative field are excluded as irrelevant or as mere "context." Spoken and written language are not the *source* of the understanding's discursivity and the materiality of whatever it handles, but they also share in it and express it: as Kant mentions in the *Anthropology*, topics of discussion must be taken up one after the other and not roam too widely from their central topic if a conversation is to be successful (*AN* 7:281). This restriction on the flow and scope of imagination affects the *subject* of understanding, as demonstrated by Kant's account of beauty and sublimity in the *Critique of Judgment* and by the mental hygiene he promotes in *Conflict of the Faculties* (which also regulates the self-affection of temporal synthesis). Pure intuitions like space and time can be broken into ever smaller units; pure concepts like mathematical number can be recombined in practically infinite ways; empirical concepts can be conjoined in a variety of discourses that allow for translation but not transparency; emotions and memories alter in quality under the introspective gaze; and phenomenal objects, of course, resist or grate or crumble when struck.

Thus subject and object are equally "material" insofar as they are affected by the work of finite understanding, which will be important for our later efforts to understand what it means for the human body to be shaped through discourse without losing materiality. The inevitability of *scale* and the persistence of *resemblance* in representation are two ways of approaching this materiality, both of which can be found in Kant's *Critique of Pure Reason* and *Critique of Judgment*. We will also discuss them in the context of Kant's reaction to the linguistic research of his own time and to the theory of hylozoism (generation of living beings from inert matter), as well as in Kant's own *non-naturalist* explanation of aesthetic judgment and human creativity.

Imagination has an ambiguous status in the history of modern philosophy. For Descartes, imagination was an activity of the *res cogitans*, but was also bound up with the sensory and motor activity of the body. It exhibited not only those aspects of sense perception that go beyond the mathematically and conceptually determinable categories of pure thought but the objects of pure thought themselves in sensory terms. Indeed, in his opinion, without an extended, material body, imagination would neither be necessary nor possible. In an important sense, all rationalists of the early modern period believed that humans were condemned to apprehend the world in a confused and largely imaginative manner, although pure thought could still grasp essences and certain scientific relationships between the essences of material things. But they disagreed over whether imagination was an essentially *positive* protocognitive ability of a material body or the defective activity (illusion) of an essentially immaterial mind (Ricken 1994, 49–50).

For one thing, the status of imagination was linked to the status of language. Language is the *sensible* form in which individual minds express pure thought to one another—and inevitably betray its simultaneity and conceptual unity. Like debates over the relative merits of rhetoric in ancient and modern European languages, the critique of language (in Spinoza's interpretation of the Bible, for example, or Leibniz's project for an artificial language) resulted from scholarly efforts to better *communicate* the objects of thought (representations or ideas). The demands of communication differed if such objects were believed to reflect a metaphysically absolute reality in itself or to arise in the passionate, temporal course of human affairs. Empiricists (such as Hobbes and Locke) believed the elements of language were drawn directly from representations with a sensory origin, then elaborated and clarified in the course of reflection. Condillac and Rousseau argued that sensory representations shaped human sociability but also threatened to fill the imagination with "empty" signs, such as demonstrations of vanity, lacking any concrete reference (Ricken 1994, 82–83). In his comments on genius and poetry, Kant recognized an association between the capacity for thought and the development of language; he even believed that the deaf, who could not hear language, lacked genuine reason (*AN* 7:155). But he insisted that concepts of the understanding were irreducible to words and prided himself for having deduced the categories from *logical relations* rather than, like Aristotle, a rhapsody of things said (*PR* 4:323).

In the context of debates between these schools, the origin of language became an object of intense inquiry. The metaphysical understanding of the cosmos as a nonspatial, intemporal whole, represented more or less accurately by language, rested on the biblical myth of Adam's receiving language from God along with mastery of the world. The naturalist, empiricist perspective, however, considered language a human response to human needs. For such thinkers, language did not distort "pure thought" but permitted it to develop in the first place, through reflection on simpler signs. Ideas about the divine or natural origin of language were closely tied to opinions regarding the legitimacy of social hierarchy and the human capacity or incapacity for self-government (Ricken 1994, 138–40).

Pietist and anthropological strands of early German romanticism were just as unable to conceive of human thought in the absence of language. But instead of associating language with the finitude of human existence, as did the rationalists, they grounded it either in a divine or a historical/national origin. Thus for Hamann, the phenomenal world *was* God's language, not a system of hierarchically determined essences that could only be imperfectly conveyed through language. Kant's irritation at these thinkers' claims to possess direct insight regarding the absolute was matched by his frustration with their unwillingness to explore the *positive, generative* potential of finite reason's specific powers—such as sensibility and imagination.

Kant compares the discursive understanding to a hypothetical "intuitive" understanding. The principle that nothing can be an object for me without being subjected to the unity of apperception "is not one for every possible understanding as such, but is a principle only for that [kind of] understanding through whose pure apperception, in the presentation *I think*, nothing manifold whatever is yet given" (*CPR* B138). By contrast, we can imagine another kind of understanding that would grasp the full complexity and existence of the manifold of intuition in every act of self-consciousness. Kant admits that the very notion of such an alternative understanding may be incoherent. The unity of apperception is so central to what it *means* for humans to apprehend an object, that "our understanding cannot even frame the slightest concept of a different possible understanding—whether of an understanding that of itself would intuit; or of an understanding that would indeed have lying at its basis a sensible intuition, yet one of a different kind of that in space and time" (B139). Because we are finite, the realm of human experience is limited in advance by forms of intuition and by the unity imposed on concepts by transcendental apperception (ultimately under the guidance of reason). The complete comprehension and determination of psychological or cosmological appearances (or the justification of their ontological necessity) is frustrated by the fact that such appearances are never *given* as a whole but must be synthesized over time. As the Transcendental Dialectic of the

Critique of Pure Reason demonstrates, the task of completely determining inner or outer nature as a whole is infinite—a task provoked as much by the limitations of our *concepts* as by the wealth of *material* presented to our thought by experience (A508–27 / B536–55). "Discursivity" is this intrinsically "successive" quality of human thought: the fact that not every aspect of a phenomenon, idea, or problem can be thought at once but takes *time* to unfold.

In "On the Amphiboly of Concepts of Reflection," Kant criticizes Leibniz for failing to distinguish between the sensibly, spatiotemporally distinct objects of experience and the logical universals or singularizing essences of pure thought. Leibniz would contend that God is capable of knowing the differences between individual drops of water which *result* in their apparent spatial difference for human perception. The difference between two drops of water is a *real* and not a merely "apparent" difference, however, when one accepts that knowledge is always human knowledge, "knowledge of appearances." But this is not to say that Kant's thought lacks a place for imagination; namely, applying concepts to spatiotemporally distinct particulars of intuition, without which regular knowledge of nature would be impossible. In the *Critique of Pure Reason*, Kant defines imagination as "the power of presenting an object in intuition even *without the object's being present*" (CPR B151). Through imagination, concepts of the understanding are used to connect particular aspects of the sensory manifold with other remembered or possible experiences they resemble. As an element of intuition, imagination is passive and belongs to sensibility, but inasmuch as it *anticipates* or *recalls* absent objects, "the synthesis of imagination is an exercise of spontaneity, which is determinative, rather than merely determinable, as is sense. . . . This synthesis is an action of the understanding upon sensibility, and is the understanding's first application (and at the same time the basis of all its other applications) to objects of the intuition that is possible for us" (B152).

The *schematism* is an employment of the imagination that correlates categories or *pure* concepts of the understanding, through which objects in general are logically determined, with temporal relationships holding between elements of the intuited manifold. Empirical concepts are like schematic images of sensory objects. But the transcendental schematism differs in being not an "image" but the set of pure forms of time (inner sense) through which any object can be given in successive and interactive order. "Hence a schema is, properly speaking, only the phenomenon of an object, or the sensible concept of an object, in harmony with the category" (A146/B186). Through the schemata of sensibility, the categories of the understanding (determinations of the purely logical unity of transcendental apperception), can be applied to sensible objects. This work of the imagination is genuine *activity*, not just a symptom of human ignorance.

However, the schemata also *restrict* concepts from applying to presentations that do not conform to the pure intuitions of space and time (*CPR* A147/B186). Concepts without intuitions are limited to a logical employment; and "leftover" presentations, which do not cohere with the manifold as a whole, like the pieces of paper which do not fit in any of the established office files, are treated as imaginative in the everyday sense—as "fictions" or "dreams," and left in the individual office worker's private space (A156–58 / B195–97, B246–47). But in a given manifold, only certain wholes "leap to the eye" and are recognized in concepts, and these concepts limit the range of further concepts that will be applied to the manifold. If I believe that I am dealing with a technical problem in medicine, I will not consider the religious meaning of illness or the epistemological "subject" of medical knowledge; my approach will be limited by the concepts I have already brought to the problem—although of course I can choose to expand the range of approaches, which will take time, demonstrate the discursivity of human understanding, and probably irritate the research supervisor. In relation to each of these recognizable intuited forms (and the conceptual forms through which they are recognized) corresponds a "matter" implying that they could be apprehended "otherwise," from a different perspective.

In the "Amphiboly," Kant argues that the concepts of "matter" and "form" (*Materie* and *Form*) "are two concepts on which all other reflection is based, so very inseparably are they linked with any use of understanding. Matter signifies the determinable as such; form signifies its determination (both in the transcendental meanings of these terms, where we abstract from all difference in what is given and from the way in which it is determined)" (*CPR* A266/B322). In other words, matter does not just refer to iron, flesh, or silicon—but it refers to them insofar as they have been *determined* by some form (and could therefore be determined in some other way).[35] Nothing material is *forever* determined. No material

---

35. My account of materiality in this essay is strongly indebted to debates concerning "prime matter" in Aristotle's *Metaphysics*, *Physics*, and *On Generation and Corruption*. Historically, the tendency has been to understand "prime matter" as a metaphysical "ur-matter" in which all qualities are indifferently present (the view taken by Descartes and Locke); however, more recent commentators have related Aristotle's comment that *hyle* is *analogically*

can be found, moreover, without *some* relation to forms of thought and to forms that seem independent of or resistant to thought.

As we have already seen, Kant's goal in this section is to argue that Leibnizian rationalism commits a fatal error in believing that judgments concerning concepts have the same implications as judgments concerning objects of experience. Just because the single concept "rain" applies equally to all drops does not mean that a storm involves only one "essential" drop of rain. "Transcendental deliberation" is required to determine whether the objects of a possible judgment are merely *thought* or sensibly *experienced* (CPR A262–63 / B319); sensible raindrops differ in their spatial relationship to the subject, and this means that Kant regards their "matter" (along with the "matter" of the air between them") as negligible to their being recognized as "raindrops." One might, however, put raindrops under the microscope in order to identify their chemical composition or impurities, "see through" a veil of drops in focusing on a rainbow (A45–46 / B62–63)—or gather the water in a barrel if one's operational concept is "cooking" rather than science or the beautiful.

Water is material because it can turn into steam or run a mill; the concept "water" is "material" because it can be compared to "earth," "air," and "fire" in the ancient theory of elements, analyzed into modern components like "hydrogen" and "oxygen," or used as a symbol of purity in a baptismal ritual. Leibniz cannot rightly conflate objects of thought with objects of sensibility because thought, being *finite*, depends on a limiting *form* for its sense—just as intuition cannot cohere without the limitations imposed by space and time. But every instance of form, by definition, can be further investigated or synthesized in relation to *other* instances of form, ad infinitum. Form thus carries with it an inescapable potential for self-limitation and self-expansion, a potential associated with but never exhausted in "matter." In fact, it seems that strictly speaking Kant must deny that even *form* (a reflective concept, after all) is any "thing-in-itself" (a kind of metaphysical matter) and conceive it as the possibility of further determining *appearances* through the understanding.[36]

---

defined in relation to any given form (*Physics* 2.2.194b9) to the indefinite potential for further discovery of form within any given "proximate matter." See Lear 1988, 21, 39, and Polis 1991, 240, for discussion and the latter for a historical overview of the debate.

36. In his commentary on the *Anthropology*, Foucault suggests that anthropology is the study in which Kant confronts the real relations between *thought* (i.e., reason and understanding) and *inner sense* (i.e., the materiality or discursivity of understanding). Inner sense is not *physiological*, but the a priori "insofar as it is temporal" and cannot be separated from reflection on language or communication (1961, 58, 89–91).

Kant uses "form" and "matter" in this way several times. Sensation is "matter" for intuition or sensible cognition because it allows intuition to be further determined in grasping the object. Concepts are "logical matter" for judgment (CPR A20/B34, A50/B74, A266/B322). In the A edition, Kant proposes that "matter does not signify a kind of substance that is so very distinct and heterogenous from the object of inner sense (the soul). It signifies, rather, the heterogeneity of the appearances of objects (with which we are unacquainted as they are in themselves) whose presentations we call outer, by comparison with those we class with inner sense" (A385). This formulation, which may have led readers to suspect Kant of Berkeleyan idealism, was abandoned in the B edition. But it gives further support to the idea that Kant believed that matter is "analogical" rather than "prime." Any logical determination of a real intuition will always leave "more" intuition to be determined, and every intuited exemplar of a concept will always permit further conceptual determination. This indicates that the very idea of form involves *open-endedness* as well as (temporary) restriction on variation.[37] For example, in his explanation of how reason can employ the cosmological principle regulatively, Kant comments that "every experience is enclosed (in accordance with the given intuition) within its bounds," but also that "no empirical boundary" must count as absolute (CPR A509/B537). In discussing the regulative use of Ideas more generally, Kant notes that reason commands that "under every species we encounter [the understanding] must seek subspecies, and for every difference must seek smaller differences. For if there were no *lower* concepts, then there would also be no *higher* ones" (B684). While each smaller difference is "matter" for the form that determines it, this open-endedness is the materiality *of form* itself.

So long as we accept that human understanding cannot capture or convey things-in-themselves, we must agree that the mind's efforts to form and find form in the world will be mediated, incomplete, and unstable. Judgments divide, combine, and subordinate those partial forms. Put in everyday terms, the distance from which one perceives, the aspect with which one approaches, or the medium through which one apprehends a form intimately affects what we see, not just *how* we see it. Anthropologists and historians of science have noted the effect of perceptual and imaginative scope on our ability to achieve descriptions and explanations that are adequately complex. Different kinds of comparisons and causal analyses are possible at different levels or "scales" of apprehension:

---

37. See Rehberg and Jones 2000 and Focillon 1989.

> The one simple constant that holds across these scales—whether one is dealing with entire regions or tiny populations, with a complex model of interrelated variables or analysis of a single work process—is that very capacity to differentiate. The intensity of the perception of similarity and difference plays an *equally* significant part in the anthropologists' account whatever the scale. It also appears to play an equally significant part in the actors' orientations. (Strathern 1991, xxi)[38]

At first glance, the "matter" of a tree's form might seem to be cellulose. But the "matter" also reflects our distance from the tree, our knowledge of botany, or the interests that cause us to ask about its species or age, such as a belief in natural conservation or a quest for timber profits. This means that forming or seeing form establishes an intimate relation of resemblance between subject and object. When this resemblance is taken for granted, subject and object can seem more rigid, necessary, or ahistorical than they otherwise might. Thomas Kuhn argues that science textbooks, for example, understate the role of present-day paradigms in the interpretation and selection of historical material, thereby making the path of scientific inquiry seem more fated and self-evident than it actually was (1996, 139–41). Heidegger makes special note of this property of the understanding when he states that the transcendental imagination (which he refuses to treat as an aspect of understanding) imitates the being it is prepared to apprehend and limits or expands itself appropriately (1990, 62). Above and beyond the indifferent "matter" attributed to any recognized form, "materiality" also refers to this resemblance between perceiver and perceived and this potential for transforming perspectives intrinsic to the apprehension of form.

In the *Critique of Pure Reason*, Kant noted that the schematism of imagination had the dual function of restricting application of concepts and realizing them in specific objects. The *Critique of Judgment* goes further in explaining how understanding and reason would be unable to find lawfulness in nature and communicate its findings without this relation between restriction and realization. "All form of objects of the senses," objects of outer as well as (indirectly) inner sense, "is either *shape* [*Ge-*

---

38. See also Lévi-Strauss 1968, 22–29, for a discussion of how scale defines the artist's or bricoleur's relation to his or her work, materials, and audience or user. Henri Lefebvre also talks about the multiple levels of spatial perception and production as resembling a "millefeuille" pastry (1991, 86–88).

stalt] or *play* [*Spiel*]" (*CJ* 5:225). Natural forms are commensurate with our judgment when "their diversity and unity allow them to serve to invigorate and entertain our mental powers . . . and hence are called *beautiful* forms" (359). Makkreel therefore suggests that aesthetic form is not about the *object's* form but about an *abstract relation* between two aspects of the faculty of imagination: its unifying and schematizing functions (1990, 59–61; see also Caygill 1989, 249–51). Rather than being "a differentiating form which would serve to specify an object cognitively," instantiated in the matter of some real object, aesthetic form would be an *internal* structure of the imagination that allows it to judge purposiveness directly. In other words, it would already be a mode of *reflection* in the presence of a sensory particular that enabled one to review the forms and purposes, the scales and resemblances, with which it could be associated.

Beauty and sublimity are two instances of this reflective mode—the first emphasizing the potential for *play* in the presence of the formed, the latter drawing attention to the need for *figure* in the presence of the formless. The kaleidoscopic play of forms that emerge when focusing a microscope or camera, especially when looking at system that "makes sense" at different levels of detail but boggles the understanding during the intermediate stages, offers a simple analogy for this process. A book, a symphony, a life have form at different moments and are experienced as overwhelming or boring during the intervals when no form emerges or when the possibilities are too numerous and arbitrary to allow for decisive action.

In the experience of the beautiful, imagination and understanding apprehend an aesthetic form without settling on a particular purpose or concept through which it must be determined as an "example of X." In some cases (free beauty), this experience arises from an encounter with natural forms whose purposiveness could never have been anticipated by humans; in others (accessory beauty), the experience arises from an encounter with man-made forms whose purpose is recognizable but whose function does not fully determine their design, or with a natural form that has become an element of everyday human purposiveness, like the horse (*CJ* 5:229–30). The pleasure that provokes a judgment of beauty "is the quickening of the two powers (imagination and understanding) to an activity that is indeterminate but, as a result of the prompting of the given presentation, nonetheless accordant; the activity required for cognition in general" (219). This pleasure does not arise from any particular bodily need or moral esteem for the existence of the form in question. Kant's original introduction to the *Critique of Judgment* describes it as "a mental state in which a presentation is in harmony with itself [and] which is the basis . . . for merely preserving this state itself," although he acknowledges that a viewer might be tempted, for reasons of physical satisfaction or morality, to "produce" "the object of this presentation" (*CJ* 20:230').

In the experience of the sublime, by contrast, the subject confronts the potentially *formless* aspect of some phenomenon, again either natural or human-made. This encounter with the formless forces her to become aware of the viewer's role in shaping and limiting the given to produce sensible experience. An experience of the sublime does not result, like beauty, directly from an encounter with a pleasing sensible form. It is a *feeling* produced when someone encounters a phenomenon, even a formless object, that appears to be boundless at the same time that it appears to present or evoke a *totality*—which, of course, implies boundaries of some kind. (*CJ* 5:244).[39] This feeling registers conflict rather than harmony between powers of thought; as well as analogizing between an impressive natural or architectural phenomenon and the scope of reason's cognitive and moral abilities. When the enormity or power of a "sublime" presentation suggests itself as an imaginative representation for Ideas of pure reason (which represent the anticipated totality), reason politely declares it inadequate. This internal relationship between two aspects of imagination differs from the one encountered in aesthetic form, but it is *felt* rather than cognized, just like beauty (*CJ* 5:269).[40]

Jacob Rogozinski argues that in the sublime this restricting self-relation is experienced as violence, as a loss. But this loss enables theoretical and practical reason to pursue their regulative and creative roles and encourages the subject to identify with reason rather than imagination (1996, 132). By demanding that imagination present the unpresentable, "put up or shut up," reason "does violence" to imagination. But it also reveals that imagination had limits all along, insofar as it had to schema-

39. This does not mean *only* formless phenomena can provoke sublime experience (although properly speaking only a state of mind can be sublime, Kant's reference to St. Peter's and to the Great Pyramid indicate that even crafted objects, if they are sufficiently imposing, can provoke the viewer to confront the limits of her own ability to impose form upon the given (*CJ* 5:252). However, objects with recognizable purposes and contours can only provoke a feeling of sublimity if we *abstract* from these purposes; in order to regard the sky, the ocean, or the human form as sublime, therefore, we must be able to view them "as poets do, merely in terms of what manifests itself to the eye" (*CJ* 5:270; see de Man 1984 for analysis).

40. Scholars disagree on whether this feeling is only comparable to the feeling of respect occasioned by the moral law (see, for example, Lyotard 1994) or constitutes the sensible experience of the law's freeing power in and of itself (for example, Rogozinski 1996).

tize the enormous complexity and scope of possible forms according to a single form of time. By revealing the limits of imagination, the sublime shows that the unity of "experience" only arose through a *selection* from possible forms of imaginative appresentation.[41] Thus in some respects this loss is a restoration of perspective, however brief. In other words, lest we forget, there is more to life than experience, and we are responsible for the shape of experience as well as for acting therein. Broadening the mind's relationship to time and totality beyond the limits of schematizing imagination is *purposive for its (moral) vocation,* just as the invitation to play with beautiful form was purposive for its (theoretical) vocation. Together, the two poles of this tension make humans into creatures that Kant hopes will be capable of moral autonomy in a physically limiting world.

The *Critique of Judgment* is alert to misinterpretations of the sublime that might seem to grant the witness special insight into the unpresentable unconditioned. Kant is anxious on this point, for it *is*, to a certain extent, in sublime feeling that the series of natural conditions determining the human being coincide with those determining him or her as free (*CJ* 5:272). Kant takes pains to distinguish the pure conviction of the moral law which is "negatively" presented by the failure of imagination in the sublime from the enthusiast's conviction of knowledge through "imagination" alone. "This pure, elevating, and merely negative exhibition of morality," he writes, "involves no danger of *fanaticism,* which is *the delusion [Wahn] of wanting to SEE something beyond all bounds of sensibility,* i.e., of dreaming according to principles (raving with reason)" (275). Claims regarding the ultimate purposes of organic nature, like claims that all faculties of thought can be reduced to one "common root," constitute just such problematic claims of insight into the unconditioned, which Kant's Transcendental Dialectic demonstrated inaccessible to finite reason. In terms of our earlier analogy with the painting *Las Meninas,* Kant wants his readers to be aware that even if the limitations of pictorial representation force viewers to confront properly sublime questions regarding the purpose and limits of gallery art, representation, and historical continuity, this does not give them the right to imagine some

---

41. A significant treatment of the violence imagination does to its own capacity for *play* in imposing *shape* is given, albeit implicitly, in the essay "Ousia and Gramme" (in Derrida 1982). There, as in Rogozinski's discussion, the question concerns the relationship between our traditional understanding of temporality on the basis of the present (analogous to the movement of points along a line) and the possibility of understanding temporality *other* than on the model of the presence of points, lines, or indeed any spatial figure held as a representation within the self-present interiority of consciousness.

particular content on the other side of the canvas whose back is depicted *in* the painting.

Clearly, as indicated by Kant's complaints about fanaticism and dogmatism in the Transcendental Dialectic and "Critique of Teleological Judgment," finite reason makes errors concerning the scope of the understanding's employment. While he affirms that human understanding is dependent on sensible passivity for its field of application and that imagination makes a positive contribution to knowledge, Kant also fears that imagination may piggyback on the aspirations of pure reason and invent ersatz "intuitions" for its Ideas. In the *Critique of Judgment*, Kant suggests that poetic imagination is capable of giving sensible form to metaphysical and moral ideals, producing analogues to the Ideas of reason. Still, in the sublime it is incapable of meeting reason's demand for a direct presentation of the unconditioned (*CJ* 5:314). The purpose of critique is to hold reason strictly to thinking, despite the imagination's desire to generate analogues for Ideas, lest reason cease to unify empirical knowledge, bypass the slow progress of a discursive understanding, and speculate directly on things-in-themselves. In other words, critique insists on the *materiality* of the cognitive powers, in the special sense of "open-endedness" that I have just described.

This complex self-relation comes to the fore in Kant's confrontation with hylozoism. Although he once flirted with the idea of "living force" occupying the same space as physical matter and accounting for real interaction and resistance between bodies, he was deeply disturbed by the sensualists' and romantics' attribution of living, creative, or reflective force to "mere matter." Epicurus and Spinoza, he contended, reduced all mental processes to the blind and accidental congruence of matter (*CJ* 5:393–94). To credit matter with a purposiveness that can only be found in human action is to undercut the very possibility of reason's reflection—or action—on itself. "We cannot even think of living matter . . . as possible. The [very] concept of it involves a contradiction, since the essential character of matter is lifelessness, [in Latin] *inertia*" (394). Kant recognized that the temptation of hylozoism arises from reason's own desire to discover ever more general principles of structure and descent in the natural world:

> This analogy among [forms] reinforces our suspicion that they are actually akin, produced by a common original mother. For the different animal genera approach one another gradually: from the

> genus where the principle of purposes seems to be borne out most, namely, man, all the way down to the polyp, and from it even to mosses and lichens and finally to the lowest stage of nature discernible to us, crude matter. From this matter, and its forces governed by mechanical laws (like those it follows in crystal formations), seems to stem all the technic that nature displays in organized beings and that we find so far beyond our grasp that we believe that we have to think a different principle [to account] for it. (*CJ* 5:418–19)

But the temptation must be resisted. Not only is it contradictory to claim that matter (defined through its lack of purpose) can be purposive. To claim insight into the purposiveness of nature *in itself* is to transgress reason's boundaries (all the more since we have no empirical experience of organized beings emerging directly from "crude, unorganized matter"). More fundamentally, however, there is *no matter* from which life can emerge apart from the indeterminacy and open-endedness of form. Matter is the "determinable as such," even in thought—not a kind of *stuff*. Kant rejects the hylozoist tendencies of sensualists like Herder along with their efforts to reduce the various powers of thought to fundamental forces, especially those, like genius, thought to inhere in language (Zammito 1992, 210–12). Concepts like force and purpose have no application outside the realm of human experience and intention, and even divine creation is beyond the comprehension of finite minds.

The Virtues of Communicability

Genius is a crucial term by which Kant's "enthusiastic" naturalist and religious opponents explain the relationship between speaking, artistically creative humans and nature itself. Although gifted individuals may have contributed to human knowledge and culture, Kant protests in the *Anthropology* that "one type of them, called *men of genius* [*Geniemänner*] (they are better called apes of genius), have forced their way in under this sign-board which bears the language 'minds extraordinarily favored by Nature,' declaring that difficult study and research are dilettantish and that they have snatched the spirit of all science [*den Geist aller Wissenschaft*] in one grasp" (*AN* 7:226). Such *Schwärmer* have denied that discursivity is a *productive limitation* and identified their own ideas directly with the spirit of nature or the spirit of language.

Kant, by contrast, suggests that we should understand *genius* as the ability to create forms in which the life-enhancing tension between shape and play is communicable. Genius produces works of art that have the originality and noncognitive appeal, but also the indefinite complexity, of nature; Kant defines it as the "innate mental predisposition (*ingenium*) *through which* nature gives the rule to art" (*CJ* 5:307). On the one hand, it involves "a *talent* for producing something for which no determinate rule can be given," namely, *aesthetic ideas* which suggest more concepts than could ever be unified by a discursive understanding, just as *ideas of reason* correspond to objects that could never be grasped in a single finite intuition (308, 314, 316). On the other hand, it involves *spirit* (*Geist*); the ability to *express* these ideas "that enables us to communicate to others the mental attunement that those ideas produce," as if these moods were associated with (necessarily communicable) concepts (317). Although genius does not directly imitate the works of an artistic or poetic tradition, its products are *exemplary* for the continuation of tradition by others. In order to communicate in this exemplary manner, the creative outpouring of genius must be tempered with discipline and taste. In other words, genius makes *discursivity* and *materiality* purposive for human understanding. Otherwise, the artist will make false claims to have grasped things-in-themselves through an idea in which no viewer can find order.

In genius, imagination aspires to communicability, but the poet/artist also recognizes that communicability (like schematization) is a *limitation* on imagination. Humanity itself requires us to demand "a concern for universal communication [*allgemeine Mitteilung*]" from one another (*CJ* 5:297). Kant analogizes this requirement to a contract, but this almost seems to be a contract promising "to be human" rather than a contract between those who are already human. Here we see the resurgent question of whether a *sensus communis* already exists or has yet to be created.

In the *Critique of Judgment*, one of Kant's more successful descriptions of communicability is associated with music (*Tonkunst*): "Its charm, so generally [*allgemein*] communicable, seems to rest on this: Every linguistic expression has in its context a tone appropriate to its meaning. This tone indicates, more or less, an affect of the speaker and in turn induces the same affect in the listener too, where it then conversely arouses the idea which in language we express in that tone" (*CJ* 5:328). This tone draws upon the same emotional resources as language, and counts as a "language of the affects." It consists in a *formal* relation between harmony and melody, or between distinct vibrations, and expresses aesthetic ideas of a "coherent whole of an unspeakable wealth of thought." Like music, dinner conversation can be a sensual context for the ripening of intelligence and social sensitivity.[42] According to the *Anthropology*, what matters more in a conflict over dinner "is the *tone* (which must be neither noisy nor arrogant) of the conversation," rather than the content (*AN* 7:281).[43]

Madness and *Schwärmerei* threaten the unity that communicable concepts and their organization through reasoned judgments promise to lend the world. Monique David-Ménard argues that the effort to distinguish critical from dogmatic thought, and to identify the former and only the former as "reason," is the overriding motivation of Kant's *Critique of Pure Reason*. Where Foucault points to Descartes's self-confident dismissal of his own capacity for madness, David-Ménard points out that "Dreams of a Spirit-Seer," which contains many of the themes of Kant's mature thought, is driven by the desire to distinguish metaphysical philosophizing from the extravagant claims of Swedenborg's spiritualism (1990, 79–84). "I find myself in the following unfortunate predicament," Kant confesses after describing the contents of Swedenborg's writings, to which he has given an eminently philosophical introduction, "the testimony, upon which I have stumbled, and which bears such an uncommon likeness to the philosophical figment of my imagination, looks so desperately deformed and foolish, that I must suppose that the reader will be much more likely to regard my arguments as preposterous because of their affinity with such testimonies than he will be to regard these testimonies as reasonable because of my arguments" (1992c, 346). The analysis of illusions endemic to reason's own effort to organize its activity permitted Kant to organize his *own* thought without sliding into delusion,

---

42. This aspect of the Kantian understanding was developed by Arendt into an implicit *political* imperative to imagine occupying the standpoint of others whenever rendering judgment. See "The Crisis in Culture" and "Truth and Politics" (in Arendt 1968) and *Lectures on Kant's Political Philosophy* (Arendt 1982). Woodmansee (1994) situates Kant's theory of aesthetics with respect to eighteenth-century elite concerns regarding the spread of literacy, the rise of "popular" tastes, and the structure of the publishing market, whose treatment of authors shaped our contemporary understanding of discourse as "intellectual property."

43. On the significance of *tone* in limiting or broadening the possible audience of a discourse, insofar as it differentiates a fanatical from a reasonable discourse or implies that the intuitions giving rise to knowledge are only available to a few, see the 1796 essay "On a Newly Arisen Superior Tone in Philosophy," including the introductory essay by Peter Fenves (Fenves 1993).

yet it forced him to define himself by contrast to an image of the "uncritical" fanatic, one who fulfilled a structural role for Kant's thought whether or not he existed historically.[44]

This problem is raised again in the *Prolegomena*, where Kant excuses the imagination for going beyond the confines of experience in the interests of imparting liveliness to thought, but declares it "unforgivable" for the understanding to "daydream," for "all assistance in setting bounds . . . to the fanatical dreams of the imagination depend on it alone." Here Kant offers a short phenomenology of the understanding's gradual conversion to the drift of fantasy; from ordering the elements of experience in keeping with the laws that constitute possible experience it leaps to cognizing "newly invented forces in nature, soon thereafter to beings outside of nature, in a word, to a world for the furnishing of which building materials cannot fail us, because they are abundantly supplied through fertile invention" (PR 4:317). In his treatment of the sublime, finally, Kant suggests that the fertility of imagination can lead to a repulsive feeling of vertigo, although reason regards its expansion as an attractive representation for moral self-sufficiency" (CJ 5:258; see Shell 1996, 333 n. 71). Publicity protects against these dangers: in "What Is Orientation in Thinking?" Kant suggests we think *more* and *more accurately* when we are "in community with others to whom we *communicate* our thoughts and who communicate their thoughts to us" (KPW 247).

Kant's *Anthropology* distinguishes between four kinds of mental illness conducing to the breakdown of common sense. "*Amentia* [*Unsinnigkeit*]" he writes, "is the inability to bring one's representations into even the coherence necessary for the possibility of experience," due to the constant introduction of imaginative material which is irrelevant to the topic under discussion (AN 7:214). In fact, such a diagnosis says nothing about the coherence of the patient's experience, but a great deal about how the patient's imagination prevents her from communicating a coherent experience to the listener. Such a person, one might surmise, attempts to elaborate an experience whose unity forever escapes her precisely because of the infinite divisibility and diversity of the spatial and temporal manifold and the range of combinations through which the understanding might present it discursively. A communicable experience can unravel in the

---

44. In "Le laboratoire de l'oeuvre" (1992), David-Ménard discusses Foucault's neglect of this essay and the potential problems it poses for his periodization of the philosophical, if not juridical, "internment" of madness.

act of discursive communication, especially if the speaker imagines that listeners are unwilling to invest themselves in the act of understanding or unable to relate to his or her content. Rape survivors and other victims of injustice know how difficult it is to describe experiences of powerlessness without feeling obliged to explain themselves endlessly and fruitlessly, because experience has taught them to imagine the listener's hostility or indifference so much more easily than their own dignity and persuasiveness.

In "dementia" (*Wahnsinn*), the second type of mental illness, the patient appears to have coherent experience and to communicate effectively—in short her understanding is intact—but includes unreal objects (such as persecutors) among the manifold of her impressions. "Insania" (*Wahnwitz*), the third disorder, affects the *judgment* such that it can no longer distinguish between analogies and similarities: "The mind is held in suspense by means of analogies that are confused with concepts of similar things" (AN 7:215). It is as if the delirious patient recognized the necessity for a common and unified experience but set about collecting the wrong elements, or too many incompatible elements, as the touchstones of a potentially communicable world. And in "vesania" (*Aberwitz/vesania*) the last and most serious of these ailments, "the mental patient flies over the entire guidance of experience and chases after principles that can be completely exempted from its touchstone" (AN 7:215). Such a patient, it appears, systematically and deliberately cognizes outside the limit conditions for possible experience in precisely the fashion debunked by Kant's own critique of pure reason. He or she lives both the thesis and antithesis of the transcendental dialectic; incarnating the inevitable self-deceptions of reason. "With complete self-sufficiency," Kant notes, "he shuts his eyes to [*wegsieht über*] all the difficulties of inquiry" (AN 7:216).

These "observed" forms of madness, while they involve failures of communicability, may or may not involve *schwärmerisch* claims of special insight into the nature of things-in-themselves. But the two categories are clearly permeable. "The only universal characteristic of madness [*Verrücktheit*]," Kant writes in the *Anthropology*, "is the loss of *common sense (sensus communis)*, and its replacement with *logical private sense (sensus privatus)*" (AN 7:219). The touchstone of sanity and the test of common sense are an ability or willingness to submit one's perceptions and opinions to the confirmation of others. Kant observes that "incommunicable" terminology for common experiences may give the misimpression of insanity, but he does imply a kind of duty to engage in communication for the sake of mental as well as political health.[45] Therefore the free press, which guarantees that citizens will have the opportunity to communicate their ideas to all and discover the limitations on such communicability, is an instrument of political as well as psychological sanity. Publicity is also an important check on the errors or paternalistic pretensions of government (*KPW* 85-86). Herder's valorization of language as the *expression* of a "people" rather than the *form* of any possible public exchange provides him with a national basis for revolution; but the language of Kantian publicity posits no "people" as its substrate.

Kant's fear regarding the violent or tyrannical consequences of adopting an "exalted tone" cannot be underestimated. While governments must permit antagonism over scholarly matters to flourish, and while Kant hoped above all else for historical progress toward a cosmopolitan society among republics, the *virtual war* or *dissensus* that lies hidden within such controversies must not be actualized by taking its real sources or stakes as possible objects of knowledge "in themselves." Kant's "Announcement of the Near Conclusion of a Treaty for Eternal Peace in Philosophy" tries to show that enthusiast Neoplatonists and critical philosophers share a common interest in morality and that philosophical conflict maintains rather than diminishes its visceral importance (Fenves 1993, 83–100). This is an effort to defuse the tension his ideas have done the most to sharpen. Critical philosophy itself, written communicably and striving to delimit the conditions for communicable experience to which intuitions and concepts contribute as *general types* rather than specific intentional objects, is the work of what Françoise Proust calls an armed peace.[46]

Kant's seemingly contradictory comments regarding revolution in general and the French Revolution in particular result from this commitment to *postpone* or prevent the outbreak of war. These comments reflect his

---

45. In a polemic against the reactionary Neoplatonist Johann Schlosser and certain other Christian sentimentalists, Kant argued that partisans of "intellectual intuition" pretended to have incommunicable insights into the nature of things (1) in order to avoid the hard work of communication and (2) in order to exert power over the many by means of a presumed secret wisdom, thereby limiting the public's scope and ability (Fenves 1993, 57, 63–64). The only secret, he insisted, was freedom and the act of self-determination through reason.

46. See Proust's introduction to "Vers la Paix Perpétuelle" (Proust 1991) for a very interesting comparison of Kant's critical "deflation" of intellectual conflicts in the *Critique of Pure Reason* and the privilege given to resistance rather than revolution in Kant's political theory, and Rogozinski 1996 for consideration of how Kant's critical method protects him against "unthinkable objects" such as revolution.

general adherence to a governmental practice in which a state is legitimated by its ability to enhance the trade and fertility of the domestic population, while maintaining a balance of power with neighboring states through diplomacy and, much to Kant's frustration, expensive standing armies. This governmental practice, *raison d'état*, will be further discussed in Part 3. On the one hand, Kant firmly considered constitutional self-government to be the only form of government that could both incarnate a people's will to moral self-governance in its political form and implement a peaceful rather than belligerent foreign policy in its international practice. When considering the question whether any evidence exists for moral progress in the human race, Kant identifies international European solidarity with the actors of the French Revolution as the surest sign that humans in general are disposed to the republican government he believes is necessary for national, and ultimately international justice.

> The revolution of a gifted people which we have seen unfolding in our day may succeed or miscarry; it may be filled with misery and atrocities to the point that a sensible man, were he boldly to hope to execute it successfully the second time, would never resolve to make the experiment at such cost—this revolution, I say, nonetheless finds in the hearts of all spectators (who are not engaged in this game themselves) a wishful participation that borders closely on enthusiasm, the very expression of which is fraught with danger. (Kant 1992a, 153)

It was with a similar passion, perhaps, that many spectators worldwide celebrated the breakup of the Soviet empire, its repressive, bureaucratic hold on Eastern Europe and the demise of the East-West balance of nuclear terror justified by Western capitalism's fear of communist expropriation. This event was a "revolution" against a particular international arrangement of forces. But revolution cannot be universalized as a principle of political change within any one regime, and therefore gives rise to a morally repugnant contradiction, even if the result is constitutional self-government. This is because (in Kant's experience) it pursues the art of government using conspiratorial means, creating a vacuum of authority between the prince and his people or between various factions of the people (*MM* 6:340). In other words, it is incapable of bringing about a state of "right" in a "rightful" way. In the *Metaphysics of Morals*, Kant declared *regicide*, the physical act of aggression against even a despotic monarch, "a formally evil (wholly pointless) crime" that no moral system should fail to consider, "although it is only the idea of the most extreme evil" (320).

Kant appears to distinguish "good" enthusiasm (*Enthusiasmus*) from "fanaticism" (*Schwärmerei*) as he distinguishes disinterested sympathy for the actors and victims of a political drama from the desire to become entangled in their passions and violence at a bodily level. Although the revolutionaries themselves cannot be sure that their goal is a just government rather than their own economic relief (especially when the two motives were combined in resistance to Louis XVI's expenditures on offensive war), the existence of sympathetic observers "can have no other cause than a moral disposition in the human race" (1992a, 153). However, claiming to know which specific policies or communities should be represented or replaced in a nation's constitution and using force to back up such claims is a form of fanaticism that can only have disastrous repercussions for the value of *rightfulness* as well as *communicability*. Kantian sovereignty, like Kantian experience, is built upon the public's recognition of its finitude—upon knowledge that the "manifold of historical memory and documentation," so to speak, can always be further determined. We will also return to Kant's attitudes on revolution in Part 3. In the meantime, perhaps one might say that Kantian sovereignty consists less in having a "final say" over policy than in being willing to take responsibility for the fact that we cannot help construing history in partisan terms, a situation that always threatens to erupt in violence and must be defused if constitutional self-government is to develop. Thus, if the "outer limit" of Kant's political thought is the imperative that "there is to be no war" (*MM* 6:354), the "inner limit" would seem to be the act of revolution and the *rational* delegitimation of existing regimes.

The moral unacceptability of revolution, which rests in part on the sanctity of *publicity* for all political action, pushes Kant to suggest that scholars should limit their own curiosity regarding the *actual* historical origins of the state in which they reside. Although the state, whatever its present form and the extent of self-governance it permits, is legitimated through the rational *idea* of an "original contract,"

> A people should not *inquire* with any practical aim in view into the origin of the supreme authority to which it is subject; that is, a subject *ought not* to *reason subtly* for the sake of action about the origin of this authority, as a right that can still be called into

> question (*ius controversum*). . . . Whether a state began with an actual contract of submission (*pacta subiectionis civilis*) as a fact, or whether power came first and law arrived only afterwards, or even whether they should have followed in this order: for a people already subject to civil law these subtle reasonings are altogether pointless and, moreover, threaten a state with danger. (*MM* 6:318–25, 339–40; also "Theory and Practice," in *KPW* 79–82)

Kant's valorization of the social contract as a *regulative ideal*, rather than a verifiable historical fact that must either be discovered or implemented, is comparable to his refusal to inquire about the "common root" underlying the exercise of human mental faculties. If Kant "shrinks back," as Heidegger claims, it is because he fears the political implications of such efforts to go beyond the limits of *communicable* reason and either reveal an actual dissensus among perspectives or violently impose a particular aesthetic experience as universally binding. Regicide *actualizes* the civil and international wars in which historians suspect the state may have originated. By touching the body of the king, regicide challenges the ways in which "peacetime" bodies have been shaped by violence even in their apparent Enlightenment, scholarly communication, and economic exchange. But Kant's reluctance to grapple with the history of "forgotten struggles" also *limits the concept of the public* in a way that helps explain why so many of today's citizens are frustrated at commercially overpublicized and relentlessly privatized or "psychologized" accounts of their own body politic.

Rejecting the madman and the *Schwärmer* at one blow, and for sharing many of the same attributes, permits Kant to identify the intellectual forms of "possible experience" with the aesthetic sensibilities of "politically reasonable" creatures. The limitations on imagination rendering experience communicable are among the most important factors discriminating the possibility of political community and psychological coherence from the fragmentary and antagonistic spaces of madness. In the process of rejecting fanatics, Kant's Enlightenment inevitably rejects critics whose sensibilities and experiences remain aesthetically "traumatic" because they are not universally communicable. Although Heidegger is right to note that imagination plays a central role in the unification of individual experience, it also gives rise to illusions and errors, to claims regarding the nature of metaphysical objects which have heuristic but not experiential validity. However, if we understand critique as a general ethos which seeks in human finitude the orienting grounds for resistance to authority, as well as the grounds for a peaceful coexistence with political and religious power, this troubles the distinction between *Aufklärer* and *Schwärmer* made possible (but also enforced) by Kantian critique.

Kant regards communicability as an *obligation* for knowers and actors, not just the *condition* of their aesthetic and scientific judgments. Madness, fanaticism, and civil war are among the demons he hopes to exorcise by committing himself to experience that has been well-grounded in critique. However, excluding the *Schwärmer* from the community of reason also justifies aggression and anxiety toward people whose failure to communicate exemplifies the finitude, fragility and susceptibility to conflict of human understanding.

The Kantian Body—Missing in Action

Where is the body among all these spaces? Especially now that materiality has been redefined in relation to form, rather than to things-in-themselves? Clearly transcendental idealism cannot speak of the body as thing-in-itself, only of the body as appearance—both to the self and to others. "Aspects" or "facets" of the body are assumed by aspects or facets of transcendental idealism.

On the empirical side, anthropology and the philosophy of physical nature deal with the physical and social facticity of embodied humans. Empirical psychology, which explores the mechanical processes of human association and motivation, must take account of the body as a content of inner sense. In the B edition's "On the Paralogisms of Pure Reason," as I mentioned earlier, Kant suggests that "body" and "mind" are both arbitrary abstractions from a whole, related in ways that cannot be known in themselves (B427–28). Aesthetically, one can speak of bodies as beautiful forms and purposive natural products (organisms) that contribute to our sense of a feasible supersensible purpose for humanity.

On the transcendental side, however, Kant takes advantage of the fact that faculties of sensibility, feeling, and imagination have historical associations with the body. The pure forms of intuition, space and time, precede all experience. But we only perceive bodies in space and time through the senses and through the capacity for motion that discovers spatial contours and measures the passage of time. *My body*, which has suffered particular accidents and developed particular motor skills, is an empirical psychological content of inner sense. But it seems valid to say that "personal embodiment" is associated in some way with all external experiences confirming and exhibiting the unity of transcendental perception, for "embodiment" is precisely how forms of intuition contribute to a *subject's* experience, rather than merely register signals like motion detectors built into a doorway. Kant accepts Epicurus's contention that "all presentations in us, no matter whether their object is merely sensible or instead wholly intellectual, can in the subject still be connected with gratification or pain, however unnoticeable these may be" (*CJ* 5:277). Although the connection is never spelled out, Kant makes no effort to deny that gratification and pain are bodily phenomena. Finally, although bodies are not empirically required for the "communicability" of the mental state that a subject of aesthetic judgment demands from his or her neighbors, something *like* a difference of bodily boundaries is required if the subject is to be numerically distinct from neighbors at all.

In the simplest empiricist sense, the human body is experienced as an external spatial object among other spatial objects. But one cannot say that the body is strictly "spatial" or "temporal," for it is apprehended through both forms of sensibility. Moreover, the division between experiences involving bodily interaction and experiences that take place "in time" while the body is at rest (including dreams and the empirical event of philosophical reflection) seems to distinguish "space" from "time" as media of outer and inner sense, respectively. To be sure, Kant would have believed himself able to distinguish space from time without having empirical experience of embodiment. But insofar as he manages to differentiate between forms of intuition at all, *that unknown feature with respect to which* he makes this distinction serves the *function* that we attribute to the body in everyday experience. This feature or function is an *internal, cognitively inaccessible* difference associated with feeling, which Kant explored in several pre-critical essays on incongruent counterparts (see Kant 1992c). We might call it a proto-body or the body as problematic object.

At the beginning of his philosophical career, Kant believed that space was neither a *property* of things-in-themselves nor a *form* of pure human intuition. Rather, he accepted the Newtonian thesis that space was an absolute manifold, a single system of relations between things-in-themselves. This system of relations guaranteed, for example, the numerical distinctness of otherwise identical objects like water droplets or right- and left-handed gloves. In the pre-critical work "True Estimation of Living

Forces" (1747) Kant argued that only substances capable of appearing in a three-dimensional space could interact, or even represent the internal relations of a substance in imaginatively exteriorized form (Shell 1996, 29; Carpenter 1998). Another text from this period, "Concerning the Ultimate Ground of the Differentiation Between Regions in Space" (1768) appealed to the felt difference between right and left sides of the body as a refutation of Leibniz's claim that spatial distinctions were purely *conceptual* distinctions. This feeling (*Gefühl*) oriented a subject in the field of experience, he wrote, even before encountering the objects of experience. But because it cannot be conceptualized, only *felt*, the greater strength of a right hand or foot is an *internal* difference between sides of the body that relates it to the *whole* of space (Kant 1992c, 1:371).[47] Finally, "Dreams of a Spirit-Seer" considered the noncognitive certainty of bodily boundaries to be a property *differentiating* empirical experience from the speculation of daydreams or metaphysics, rather than a fact for which metaphysics itself must account:

> For in this case [waking experience], everything depends on the relation in which the objects are thought as standing relatively to himself as a human being, and, thus, relatively to his body. Hence, the images in question may very well occupy him greatly while he is awake, but, no matter how clear the images may be, they will not deceive him. For although, in this case, he also has a representation of himself and of his body in his brain, and although he relates his fantastical images to that representation, nonetheless, the real sensation of his body creates, by means of the outer senses, a contrast or distinction with respect to those chimaeras. (1992c, 2:343)

These early essays give feeling and embodiment an important role in the philosopher's effort to overcome metaphysical disorientation. With his critical transformation, however, Kant asserted that substances and spatializing points of view on the whole of reality are linked by the *form* of intuition that relates them externally (Kant 1992c, 2:402–3). In other

---

47. Gil (1985) intended to use this inner difference as the basis for a course in the anthropology of the body at the Collège International de Philosophie. He presented some of his ideas in *Metamorphoses of the Body* (1998), where he describes bodies as systems of forces/signs that open onto a plurality of discontinuous spaces only superficially identifiable with a single "body image," focusing, however, on Deleuze rather than Kant.

words, after "Dreams of a Spirit-Seer" Kant no longer referred to the body for counterexamples to metaphysics. The body became an element of his new strategy to contain and ground metaphysics. Kant established a *finite* but *single* sphere of interactive appearances by rejecting the noumenon as an object of a purely problematic concept, accessible only to an intuitive understanding (if such exists). He also accepted the *phenomenon* as the prototypical, though non-noumenal, object of finite understanding and intuition. While he was still formulating transcendental idealism, Kant seemed to take the thinker's body as an exemplary phenomenon modeling the relationship between the limited and intrinsically structured sphere of appearances and the potentially exorbitant claims of reason and imagination. But once he oriented himself, so to speak, bodies dropped out of the picture and became one more example of the "phenomenal" in general. The body thus served as what Žižek calls a vanishing mediator between pre-critical and post-critical works, disappearing when Kant's "epistemological break" was successfully accomplished.[48]

It is as if Kant had managed to detach himself from the space in which his nature was externally determined as a thing-in-itself determined by other things-in-themselves, and to use his own bodily boundaries as the horizon within which to unify the world so that his own body no longer posed an *obstacle* to autonomy—even if it did not positively facilitate autonomy. In effect, he limits his awareness of the empirical body for the sake of forces and faculties *vested in the body* and *capable of transforming its meaning*. The consequences of this selection are enormous and quite positive, for against the backdrop of the inaccessible noumenon, the infinite sphere of phenomenal appearances can now be rationally and socially determined under the regulative guidance of Ideas. By posing questions to nature, rather than attempting to accumulate evidence in light of a totality known only to God, progress could be expected in metaphysics just as in natural science. "For otherwise our observations, made without following any plan outlined in advance, are contingent; i.e., they have no coherence at all in terms of a necessary law—even thought such a law is what reason seeks and requires" (*CPR* Bxiii).

---

48. Slavoj Žižek adapts the phrase "vanishing mediator" from Frederic Jameson to refer to a temporary phenomenon or group which makes possible the transition from one (failing or fragmenting) "problematic" to another. "A system," Žižek explains, "reaches its equilibrium, i.e., it establishes itself as a synchronous totality, when—in Hegelese—it 'posits' its external presuppositions as its inherent moments and thus obliterates the traces of its traumatic origins." During the transition, however, the soon-to-vanish mediator demonstrates or reveals the antagonisms implicit within the changing situation (1993, 227–29).

Henceforth, things which may be ontologically diverse share a common form, and this form is bound up with the form of *finite* human being rather than the omniscient and omnipotent form of God. This strategy enabled Kant to explain community among substances in terms of the regularity of human understanding rather than the divine, if benevolent, contingency of God's will (Shell 1996, 44; Carpenter 1998). By sidestepping references to divine knowledge or will, Kant could refer community, "the universal connection of things—insofar as the world of appearances is concerned—to a ground that is human rather than divine"; i.e., to the unifying functions of space and time (Shell 1996, 134, 141). It is as if the skin linking us to God, and thereby to other beings, ceased to be a "divine sensorium" (Newton's theory) and became our own, enabling us to touch one another, but rendering any other form of interaction unthinkable.

On the other hand, the fact that the mediator vanished upon hearing that critical reason had unified its domain deprived Kant of a metaphorical, that is, "noncognitive" way of conceptualizing *other* problems remaining from the pre-critical period—namely, the balance between form and play necessary for coherent experience and the community of individuals who disagree about how that experience looks and should look. As the *Critique of Judgment* shows, this is a problem with political as well as epistemological implications. Before humans can engage in autonomy or obedience, they must at least be able to guarantee that they occupy a common *physical* space of perception and action, nurturing or violence. In *The Critique of Pure Reason,* Kant explains that traditional metaphysics conceived of community or "reciprocity" among substances as both a *logical* attribute of subjects and as a *real* attribute of things-in-themselves. Dogmatic rationalism attributed the logical meaning of these categories to things-in-themselves. Instead, Kant proposes, community should only be understood as a way in which the categories of logic orient us in a world of *appearances.* These appearances are real for us because they present a permanent aspect to viewers and exhibit *irreversible* causal relations.

"All substances, insofar as they can be perceived in space as simultaneous, are in thoroughgoing interaction," Kant states in the Third Analogy. "Community" refers to the condition through which entities can be perceived as simultaneous or reciprocally active within a manifold of intuition, not to their intrinsic nature. "The relation of substances wherein the one substance contains determinations whose basis is contained in the other substance is the relation of influence; and if this latter thing reciprocally contains the basics of the determinations in the former thing, then the relation is that of community or interaction" (*CPR* B258). We could not recognize things as simultaneously coexistent if we could not presuppose that they were capable of exerting influences on one another, influences perceived by means of irreversible effects that some have on others. Put in social rather than physical terms, we could not know that we inhabited the same world as our neighbors if we never saw evidence of our influence on their actions (if only through brute coercion) and never felt ourselves to be affected in turn.

Time, as pure form of inner sense, is an "enduring" structure associated with the mere sensation of intensity or reality (*CPR* A182/B225). This permanence need not be associated with the appearance of any *particular* substance, but is rather a feature of the *form* which structures sensation as an "object," a more or less complex manifold. Against the backdrop of this permanence, changes in appearances prove either irreversible and explainable through rules of cause and effect, or contingent and reversible, in which case we recognize them as simultaneous rather than successive states (A190/B235). If a ground can be found for an irreversible change—a ground in the original state of affairs that would produce a similar change under similar circumstances, as when a falling weight creates a hollow in a pillow (but a pillow on its own fails to exhibit any change), the succession can be considered "objective," that is, in the manifold as "object." If not, then the succession is considered subjective, as when the appearances encountered in moving through a house could change if one entered rooms in a different order, but the structure of the house itself remained stable (A195/B240). Time would not be a register of simultaneity and change if we were unable to distinguish the changes caused by real interaction among appearances from the subjective changes of appearance provoked by our perception of what is simultaneous.

Thus the very fact that we perceive succession and simultaneity demonstrates that we are required by the very *form* of our experience to conceive substances as *really interactive*, rather than merely imaginatively or logically dependent on one another. According to Béatrice Longuenesse, the rules of cause and effect engaging the human body in reciprocal relations with its surroundings play a privileged role in organizing the simultaneous manifold as well as detecting its alterations in time. "Our body," she writes, "appears to us as one substance among other substances, with which it is in a relation of dynamical community (*commercium*), simultaneity, and thereby spatial community (*communio*)." We learn the stable contours of the world, in other words, only through bodily experimentation with what can change. "Our experiencing the coexistence of other substances with our own body is the condition for our experiencing their respective relations of community" (1998, 391). The passage to which she refers is the following:

> We can easily tell by our experiences: that only the continuous influences in all positions of space can lead our sense from one object to another; that the light playing between our eye and the celestial bodies can bring about an indirect community [*eine Mittelbare Gemeinschaft*] between us and them and can thereby prove their simultaneity [*Zugleichsein*]; that we cannot empirically change place (and perceive this change) unless matter everywhere makes possible the perception of our position; and that only by means of matter's reciprocal influence can matter establish its simultaneity and thereby establish (although only indirectly) the coexistence of objects, down to the most remote ones. (*CPR* A213/B260)

But what does it mean to talk about "our" body? Is that body a "generic" body, or does it refer to the ensemble of bodies in their social, i.e., anthropological, coexistence? (see, for example, Gatens 1996, 131, and Lefebvre 1991, 194–99). Are bodies the phenomenal entities with which each of us can identify, thereby entering into a common social space, despite the obvious external differences of appearance and performance and their qualitative differences of affect or sensation? Is one kind of body among many (white, bourgeois, male) being taken as the norm, along with its opportunities for physical interaction with the natural world, and the social rights that protect it *as an object in space* against threats of starvation or violence? Or should we surmise that differential racial, gendered, class, and cultural marks are imposed upon or educed from originally *indistinguishable* spatial "objects" to explain the differences in their qualitative and affective experiences—differences which, even more obscure than the incongruence of left and right hands, are otherwise very difficult to represent in "space"?

"Inner" differences among entities are incomparable. "In an object of pure understanding, only what has (as regards its existence) no reference whatever to anything else different from itself is intrinsic" (*CPR* A265/B321). Thus "inner sense," for example, is not inner in the sense of being "inside" a person's head, but inner insofar as it involves relations among elements of someone's experience that cannot be compared to those of someone else (except, that is, through externalization in space and time—speaking, writing, and perhaps pantomime). It also "individuates" a person, in the sense that we feel sure two cloned persons would have "different" psyches and experiences, even if they shared the same genetic makeup. Leibniz believed that two physical entities could be truly differentiated from each other only if they possessed internal, incomparable properties analogous to "inner sense"; thus two drops of water ought to be distinguished by their tiny minds (internal differences) rather than by location in space or observable difference. At the same time, inner relations that differentiate entities from one another are positive *relations* the entities hold with themselves. Kant's goal is to make self-relation compatible with external comparability; and in the political realm, autonomy with social interaction and hierarchy, according to a common form of self-limiting imaginative reason.

"On the Amphiboly of Concepts of Reflection" in the *Critique of Pure Reason* states that reason is capable of distinguishing the mind's faculties from one another (examining the properties of their characteristic representations) on the basis of *feeling (Gefühl)*. This feeling is not sensibility (*Sinnlichkeit*), for sensibility is one source of the representations being compared. Lyotard contends that there are several types of sensibility at work in transcendental deliberation (1994, 10). While the feeling of pleasure and displeasure (*Gefühl*) is a present state, sensibility (*Empfindlichkeit*) is generally used to refer to those aspects of experience that contribute to cognition by providing the empirical material for intuitions. Feeling, by contrast to sensibility, encompasses aspects of experience that cannot be captured in cognition, but it may still be empirical. Apart from respect (*Achtung*), *pure* feelings only become important for Kant in the *Critique of Judgment*.[49]

49. In the *Anthropology*, Kant classifies boredom, joy, grief, and aesthetic pleasure as varieties of *feeling*, while emotion and passion are examples of *desire*, that is, ways of imaginatively orienting oneself toward the future (*AN* 7:230–50, 252). Emotion (*Affekt*) responds to a particular situation; Kant's examples include hope, cowardice, bravery, anger, and shame. The lust for freedom, vengefulness, ambition, and avarice are among the passions (*Leidenschaften*), which endure over time, govern the subject as if according to a maxim, and make use of the same impulses and capacities as morality. This dangerous property enables them to coexist with and compete with reason (266). Emotional intensity or attentiveness seems difficult to classify as either feeling or desire, though Kant believes that lack of progressively changing stimulation causes boredom and the intellectual equivalent of death (235).

Kant differentiates between empirical and pure feelings. In his moral writings, all empirical feelings are assimilated to sensible motivations of varying intensities (whether they are produced by intellectual or physical activity). Such feelings are thereby opposed to reason (see,

What is important here is that insofar as the representations/faculties are noticeably distinct from one another, they have *external* relations, although they are not situated in space. Space is a pure form of intuition, and thus a property of the representations that go to make up intuition. *On the other hand, insofar as these representations/faculties are indistinguishable from one another, they are associated with a feeling tone.* As explained above, Kant refers to the "representational space" (to borrow Lefebvre's phrase), or "external relations" in which these different tones are distinguished as a "transcendental topic." Each power of thought is a different way in which reason relates to, and limits, itself. We may infer, from the overall purpose of the critique, that Kant hopes that however diverse our individual experiences and thoughts, we will all distinguish our faculties and limit our reason in the same way. But attempting to follow a common model will invariably produce distinctive results for each person, since reason is not the *only* self-relation we possess.

The *Critique of Judgment* distinguishes *feeling* from sensation (*Empfindung*) in this way: feeling is an effect of a presentation that "is referred solely to the subject and is not used for cognition at all, not even that for which the subject *cognizes* himself," whereas sensation is an effect of a presentation on the senses and admits of a determinate cognition (*CJ* 5:206). We sense that a field is green, he offers by way of example, and surely some people find green fields agreeable; but the "agreeableness" of the color rests on the *beauty* of a certain proportion which has nothing to do with the color itself. "A judgment of taste," he adds a few pages later, "is merely *contemplative* . . . it [considers] the character of the object only by holding it up to our feeling of pleasure and displeasure" (209). Thus feeling is an *internal* relation which we might analogize to a "frequency," while sensation is an *external* relation (that may, like certain visual signals or radio transmissions, only make sense at that frequency). Elsewhere in the text, Kant reminds the reader that pleasure encourages us to *linger* over an aesthetic image "because this contemplation reinforces and reproduces itself" (222); sublimity is even more explicitly an *internal* conflict between mental powers. What is properly beautiful is not the

---

for example, *GR* 4:399, 418, and *CPrR* 5:23). Respect for the moral law (*Achtung*) as a motivation may be a cover for other emotions or sensible incentives (*CPrR* 5:116–17). The calmness that follows genuine moral action, however, is a *feeling*, not a motivation (*CPrR* 5:76); like *disinterested aesthetic pleasure*, it instructs the subject about a priori relationships holding between feeling and other faculties. See Schrader 1976 for an overview of the development in Kant's understanding of emotion.

object, but the *relation* we have with *ourselves* as a result of our *relation to the object* (two relations which are thus internally linked); if we consider it a feature of the object, this *relation to the object* is something occasioned by, but exceeding, the object.

The term *feeling*, therefore, indicates a relationship between faculties or between one aspect of imagination and another. Such a relationship mirrors the judging self's relationship to all other judging subjects, who may or may not "resonate" at the same frequency. Kant sharply distinguishes pure aesthetic pleasure, which indicates that a perceiver is in the presence of a form provoking a free, lawful play of imagination and understanding (or, in the case of the sublime, confronted with a formlessness which causes imagination to subordinate itself to reason's desire for self-legislation), from empirical pleasures that are unique to the individual and have no heuristic import for critical philosophy *or* the transcendental grounding of the sciences. Pure aesthetic pleasure is *disinterested* (*ohne alle Interesse*), a state in which one's self-identification is difficult to distinguish from identification with others. In "Analytic of Aesthetic Judgment," in the *Critique of Judgment*, Kant declares that "*taste is the ability to judge an object, or a way of presenting it, by means of a liking or disliking [Wohlgefallen oder Misfallen] devoid of all interest*" (CJ 5:211). We have a practical or moral interest in things which are *good*, and an empirical or sensible interest in things that are *agreeable* (*angenehm*)—that is, we have an interest in their existence as opposed to their mere form. But objects of taste are forms that please even if they are only imaginative.

"In order to play the judge in matters of taste, we must not be in the least biased in favor of the thing's existence but must be wholly indifferent [*gleichgültig*] about it" (CJ 5:205). But this does not mean that the thing provokes no pleasure; rather the pleasure it provokes is one which lingers and animates the mind even if it does not correspond to a sensible reality or impose upon us a moral obligation. Individually, we may find different things agreeable in a sensual manner; but we consider beautiful only those things which provoke a certain *kind* of pleasure, a pleasure *in* our pleasure, a pleasure in our *capacity for pleasure,* which is occasioned by, but not dependent upon, the state of our senses and the world. We might translate as follows: agreeable things reflect or symptomatize inner, individuating differences that remain idiosyncratic. Beautiful forms, by contrast, reflect or symptomatize inner, individuating differences that promise to alter or elaborate upon the *form of exteriority* through which individuals relate to one another.

The pleasurable relationship to oneself in aesthetic pleasure is, therefore, inevitably a kind of relationship with others as well. "There is no disputing [*disputieren*] about taste," runs one half of the antinomy in Kant's "Dialectic of Aesthetic Judgment," since taste does not refer the pleasurable form to a concept whose communicability we can debate. Still, we are prone to *quarrel* (*streiten*) endlessly about taste, as if there were hope of agreeing about it, which implies that such judgments do ultimately have a conceptual basis (*CJ* 5:338). Kant's solution (as discussed above) was to make judgments of taste refer to an *indeterminate* concept. It is over the interpretation of this indeterminate concept—the concept of the "supersensible substrate of humanity" (*das Übersinnliche*) understood either as a play of powers or as a *sensus communis*—that quarrels inevitably break out. But agreement on taste, as discussed earlier in the chapter, is also something that can be cultivated by culture, through an appeal to practices that involve not our "cognitive" but our "aesthetic" response to experience.

Such responses seem instinctive, bodily. And yet taste does not concern the responses of the body as we know it empirically, governed by needs and whims, irrevocably separated from others in its biological and psychological particularity. Rather, it concerns sensibility insofar as it has not yet been schematized in relation to one particular body or in relation to a scientifically understood function of the body. Kantian beauty relates a representation to the subject's "feeling of life, under the name of the feeling of pleasure and displeasure" (*CJ* 5:204), but this feeling of life is explicitly distinguished from sensual satisfaction. It constitutes the "sensibility" of a "body" shared with others but emerging over time, through disputes in which the visual and performative aspects of empirical bodies are altered by the forms they agree are beautiful or disdain as someone's private predilection. One might conceive, though Kant does not turn in this direction, that this feeling of life could also be confused with or manifest as an *anxiety at the fact of individuation*, just as conscious anxiety may represent the effort to master a repressed pleasure or excitation. In judgments of taste concerning the public world, we reveal our similarities and dissimilarities to other perceivers—similarities and dissimilarities that are grounded in our relation to *that aspect of appearances which does not itself appear as a possible object of cognition,* and which trouble the distinction between thought and its object as well as between the thinker and her neighbors.[50]

On the basis of his analysis of aesthetic judgment, Kant wants to identify those elements of the world that provoke agreement among individuals and indicate the possible existence of a supersensible substrate in which theoretical and practical reason might tend toward the determination of a single sphere of experience and culture (*CJ* 5:355–56). Apart from the preferences of the empirical body, the pure pleasure and displeasure that accompany aesthetic judgments are the phenomena from which the existence of such a substrate is to be educed. On the one hand, these pleasures reflect a potential harmony of the faculties that must be assumed to underlie the theoretical exercise of reason and that is assumed to function similarly in all humans. On the other hand, they reflect a potential harmony among human individuals, which *cannot* be taken for granted and may turn out to be a task of education and culture. Judgments of taste are universally valid, but we will quarrel endlessly over which judgments are authentic and which are merely the expression of our individual empirical preferences and bad upbringing.

According to Arendt, Kant's *Critique of Judgment* contains an "unwritten" political philosophy addressing the significance of a necessarily plural and *public* horizon for the development of human thought (1982, 9, 13–15). The "sense" for which objects are likely to arouse pleasure in *other refined persons* (others, that is, who have themselves been trained to distinguish between interested and disinterested pleasure, in other words to take stock of their bodily and moral desires) shapes the physiognomy of a community *quite literally* inasmuch as it trains us to differentiate ourselves from one another on the basis of similar and dissimilar "capacities" for taste (Kant alludes to this when he speaks of the need for cultivation: *CJ* 5:265; 355–56). For Arendt, aesthetics has a role in politics because it is only according to taste, suitably cultivated by reference to the potential judgments of others, that events can be judged *impartially*

---

50. Along these lines, Žižek argues that what most bothers the subject of racism or ethnic hatred is the fact that the other group "enjoys differently," eats foods with a different smell, speaks the same language with a consistently different accent, or listens to jarring music (1993, 201–5). Appadurai suggests that he or she feels "betrayed" that the neighboring nation turns out to be inhabited by people with whom one's reactions and pleasures are not immediately communicable, diminishing the assumed power of the aesthetic community in which he or she does participate (1996, 154). Instead of reacting with anger that parents or the media have misrepresented the extent of international similarity, they feel that the other group has taken something from them.

by those who witness them (such as judges and historians), whereas those who engage in action are necessarily limited by the perspective that permits them to take a stand. Arendt's belief in the impartiality of taste is supported by Kant's account of the French Revolution as sublime from an aesthetic standpoint, though a morally repugnant act for any potential actor (Arendt 1982, 52).[51] But Kant's account of genuine aesthetic judgment as asserting universal validity even in the face of likely skepticism from others certainly presents the critic as an individual whose *unique way of being impartial* nevertheless renders him or her *very partial* in the eyes of others. It is this partiality whose humanity and rectitude he or she must valorize through acts of interpretation and obedience, rendering us both attractive and abject to one another and ourselves.

Kant's lectures on anthropology and physical geography suggest that although bodily appearance may only be a *sign* of differences or similarities in feeling, cultural belonging, or subjection to a certain way of life, we only ask about capacities for reason and taste in those with whom we can already communicate. The pre-critical *Observations on the Feeling of the Beautiful and the Sublime* (1764) addresses sexual and national differences in emotion and taste rather than aesthetic judgments per se, since the transcendental meaning of these feelings in Kant's thought was not clarified for decades. Kant does not ignore the role of culture in distinguishing women's interests and capacities from those of men, any more than he fails to consider the role of climate in shaping the supposed "character" of peoples. But neither does he question his affinity with the culture and the talent, or style of communicability, prevalent in the culture that makes such differentiations. European men are to be preferred on account of their supposed capacity for sublime feeling, whereas women only show their capacity for sublimity in acknowledging men's qualities (Kant 1965, 94), and non-Europeans either lack appreciation for beauty and sublimity, favor one or the other disproportionately, or give them "distorted" cultural expressions. Most extreme is Kant's approving citation of Hume to the effect that "the Negroes of Africa have by nature no feeling that rises above the trifling," although the natives of Central and South America are also characterized as possessing "an extraordinary apa-

---

51. Arendt, however, argues that Kant's seemingly contradictory response to revolution is irrevocably shaped by the assumption that revolution must proceed by coup d'etat, that is, by actors working in secret. Such secrecy conflicts with the publicity that alone justifies the laws of an emancipated state and that makes popular acclaim for such events a sign of potential progress in history (1982, 60–61).

thy" (Kant 1965, 110–12). In his lectures on physical geography, Kant admitted that the slave trade had prevented Europeans from gaining genuine knowledge about Africa but nonetheless was so indifferent to the humanity of Africans that he included in his lectures a description of how they responded to different kinds of beating or brutal punishment (Eze 1997, 58–64).

The problem is not simply that white bourgeois men may not be able to tell the difference between pure aesthetic pleasure and inclination, thereby substituting their purely anthropological preferences for a universal standard, and using their financial and media power to impose it on others. This is an inescapable problem, but Kant would say that it concerns an abusive appeal to aesthetic judgment, as well as, perhaps, an abuse of justice. The more fundamental problem is that truly beautiful form should not only allow individuals to express their conformity with a norm, but also to successfully communicate their *difference* from that norm to others without leaving the medium of communicability.[52] However, colonialism and the deployment of sexuality have resulted in a situation where the most powerful signs of such a potential commonality have come to be those associated with European culture and a racializing and/or sexualizing perception. The people we regard as white bourgeois men are the ones whose imperial and industrial adventures led their skin color and the way of seeing, hearing, and moving embodied in their artworks to become widely recognized touchstones for a pleasurable self-relation that is *both* differentiating and universalizing.

Every culture and class appreciates the power of a state of mind in which universal communicability seems plausible and necessary. Given the historical legacy of colonialism, however, other cultures and classes have to fight against the recognized image of the "aesthetic perceiver" as white and bourgeois or aristocratic if the universality of their aesthetic judgments or the communicability of their *equally* noncognitive, invisible but disinterestedly pleasurable self-relations are to be taken seriously. When the oppressed must express or "discover" that state of mind using an aesthetic language in which their bodies, language, and tastes signify the unformed, their capacity to experience such power is rendered invisible, while the oppressor's awareness of his or her moral destiny and shape-giving abilities is enhanced. Efforts to draw others into a *sensus*

---

52. Here I am using the term "norm" to indicate an anthropological manifestation of the *sensus communis*, not in the sense of "standard idea" (e.g., *CJ* 5:234–36).

communis marked in this "partial" way add inadvertently to the persuasiveness of the oppressor's media, because they must reference it to communicate widely. It is rare that they manage to reshape or retune it—as in the case of African-American music, which has forever changed any possible "white" aesthetic, and may ultimately change the "color" of the U.S. population in the imagination of global listeners. The fact that a minority literature is considered "worth translating" into one of the dominant tongues in a multilingual society generally brands its point of view as particular in the very act of valorization. It also physically reinforces the time that readers spend experiencing and imagining in the cadences and imagery of the dominant language, whether or not they speak a minority tongue of their own. To the extent that they want to train others in the aesthetic of an oppressed people, artists and scholars must make reference to the forms that signify the "aesthetic as such" to less-educated readers and viewers; thus subcultural art forms are not taken seriously or distinguished from entertainment and advertising until they are removed from their concrete life context and placed in the university.

"Whiteness" is the effect and the sign of this accumulation of universal communicability in the hands of people who look and see or feel in a certain way. Aspects of "white" attunement to form have just as much potential for universal validity (and are just as potentially "empirical") as the aesthetic attunement of any culture, although they may also be responsible for making "culture" a visible difference or object in the first place. Other aspects of "white" attunement, however, were used to give colonial and capitalist domination a transcendental ground, and cannot be separated from the way it looks and feels to inhabit a culture with these specific injustices. The potential for accumulating communicability in this way exists in every society, and is a cause for serious ethical concern.

Like shifting scales or conceptual paradigms, it takes time and effort to learn the potential universal communicability of an alien aesthetic. This is why aesthetic perception and creation are *material*—plastic, but difficult to reverse or redirect—even when Kant emphasizes the *disinterestedness* of the pleasure that follows aesthetic judgment. Very few people from a dominant culture or sexuality will learn another way of articulating aesthetic pleasures to the point where they genuinely experience the *kind* of *sensus communis* it makes possible—its particular *style* of potentially universal communicability. Indeed, one of the effects of oppression is that the potential universality of aesthetic apprehension is forgotten or becomes indistinguishable from inclination and is finally affirmed *as* need or inclination. Although every individual faces some difficulty distinguishing aesthetic judgment from individual inclination, those who physically represent the existence of an "outside" or an "incommunicable" state of mind to a dominant *sensus communis* lack the social power to be persuasive about the universal communicability of their own aesthetic pleasure, which they nevertheless can and should not abandon. After all, lacking determinate concepts, aesthetic judgment refers to an empirically and logically *incomparable* aspect of the critic's being. Such failures of communication, like the pleasure in successful communal perception, shape the critic's *inner* relations: those of inner sense. In other words, there is a socially invisible aspect to every individual's experience (just as I have argued there is a heterotopic aspect), but it took white domination to make dark skin into a concrete sign of what was incommunicable in Ralph Ellison's experience or the cause of double consciousness for W.E.B DuBois, and this signification had indelible effects on the inner sense of these extraordinarily gifted communicators. This is how racism and sexualizing domination reproduce themselves at the level of the psyche and imagination as well as overt economic or political domination. It helps us understand why the slogan "Black is beautiful" was so powerful and necessary for African-American activists in the early 1970s and yet became an artifact of identity politics even for its advocates.

The representational space in which Kant arrives at his "representation of space," to borrow Lefebvre's language again, begins to one side of things-in-themselves, and is subsequently divided by acts of transcendental deliberation into realms of sensibility, understanding, and—if we follow Lyotard's principle that *Überlegung* is the aesthetically disinterested state in which Kant develops the critical philosophy—feeling and action as well (1994, 31–32). But all these representational spaces interact and enter into a temporal manifold that *could belong to anyone* insofar as they belong to a body—specifically, to a normal, cultured body willing to exchange reasons with others. Perhaps this body is the "ideal of beauty" described in Kant's "Critique of Aesthetic Judgment"; namely, a body which exhibits "the moral ideas that govern man inwardly" and which provoke a purely disinterested admiration, by contrast to the "standard idea" of human beauty which reflects the *average* proportions and features enabling us to judge a creature as human and which, in fact, Kant expects to differ from culture to culture (*CJ* 5:234–36). But it seems unlikely that this is what Kant has in mind, for what would meet his criteria is a body exhibiting the "aesthetic" sensibilities that govern men and women publicly—and which, at his point in history, were governed by the cultural preferences and economic needs of a particular warlike culture (despite Kant's genuine desire for international peace). In fact, it is not clear that such a body belongs to anyone at all, although techniques exist for detecting and producing human types regarded as foreign, dumb, or culpable, even in peacetime. Thus Kantian reason, first unified by its exclusion of the merely problematic noumenon, was historically unified around the body whose empirical "pathology" or private deviations from the norm absorb all the incongruities created by reason's inevitable bifurcations.

One might argue that if the *sensus communis* is, in fact, a regulative idea, then a tenacious member of the community could always *invoke* the dialectic of aesthetic judgment, denouncing an apparently universal example of beauty as mere empirical preference. Indeed, there is probably a fine line between examples of contemporary art that invoke a *sublime* aesthetic rather than the experience of beauty (an all-pervasive trend, according to Lyotard's *The Inhuman*) and those that try to expose the *interested* moral or empirical quality of traditional Western aesthetic values, without proposing a "more genuine" object of aesthetic pleasure in their stead (Lyotard 1991). Kant allows and even encourages readers to exercise "veto" power in public debate against ideas that are less than rationally persuasive (A738–39 / B766–67). However, the ability to use this veto against the "empirical" or "anthropological" *sensus communis* depends in large part on whether or not an individual knows what it *is* about that sphere of communicability that feels oppressive. The Leibnizian and Wolffian image of intellectual intuition as a standard for knowledge gave Kant an image for the kind of subjection and suspension in indefinite confusion or imagination he wanted to escape. Likewise, paternalistic religious government is the clear target of Kant's project to convince each actor that only self-legislation renders humans worthy of happiness. But if no element of "the normal" stands out as a clear symptom of his or her peers' empirical inclination rather than genuine taste or moral motivation, it is extraordinarily difficult for him or her to "shift terrain." This is the task of Foucauldian problematization.

In "Dreams of a Spirit-Seer," the body's boundaries enabled Kant to distinguish between, on the one hand, elements of real experience and knowledge, and, on the other, fleeting psychological events or acts of speculation and imagination to which determinate concepts could not be applied. In the *Critique of Judgment*, Kant distinguished between judgments motivated by morality or private inclination and those that all other humans could be expected to share on the basis of a distinction between interest and dispassionate pleasure. Of course, Kant never claims that it is easy to distinguish judgments motivated by inclination from those motivated by morality. To claim full knowledge of our reasons for acting, as if we had insight into ourselves as "things-in-ourselves," is *Schwärmerisch* (GR 4:407, 451). The stakes are high, moreover, for while Kant believes that the idea of acting for the sake of the law gives him access to something of "absolute worth" (428), *failing* to act for the sake of law should lead to "contempt and inner abhorrence" (426). The universalizing criteria of the moral law can rule out some moral judgments as inevitably "pathological," and a feeling of respect (*Achtung*) indicates that morality is *among* the motives driving an action. But Kant believes that few morally acceptable judgments are likely to be motivated by the sake of the law alone; they are mixed with empirical motives, including benevolent ones. And although an actor's *solution* to a morally problematic situation may be motivated by pure duty, her ability to recognize that situation as morally problematic in the first place, or to formulate her duty in a relevant maxim, are shaped by empirical and aesthetic factors. Likewise, disputes over aesthetic judgment force each individual to divide the elements of her perceptual experience that can be credited to a *sensus communis* from those that "stick" to her as *symptoms* of her bodily interest.

What Kant refers to as "hypochondria" (*Grillenkrankheit*) may be the *individualizing* and *isolating* effect of pleasures *in* one's capacity for pleasure that remain incommunicable and therefore a sign of vulnerability rather than potentially collectivizing strength. Perhaps the fear of impending illness is what happens when "inner differences" in one's aesthetic perception of a social object become a source of worry and guilt. In the *Anthropology*, Kant defines hypochondria as the tendency to pay inordinate attention to the specific sensations of one's body as if they illustrated diseases about which one has read:

> Certain internal physical sensations do not so much disclose a real disease present in the body but rather are mere causes of anxiety about it; and . . . human nature, by virtue of a peculiar characteristic (which animals do not have), can strengthen or sustain a feeling by paying attention to certain *local impressions*. On the other hand, either intentional *abstraction,* or abstraction caused by other distracting occupations, may weaken the feeling,

and if the abstraction becomes habitual, make it stay away completely. (*AN* 7:212)

It seems that Kant is recommending that the hypochondriac concentrate on what is abstract rather than concrete—for example, philosophy rather than health. But further consideration suggests that he is actually recommending abstraction, the *method* of philosophy, as a way to alter and prevent the incommunicable aspect of our relations and differences from others from turning into worrisome symptoms. Of course, given the state of medicine in the latter part of the eighteenth century, some of these symptoms might have been empirically grounded. But the persistence of conditions whose causes remain obscure even today indicates that "empirically" invisible aspects of embodiment are philosophically interesting even when their cause remains obscure. A man may be mistaken about his actual health, because it is a question of the causes of death and therefore of the understanding, but the *feeling* of illness can neither be proven nor disproven (Kant 1992a, 181). Here the incommunicable is a potential hazard because it symptomatizes the individual's vulnerability to the effects of living according to norms he or she did not help shape.

In the third essay of "Conflict of the Faculties," Kant reiterates his belief that "*philosophizing*, in a sense that does not involve being a philosopher, is a means of warding off many disagreeable feelings and, besides, a *stimulant* to the mind that introduces an interest into its occupations" (Kant 1992a, 185). Here we see a different way of regarding both thought and symptoms; rather than being what abstracts from the particularity of feeling or returns a body unable to distinguish genuine taste from the merely agreeable to a universal state of mind, "mental work can set another kind of heightened vital feeling against the limitations that affect the body alone" (189). This is a work of resistance, rather than expansiveness, obviously preferable to anxiety, but failing to account for the circumstances under which individuals recognize their bodies as individuated in the first place. Andrew Cutrofello refers to it as a disciplinary care of the self (1994, 48, 60–61).

Freud conjectured that individuals construct a "bodily ego" or imaginative anatomy based on experiences of pain and illness in particular parts of their bodies (1961, 26). I suggest that individuals construct a similar anatomy from those experiences of pleasure or reflective fascination that *fail* to communicate or resonate with others. They live "inside" the world cognized through the dominant forms of harmony between imagination and understanding—and experience their own imagination as a byproduct of bodies that are understood in merely empirical terms. Although Kant's overall goal was to ensure that every rational being could step *outside* the empirical, anthropological world through transcendental reflection or moral self-legislation, some people's bodies stand before them as signs that these escape routes are forbidden. For example, in a media-saturated world, everyone has some knowledge of how the rich live and the aesthetic forms (activities and products) through which they represent their most disinterestedly pleasurable self-relation. Those who are poorer cannot afford such products or activities, and this may be because of overt discrimination or geographical location. But they also know that the aesthetic forms they would use to represent this universally communicable state of mind are unknown and uninteresting to the rich because they come from bodies already known to be uneducated and addicted to the "agreeable." A similar dynamic holds between older and younger generations in cultures where advertising and education favor one set of aesthetic norms over the other, which are suspected of being merely "agreeable." Those who must communicate using aesthetic forms that cast suspicion on their own capacity for genuine—that is, communicable—aesthetic pleasure become passive rather than active participants in world-historical events. How else can we explain the fact that in a world of increasingly unequal life chances and opportunities, so many of those whose bodies are endangered feel and are made to feel that disorder is "all in their heads"?

Kant's aesthetic doctrine offers an account of the transcendental conditions for detecting similarities among political spectators or scientific observers as well as for grouping and finding representative kinds among natural phenomena. These conditions include a certain range of *feelings* held to be epistemologically significant and communicable in anyone who qualifies as human. The Kantian, however, has been trained to respond to such feelings with an immediate concern for their universalizability or partiality, rather than concern for his or her *power* to enjoy or detach from them in order to combine with other humans. In a society whose economic and political structures produce a variety of recognizable body types with different rights and obligations to one another, the aspiration to *universal* aesthetic feeling can only force individuals from marginal groups to assume personal responsibility for the *non-universality* of their social and bodily situation (Warner 2002, esp. 160). At the same time, the experience of *dissensus,* of an inability to impress others with the imaginative potential of certain despised experiences or forms, enhances and develops the differentiation of those bodies, their distance from any "universal humanity." The social meaning of bodily differences, which corresponds in so many ways to a differentiation of qualitative life opportunities, health, and emotional security, is produced in relation to the *sensus communis*. These differences define an "object" which consists in ties as well as distances between people, embodying the subject's awareness that human experience has a form which is imperfectly collective, a form which both holds out the possibility that the world should be a *whole* for each individual and prevents this whole from being *completed*.

The "Third Analogy" establishes that a judgment concerning the *simultaneous* coexistence of substances presupposes their capacity for mutual interaction, their collective determination as a formal whole grounded in an experience of sensible intensity, apart from things-in-themselves. But Kant's "Analytic of the Sublime" suggests that the experience of simultaneity *also* results from commensurability between the size or force of appearances and the unit through which we attempt to judge the coexistence or successiveness of their constitutive parts. This discussion from the third *Critique* adds the *qualitative* dimension of differences among experiences of extensive magnitude and intensity to our existing understanding of how bodies can allow us to represent the "baseline" rules for distinguishing between permanence and change. Commensurability with the feelings of others is precisely what "pathological" judgments of private preference fail to achieve. But the sublime also complicates our assumption that the unity of social space, space as pure form of appearances, and the "representational space" of Kantian reason are homologous.

Ordinarily, Kant explains, measuring appearances and comprehending them as units are commensurate: my desk is four feet long, and both the foot and the length of the desk can be grasped in a single intuitive act (*CJ* 5:259). When the appearance I am measuring not only exceeds what I can grasp in a single intuition, and requires a measure that also goes beyond my comprehension (or suggests itself as a potential measure for something even more inconceivably colossal), then this "subjective movement of the imagination" "does violence to the inner sense." In relativizing the scope of spatial measures, the sublime casts doubt upon the scale of temporal determinations used to schematize ordinary experience, suggesting that experience can comprise a potential plurality of simultaneous durations. The scale of space is always related to the amount of time we have to attempt its comprehension, or to contrast the apprehension and imagination of unboundedness with reason's Idea of totality. Once we become aware of the link between finitude and the multiple time frames required for aesthetic comprehension, something we might call historical shows itself within the temporal. Finite human reason becomes historical when it discovers that its psychological or phenomenological temporality is implicated in a world involving objects and durations that are longer than itself, both older and extending into the unforeseeable future. In "What Is Enlightenment?" "Perpetual Peace," and other late essays, the historical dimension eventually becomes the *topos* within which scholarly dialogue and law will progressively increase the scope of human communicability.

Because it works with the concepts and forms of judgment, rather than an intuitive understanding, the unity of reason is indefinitely multiple, involving a plurality of representational domains that cannot be reduced to the forms of pure inner and outer intuition. We might ask, however, what gives Kant reason to think that spatiality and temporality are any less complex and diverse than reason itself has proven to be. When he abandoned his pre-critical Newtonian view of space as a "divine sensorium" unified by correspondence with the unity of a single, eternal, intuitively understanding God, why did Kant assume that the space and time of finite human intuition would remain unified and uniformly divisible or comprehensible? These questions point to the real defect of Kant's account of space as "form of outer intuition" and his failure to consider time as a form of anything other than an "inner," "psychological" sense—namely, these uniform forms of intuition cannot account for the *felt* or *qualitative* differences among regions and "styles" of space and "periods" of historical culture which provoke aesthetic delight or dread.[53] Interactions between bodies and their common spaces have qualitative differences, but these cannot be accounted for except as deviations from the dispassionately pleasurable norm that is presupposed by aesthetic judgment. The universalizable experiences of feeling are those which are associated with the body inserting us into the manifold of "socially shared intuition," one among many objects in abstract space, while other feelings are relegated to the limbo of psychological interiority. Where once the body's boundaries offered Kant the very model for the act of distinguishing between private fantasy, speculation, and the reality of shared experience, now bodies appear only as objects among others, apprehended within a manifold of intuition.

---

53. See Kubler 1962 for a discussion of style and historical differentiation; also see Lefebvre 1991, 205–7.

Kant's discussion of the sublime in the *Critique of Judgment* acknowledges that in neither its scope nor its affective tensions are we capable of grasping the entirety of space at once. In the *Critique of Pure Reason*, Kant's first and second conflicts from "The Antinomy of Pure Reason" demonstrate the wrong-headedness of asking about the reality of the world's spatial and temporal unity, infinity, and limitation. In discussing the first antinomy, Kant argues that it is mistaken to regard the world as either limited or infinitely extended in time and space; in resolving the second antinomy, that it is mistaken to presume that the world is composed of simple or infinitely complex elements. It is always possible to change the level of detail or scale at which form may be found in the world, as I discussed in the context of materiality; neither space nor time are properties of things-in-themselves. And yet Kant never considers the possibility that space could be unified but striated, or that its uniform and uniformly divisible expanses could involve a variety of overlapping and conflicting metrics. In other words, the unity he hopes to give experience, social consensus, and reason remain *ideals*, albeit mutually enabling ideals. According to Rogozinski, what provokes the feeling of sublimity is not the size of the *measured* object (pyramid, basilica) but the *magnitude* or "size" of the *measure*, which is capable of infinite differentiation and determination and *in which* objects of greater and greater size can be successively determined and swallowed (1996, 133–35). Its enormity is not the enormity of a large rigid object but that of a stretchy fabric whose weave can allow immeasurably large (or small) objects to pass through and be perfectly measured.[54] Recall that *scale* is one of the crucial factors in giving an entity resistance to apprehension, resistance to other objects, but also the potential for conceptual repositioning and physical manipulation. Stretching an object or one's attention span, like expanding or con-

---

54. The *elasticity* of relations between substances, as Shell notes, was in fact a theme of Kant's early *Physical Monadology* (Shell 1996, 76). "Savage space" is the Merleau-Pontyan name Rogozinski gives to this form of pre-intuition: "prior to phenomena, but already differentiated, asymmetrical, heterogenous, traversed by fractures and wrinkled by tracks," and "situated in the 'personal space [*éspace propre*] of the body,' its primordial *orientation* structured by the opposition of high and low, before and behind, left side and right side" (1996, 137). Gibbons (1994, 144–46) contends that the body is the fundamental "measure" of spatial magnitude or living force exposed by both the mathematical and the dynamic forms of the sublime. This makes best sense, then, if we understand by "body" the most fundamental rhythm or style of spatializing and apprehending time that a body performs. Rogozinski's reading helps us to see why the "body" must be rethought in terms of the role played by pure forms of intuition in guaranteeing the communicability and unity of the world, rather than the pure forms of intuition in terms of the body understood as a naturalistic entity.

tracting a discursive viewpoint, reveals different points of contact and "noise" or smoothness and intelligibility; even boredom (Strathern 1991, xxiii–iv). In Rogozinski's view, Kant ought to have conceived both space *and* time as infinitely stretchy measures, whose internal differentiability constitutes all possible "community" of persons and objects in the world.

But is it meaningful to talk about space as being stretchy or fractured in everyday experience, rather than in the special context of aesthetic apprehension? Kant was wrong, we now know, to presume that Euclidean space is the only possible way to conceive of the space that defines our existence.[55] In fact, the everyday expectations governing our experience of distance and movement can be radically redescribed from the hypothetical standpoint of spaces having more than three dimensions, such as, for example, a space-time which explains the relations between energy and matter better than Kant's form of outer intuition or Newton's absolute Euclidean space. The act of displacement through which we understand ourselves as inhabiting space-time, like that which gives rise to Kant's transcendental topic, cannot even be considered a "hypothetical" standpoint, although it takes a special act of reflection to recognize the temporal aspects of an experience that, outside the lab, makes perfect sense in terms of Euclid. Yet there is no reason, topologists and physicists argue, to assume that space or space-time has a uniform character; mass, for example, alters the way in which objects embedded in such a space behave. Although a figure drawn on a Möbius strip retains all its proportional features in sliding around this two-dimensional surface, the same figure might be forced to stretch when sliding around a torus, because the degree of a torus's curvature is constant on the flat side but steep if one goes toward the center.

Graham Nerlich offers several descriptions of how three-dimensional humans might experience the process of "sliding around" in a space that had more dimensions than our own:

> Squash a flat piece of paper onto the surface of a sphere and it wrinkles up and overlaps itself; if it were elastic, some parts of it would have to stretch or shrink. Since we ourselves are reason-

---

55. In the A edition of the *Critique of Pure Reason*, Kant asserted that his conception of Euclidean space as a "necessary a priori presentation" *ought* to owe nothing to experience (otherwise geometry would be contingent), but admitted that he could not absolutely preclude its supersession: "As far as we have been able to tell until now, no space has been found that has more than three dimensions" (*CPR* A24).

ably elastic we could move about in a space of variable curvature, but only by means of distorting our body shapes into non-Euclidean forms. We would have to push to get our bodies into these regions, for only forces will distort our shapes. If the curvature were slight, the rheumatism might be easy and bearable; if acute, fatally destructive, just as if you fell into a black hole too. (1994, 39)

A figure squashed onto a sphere exists in two-dimensional space, but it can also exist in three-dimensional space, as can be surmised from the fact that a human from outside the strip or the sphere put them there in the first place. Humans, likewise, may exist in a variety of n-dimensional spaces in addition to the one that makes sense to us as a "form of outer intuition." This is what I tried to show with the earlier reflection on transcendental deliberation and *Las Meninas. Even the space of "real interaction" may be thought of as multiple, as well as stretchy.*

Nerlich is convinced that Kant was wrong to abandon the hypothesis that space is both real and absolute, which makes sense of Einstein's space-time far better than Newton's Euclidean world. But Kant, unlike philosophers of mathematics, was interested in explaining the social impact and causes of our "representations of space" along with the topological impact and causes. Nerlich assumes that bodies develop naturally "in" space but may be distorted by its curvatures. Kant assumed that bodies would be most "at home" in a space materially and interpersonally unified by a common "transcendental aesthetic." I suggest that bodies are historically recognized and fashioned in ways that enable the multiplicity of representational spaces and the variable metrics of perceptual space to preserve the ideal unity of a "manifold." The social space in which we interact with one another, with nature, and with technology can, like the rheumatic space of Nerlich's non-Euclidean holes, be incredibly painful or pleasurable, depending on how communicable one's organization of these spaces turns out to be. Its curvatures are measured by the greater or lesser universality of the qualitative experiences or feeling that we attribute to the bodies of fellow human beings.

We cannot know, according to Kant, what bodies are in themselves, or to whatever "thing-in-itself" our bodies may give access. The appearances and abilities of bodies are mere *forms* for the exercise of sensibility and understanding, in conjunction with imaginative reason. But they are *material* forms, that is, forms whose possibilities for variation are limited in space and time, by the very capacities that enable them to move, reflect, and change, including change in perspective. It might be best to think of them as *symptoms* of a certain historical configuration of imaginative reason, one put together by Kant in order to create community and knowledge for finite knowers independent of God's superior intuition or arbitrary power to preserve harmony. To get free of this configuration of imaginative reason and to apprehend the world in any other way than through these administratively and economically recognized bodies does not mean giving up the capacities for receptivity and activity associated with sensibility, understanding, or imaginative reason, any more than Kant gave them up when the body vanished from his philosophy.

By asserting that concepts and laws of the understanding apply only to appearances, not to things-in-themselves, Kant acknowledges limits on human knowledge. But he secures room for an empiricist scientific practice that owes nothing to a theological conception of God's creative or intellectual intuition regarding the essences of material and psychic beings. Without turning language into a mysterious source of appearances (natural or divine), Kant gives *discursivity* a place in the process of human knowledge. The principle of publicity and the touchstone of communicability enable him to place limits on the discourse's tendency to push understanding beyond the limits of experience, toward inadmissible knowledge claims regarding the objects of pure reason. At the same time, Kant's principled refusal to elaborate the unity of reason without recourse to fundamental powers of the mind or nature leaves each individual room to legislate over his or her actions without being haunted by an absolute vision of the human good.

The European domination of "non-standard" bodies and of physical geography in the colonial world was facilitated, in part, by the growth of positive scientific knowledge beginning from these limits, along with the experience of autonomy held in check by law. The being of language and the being of the physical human body changed in significant ways when they ceased to be, for all epistemological and political purposes, the expression and the creation of an all-knowing and self-revealing God. After Kant's Copernican revolution, the body could no longer be the essential form linking microcosm and macrocosm, monad and preestablished harmony, as knower and known. This role is now assumed by the transcendental structure of possible experience. Neither could the body be a simple source of empirical data and a datum in its own right. For, as a datum, it could not explain the organization of data into a single experiential manifold. Rather, we find it on the side of both the transcendental and the empirical, obliquely referenced in the doctrine of faculties (especially sensibility, feeling, and imagination) and overtly referenced as an object of the physical and anthropological domains.

According to Heidegger, Kant "shrank back" from considering imagination the hidden root of intuition and the conceptual powers of understanding and pure reason because he had forgotten that the finite knower is "man," not simply "any rational being." But "man" is hardly a self-evident concept or phenomenal appearance. In the reading I have proposed, Kant appealed to the anthropological sphere to resolve conflicts in pure reason associated with the divisiveness of imagination. If "God" was one way in which humans made sense of the limitations imposed by their imaginative, sensual reason; critical "man" is yet another way in which reason both employs and limits imagination. It is up to those who regard themselves as "man," in the end, to resolve the bifurcations among reason's domains by rigorously bracketing their pathological interests and incommunicable pleasures in the domain of psychology and approaching the natural and political world as if from a *sensus communis*.

PART 2

MAN AND HIS DOUBLES:
TWO WAYS TO PROBLEMATIZE

In the last few decades, scholars and activists have turned to the human body, as seat of experience and as target of power, for ideas about political resistance to a multiplicity of oppressions—sexual, racial, and economic—as well as possible strategies for reconfiguration. The body's interests seem to be "real" in a world of deception and emotional as well as economic exploitation (Harvey 2000, 14, 100–101; Lowe 1995, 14–15). At the same time, imaginative associations with the body have allowed vast numbers of people to be seduced by consumerism, intimidated by modern medicine, and subordinated to hierarchical religions or political ideologies that promise health and harmony. The body is at once a seat of self-evidence, autonomy, and resistance to power—and the most powerful reference point for people to struggle over the content of one another's experience and persuade one another to adopt this or that set of "real" interests as their own.

I have tried to show how Kant assumes that the human body exists at the crossroads of many conceptual and perceptual spaces: empirical appearances in external nature, psychological introspection, transcendental deliberation, moral self-legislation, and the anthropological/historical world. But the body at this crossroads is invisible and unspoken—it is a way of organizing phenomena and acts of reflection, of being passively affected by the given and spontaneously seeking out the given when reflection proves unsatisfying. Foucault's histories show how modern forms of administration and knowledge put "flesh" on the bones of this structure, taking advantage of the very fact that we have no access to our bodies as "things-in-themselves" in order to tell us what those bodies *really are* and what they *really need* or *really can do*. Foucault's controversial claim that we should turn away from sexuality and desire toward "bodies and pleasures" if we think that sensual experience has a liberatory dimension should be read in this light: there is no one thing such as the "body" in itself to which a confessor, doctor, or psychiatrist has privileged access (*HS1* 157). Nor was there a "natural," premodern way of living that body in unreflective harmony with society and the sacred. There are only plural, ultimately unique bodies and pleasures or pains produced by power relations that give individuals more or less access to and ability to explore the singular experience of their own bodies.

The following sections lay out Foucault's treatment of the problematic object in his histories of madness, medicine, prisons, and sexuality. We begin by comparing the unified "world" that serves phenomenologists as a horizon for meaning to the presumption of heterotopia that orients Foucault's thought. The subsequent section shows how madness, death, criminality, and sexuality function as problematic objects enabling social-scientific discourses and administrative practices to conceal the appearance of heterotopia behind their treatment of the human body and population. In doing so, has Foucault simply put a new "transcendental-empirical doublet" in place of the "man" in nineteenth- and twentieth-century humanism? The third section argues in the negative: "man and his doubles" are a system of resemblances, not an identity. This system is based on specific experiences of aesthetic pleasure and communicability, in which bodies, like statements, can function as *events* of resemblance rather than *things* or *identities*. But how can events be discursive and material? If one thinks of bodies and statements as bundles of resemblances, gathered by a schematic imagination but dissociable by reflection, I argue, then phenomenological and genealogical approaches to the body are not as incompatible as some feminists have feared. Nor, as the final section contends, does Foucault's desire to distance himself from phenomenological reference points like "man" and "world" necessarily lead to an unethical and an-aesthetic foreclosure of emotion.

Heterotopia and the Phenomenological World

In *The Order of Things*, Foucault states that his goal is "not to calculate the common denominator of men's opinions, but to define what made it possible for opinions about language [among other topics] . . . to exist at all" (*OT* 119). In other words, he seeks to identify the conditions of possibility for a specific *sensus communis*, that of seventeenth- and eighteenth-century European intellectual culture. Foucault also questions whether the "world" is an object of experience, or a mere Idea of reason made possible by the enactment of a preliminary cut between "appearances" and "things-in-themselves." *History of Madness* and *Birth of the Clinic* begin by positioning the philosopher at the dawn of a discriminating practice. For example, we must "try to recapture, in history, this degree zero of the history of madness, when it was undifferentiated experience, the still undivided experience of the division itself" (*HM* xxvii).[1]

Unlike Kant, Foucault identifies this division between the realm of experience and the "object" of a problematic concept with a series of specific historical events. Zero points, moments of historical hesitation between two or more forms of order, foreground the cognitive and perceptual acts through which thinkers and their cultures consign certain possibilities to historical *impossibility*. Confronting Borges's description of a certain Chinese encyclopedia whose categories stagger the Aristotelian mind, he muses, "But what is it impossible to think, and what kind of impossibility are we faced with here?" (*OT* xv). Again, "what does it mean, no longer being able to think a certain thought? Or to introduce a new thought?" (50) The "unthinkable" is a point of reference whose *exclusion* grounds the possibility of positive knowledge: the unthinkable can always be historicized in relation to some "other" possibility that happens to be actual for us. But positive knowledge of the world, according to Foucault, never forms a seamless reading, except by reference to some problematic object on which each discourse touches and from which each discourse recoils.

In a pseudonymous article for the *Dictionnaire des Philosophes*, Foucault described his own work as falling into the "critical tradition of Kant" (*AME* 459). But obviously he made choices in his Kantian inheritance. Some elements of this inheritance are the presumption of real conflict between objects of experience and the merely regulative use of ideas like God or the world, which may play an important role in orienting more phenomenologically or hermeneutically oriented thinkers. He is not seeking to unify pure reason, nor to criticize pure reason in light of an ideal unity, but to describe its plurality:

> I would not speak about *one* bifurcation of reason but more about an endless, multiple bifurcation—a kind of abundant ramification. . . . Everything propitious to the development of a technol-

---

1. On the importance of the division rather than the excluded object (madness), see "Sorcery and Madness" (*FL* 109).

ogy of the self can very well be analyzed, I think, and situated as a historical phenomenon—which does not constitute *the* bifurcation of reason. In this abundance of branchings, ramifications, breaks, and ruptures, [the Kantian "self-limitation" of reason] was an important event, or episode. (*AME* 442; see also *PWR* 328–29)

In *The Order of Things*, Foucault puts his project under the sign of a "general critique of reason," but insists that he learned the elements of this critique from Cuvier, Bopp, and Ricardo rather than Kant or Hegel (*OT* 342, 307).[2] One might say schematically that Foucault is Kantian insofar as he believes modern structures of power/knowledge solve problems arising from the plurality of rational discourses in the epistemological, historical, and political *episteme* inherited from Kant. More important, Foucault resembles Kant in following the philosophical *method* of beginning his historical analyses to one side of a field unified through infinite judgment.

To understand where Foucault both continues and breaks with the Kantian problematic, we must look at how his various texts describe the relationship between imagination, language, and bodies. Foucault's earliest essays addressed the topic of phenomenological imagination and its ability to project a coherent world of appearances. In "Dream, Imagination, and Existence," *Mental Illness and Personality*, and even *History of Madness*, Foucault was intensely interested in the "world" of meanings projected by the dreamer, madman, or witness to madness. He was especially intrigued by how statements and imagery that might have lacked human significance in isolation generated "meaning" when brought together in the context of a shared and potentially threatening world. His *thèse secondaire* for the *doctorat d'état*, on the genesis and structure of Kant's *Anthropology*, presented the human being as a "citizen of the world," not primarily as finite knower or moral agent. But Foucault's interest

---

2. A very thorough treatment of the "critical" theme in Foucault's work can be found in *Foucault et Kant* (Fimiani 1998). Fimiani traces the development of Foucault's work from its original effort to identify the conditions of intellectual and political "minority" decried by Kant in the essay "What Is Enlightenment?" to his late explorations of *parrhesia* as comparable to the public use of one's reason. Bernauer (1990) stresses the "human" as conceptual prison and object of critique; Simons (1995) focuses on the implicit agonism of Foucault's later work and notes how limits, for Kant as well as Foucault, are both enabling and restrictive conditions on life and thought.

quickly moved from phenomenology to problems involving political conflict over the being of language and the human body.

What does it mean to have a world? Kant's solution to "The Antinomy of Pure Reason" asserts that competing philosophical descriptions of the world (as limited or limitless in time and space) cannot be resolved while the world is regarded as accessible *in itself* to human understanding. The wholeness of the world is merely an Idea of reason, and therefore the object of a merely "problematic concept" or Idea of reason, marking out boundaries for the extension of natural science without allowing or requiring a stand on cosmology. Kant does assert that science will progress and Christians will avoid morally dangerous naturalism, if they recognize the *regulative* validity of this Idea. By this he means that reason uses the Ideas to "direct the understanding to a certain goal by reference to which the directional lines of all the understanding's concepts converge in one point" (the *focus imaginarius*), rather than imagining that they give the understanding access to an actual object (*CPR* A644/B672). The *Prolegomena* reiterates Kant's critique of the dialectical illusions resulting from constitutive, rather than regulative use of this Idea. The community among objects of experience is one kind of "world." But shared "objectivity" or the intersubjective validity of logical judgments is a world in a different sense than the community among *objects* of experience (*PR* 4:298). Finally, in the *Critique of Judgment, Anthropology*, and other later writings, the "world" connotes a sphere of society, custom, and political action. Thus the cosmopolitan right to be a "citizen of the world" is not endangered by Kant's criticism of the "world" as cosmological Idea.

Foucault's contribution to the relationship between imagination and world is mediated by twentieth-century phenomenology and by the style of historical analysis practiced by the *Annales* school in the first half of the century, in which the intelligibility of certain historical phenomena depended upon the time frame over which they were observed.[3] Edmund Husserl, the founder of phenomenology, sought to determine the basic structure of conscious experience through a reflective technique similar to Kant's (Husserl 1962). Bracketing the question of whether a particular object of thought, imagination, memory, or value really existed external to the thinker, Husserl asked about the characteristic styles or structures of the intentional acts through which memories, fantasies, acts of evalua-

---

3. Dreyfus and Rabinow (1983), Han (2002), and Lawlor (2003) devote significant space to analyses of Foucault's relationship with phenomenology.

tion, and perceptions could be distinguished from one another. One of his goals was to identify the essence (or meaning) of a typical intentional object. He was also interested in the essence of different *kinds* of intentionality (memory, fantasy, evaluation, perception, etc.).[4]

Husserl discovered that a single object, essence, or meaning could be grasped in multiple intentional attitudes or acts and often had a place in different webs of meaning and styles of intention (as a single lover can have a well-developed place in memory, fantasy, evaluation, and perception). But Husserl's focus is *consciousness*, rather than *reason*. Imagination is a privileged mode in which consciousness explores its own structure. Phenomenological reflection reveals a transcendental structure of consciousness *in which* every consciousness participates knowingly or unknowingly, and conversely every consciousness can immerse itself in this transcendental structure through reflection. But Husserl takes pains to stress that consciousness experiences its unity in a single time-flow (1962, 215–18), and its intentionality does not just aim at this or that meaning, but at the entire web of related meanings that constitute, together, a "world" (134). Although this world may only be the problematic object of a Kantian Idea, it is still necessary to the structure of Husserl's transcendental consciousness. Foucault challenges the philosophical priority of a world "for consciousness."

Unlike phenomenologists, the historians gathered around the *Annales* school gave historiographic priority to phenomena whose changes could not be observed in the time frame of ordinary human consciousness.[5] And unlike traditional historians who study battles, kingdoms, biographies, and beliefs, *Annales* historians like Marc Bloch, Fernand Braudel, and Lucien Febvre focused on rhythms in the price of basic staples, trade routes, and mentalities rooted in practices of everyday life. These phenomena, while clearly human activity, seemed self-evidently part of the natural environment, and thus outside the realm of historical reflection. Although

---

4. Husserl did not work with "faculties" in the sense of early modern philosophy, but his effort to analyze and distinguish the various acts of consciousness compares easily to Kant's effort to distinguish between powers of thought in transcendental deliberation, on the basis of felt differences in their object-representations.

5. Dosse (1994) gives an overview of the *Annales* school. Gutting (1989), Kusch (1991), Daddabbo (1999), and Flynn (2005) address Foucault's relationship to their work. Both Gutting and Kusch relate the *Annales* historians to antipositivist historians of science such as Bachelard and Canguilhem; Kusch traces changes in their use of the "event" as a primary analytic concept (169–78). Daddabbo, who reads Foucault strongly through Deleuzean eyes, is also interested in how to render structuralist theories capable of accounting for historical change.

the emphasis on social rather than political life is the most obvious contribution of the *Annales* historians to contemporary historiography, their belief that phenomena with different rates of change break up any apparently unitary historical "flow" into *multiple* time frames is equally important. Women and colonized peoples were considered by many scholars to be "without history" or "outside history" because changes in their life habits and practices were either considered "natural" or determined by outside powers. Attention to multiple levels of historical change revealed events in the life of subaltern groups. These events could now be considered qualitatively distinct historical—and existential—standpoints for the organization of heterogeneous temporality. In his emphasis on the political significance of private life as well as his exploration of history as a layering of series, Foucault was influenced by the *Annales* school, especially in *The Order of Things* and *The Archaeology of Knowledge*.

But Husserl's focus on consciousness was insufficient for later phenomenologists, leading them to apply the ideas of *imagination* and *world* in new ways. Heidegger and Merleau-Ponty address the manner in which intentionality and the anticipation of meaning characterize human *being* rather than simple *consciousness*. Language and the body, respectively, are Heidegger's and Merleau-Ponty's chief vehicles for exploring humans' ontological orientation toward past and future. For existential phenomenology, the world is an orienting horizon for language and the body, understood as modes of being rather than as essences or meanings for pure thought. But like Husserl, these thinkers assume that inasmuch as we share a single world, there is a single transcendental or ontological structure for all "normal" humans, discovered and varied instructively through imagination. Henri Lefebvre follows in this tradition when he assumes that representations of space, spatial practice, and representational space all refer to facets of a single social world, even though they may permit it to be reconfigured by slipping like tectonic plates in relation to one another. Neither Lefebvre, Heidegger, nor Merleau-Ponty consider that a single social world might not *appear* except as multiple spatializations compete for control over the actions and words of a multiplicity, a population, or a public.

In the essay "Age of the World Picture" (1977b, 115–54) Heidegger argues that modernity involves the effort to grasp all beings, including humans themselves, as representations or potential objects of representation. Modern historical awareness, according to this perspective, involves the projection of distinct representations or "world-pictures" at different moments in time, where time is conceived (as in Kant's Transcendental Aesthetic) as a uniform form of appearances. Interpreting the being of language as *representation*, as early modern thought did and many contemporary philosophies do, prefigures the project of technical mastery over nature Heidegger calls *Ge-stell* or enframing (1977b, 19–27). In *The Order of Things*, however, Foucault proposes that the idea of "man" as the historical subject and producer of representations only emerged with the *collapse* of an order of knowledge founded on representation, an order in which language and created beings were presumed homologous within the act of divine creation.

In his *thèse secondaire*, Foucault suggested that the need for anthropology emerged when Latin ceased to be the unquestioned language of international scientific and philosophical thought (1961, 95–99). Lessing's reflections had also begun to detach poetry from the representational and therefore ultimately pictorial ideal of Classical discourse (Lebrun 1970, 611–13). Literary authors sympathetic with romanticism began to reflect on the *positive* aesthetic value of time and the way the temporal flow of language could evoke emotions of hope, relief, and regret, rather than regarding it as an inevitable deviation from the presentation of a contemplative whole. When languages such as Sanskrit were discovered to be more ancient than Hebrew, and when processes of development and mutation in language comparable to the families of living beings were first identified, the finitude of human knowledge could no longer so easily be regarded as a simple limitation on God's perfect table of cosmic order. Language had once been one "version" of representation among others—as for Kant, who celebrated the discursive understanding rather than the historically linguistic dimension and was proud to have derived his table of categories from the (intemporal) functions of judgment rather than a "rhapsodic" reflection on grammar (*PR* 4:323). Now, with the advent of romanticism, more philosophers were willing to agree that representation was both made possible and limited by the specific language in which the representation was made.

In the century following Kant, European scholars no longer identified knowledge with an ideal grid of representations but with the transcendental conditions for a specifically finite, *human* experience. As domains of positive knowledge and conversations in specific languages began to break away from the model of representation, only "man" as living, speaking, and producing being could unify them in a single trajectory. All knowledge was referred to the forms through which "man" understood himself as *finite*—life, labor, and language. Where phenomenology investigates the experience of the conscious, embodied, social subject who participates in these processes, Foucault believes that this subject cannot be generalized across all historical periods and cultures. He also suggests that there are advantages to an epistemological standpoint that is neither "human" nor historical in a general sense.

Foucault suggests that the concept of "man" was a *screen* protecting the era from developing the full implications of the idea that life, labor, and language were specific, discontinuous, and "active" levels in relation to human being. These plural historical trajectories give the impression of an "originary" stratum that precedes every empirical human existence; and "it is by this means that men enter into communication and find themselves in the already constructed network of comprehension. Nevertheless, this knowledge is limited, diagonal, partial, since it is surrounded on all sides by an immense region of shadow in which labor, life, and language conceal their truth (and their own origin) from those very beings who speak, who exist, and who are at work" (*OT* 331; see also *FL* 76). He conjectures that the twentieth century's fascination with language's independence from human intentions coincides with a new willingness to accept that the historical process is no "single" flow but rather a plurality that the student of history can grasp only imperfectly at any one point of simultaneity. This fascination also provokes reflection on how training bodies to speak or be silent helps *each* person participate in collective control over the language's "independence."

"In every society," Foucault conjectures, "the production of discourse is at once controlled, selected, organized and redistributed according to a certain number of procedures, whose role is to avert its powers and its dangers, to cope with chance events, to evade its ponderous, awesome materiality [*d'en esquiver la lourde, la redoubtable materialité*]" (*AK* 216). In order for knowledge to take the form of a great text, a table of representations, or a description of lived experience and its world, resemblances and fictions must be circumscribed. "If language were as rich as existence," he observed in *Death and the Labyrinth (Raymond Roussel)*, "it would be the useless and mute duplicate of things; it would not exist. Yet without names to identify them, things would remain in darkness" (Foucault 1986a, 165). The situation is worse, however: language threatens to be *richer* than being, as shown by its potential for infinite complexification and interpretation. Every human speaker forces language to express his or her conflicting demands and even to invent new ones. What is to be done? Prohibitions, divisions between those qualified to speak and those not, and the privilege or priority of potentially verifiable speech over expressive speech enable this fertility to be brought under control. "This will to truth, like the other systems of exclusion, relies on institutional support: it is both reinforced and accompanied by whole strata of practices such as pedagogy . . . the book-system, publishing, libraries, such as the learned societies in the past, and laboratories today" (*AK* 219; see also *OT* 319). Those who appear deviant within this economy of discourse are not just those who say incommunicable things, but those who break the rules and betray their ignorance by speaking when they were expected to be silent, violating the conventions of propriety that enable speech to be meaningful (*P/K* 82–83; *SMBD* 182–84).

As Foucault's work progressed, he came to understand bodies less and less as the paradigmatic site of phenomenological self-understanding. Rather, they came to take on the same kind of materiality associated with language in his early work, and were inextricably bound up with the certification or exclusion of speakers. Like "language," the "body" is an ontologized abstraction that occurs only in historically variable examples and remains invisible while culture takes the unity of reason for granted.[6] The "human" body as targeted by modern social science and administration is therefore a solution to *problems* arising from increased awareness of the discursivity and materiality of language. By providing a stable anchor and *form of appearance* for multiple discourses that cannot be captured in a single scholarly practice (however stringent the requirements for producing communicable utterances may be), the communicable, communicating body makes the world seem naturally unified—a plausible regulative Idea, if not a "thing-in-itself." It is crucial to consider the body's role in "naturalizing" the systems of order provided by scholarly discourses and other specifically modern practices if we are to understand the significance of the body for contemporary discussions of political imagination.

Unlike Kant, therefore, who attempted to unify reason despite the plurality of "domains," Foucault regards any possible "world" as involving

6. For example, languages were only unified as "entities" by grammarians in the context of early modern projects of national unification. Before the seventeenth century, language was clearly tied to class and situation, just as people were, and communication was in no way expected between members of widely varying occupations. Humanism marks an advance for democracy, but does not mean that advocates of democracy should regard "man" as a true being rather than a desirable construction, nor that such advocates should regard "language," "life," and "labor" as ontological universals from which qualitative specificity or multiplicity have been stripped.

a plurality of spaces that are associated with various rates of historical change—including, but not limited to the economic, linguistic, and biological strata. *Heterotopia* replaces the *table* and *world* as a provisional way to think about the materiality and discursivity of historical forms. He also proposes that "man" be abandoned as an illusory object of self-understanding, much as Kant abandoned talk about the "world" as thing-in-itself. His goal is not to *return* to a Classical metaphysics of the infinite, but at the same time to *resist historically given interpretations of finitude*. Finally, he is opposed to a transcendental search for the conditions that make finite human experience possible, suspecting that they will simply ground what we *already know* ourselves to be empirically rather than uncover what we *must necessarily* be (*EST* 315–16; *FL* 252).

What Foucault preserves, despite his skepticism concerning Kant's "analytic of finitude," is a sense that the *discursivity* and *materiality* of the human understanding, as exemplified in language and embodiment, are objects of constant political antagonism (e.g., *AK* 216).[7] But none of Foucault's books approaches this problem from the same angle; it is a mistake to read *The Archaeology of Knowledge* as an elaboration of the "being of language" alluded to in *The Order of Things*, or to try and situate the "bodies and pleasures" of *The History of Sexuality* too carefully in Foucault's historical account of libertinage in *History of Madness*. Each of his books is discontinuous, beginning from a different *question, problem,* or division in the field of possible knowledge. In "What Is An Author?" (*AME* 205–22) Foucault famously suggested that the very concept of authorship was an attempt to control the proliferation of planes across which a single thought could extend itself; and just as I feared doing injustice to Kant by stressing the fragmentation of his thought, I fear imposing a false unity on Foucault's work in the name of Kant. Kant is only one name for the event "Michel Foucault"; we must remember that the name "Foucault" does not encompass it either.

---

7. Ernesto Laclau and Chantal Mouffe define antagonism as an experience of social conflict revealing the limit conditions under which a society can grasp itself as a totality (1985, 122). Antagonism corresponds to what Kant called "real conflict" as opposed to "logical contradiction." Hegemony results when this limit condition is invisible, when it is unclear which subject positions must exist for the status quo to be intelligible, or when those who suffer the limit condition go unheard. Laclau and Mouffe, unlike many Marxists, do not insist that only one antagonism can organize the social order, or that it be in the economic domain. Confinement of the mad, the indigent sick, the creation of delinquency, and marginal experiences of sexuality are occasions for antagonism in this sense (see 115–17 for their reading of *The Order of Things* in light of this theory).

In the Field of the Problematic Object

"Between the already 'encoded' eye," which has been trained to see in a way that anticipates meaning and coherence, and arises from an institution with the power to repress social disorder, "and that reflexive knowledge," which studies the elements and functioning of the world anticipated by perception, Foucault writes that

> there is a middle region which liberates order itself: it is here that it appears, according to the culture and the age in question, continuous and graduated or discontinuous and piecemeal, linked to space or constituted anew at each instant by the driving force of time, related to a series of variables or defined by separate systems of coherences, composed of resemblances which are either successive or corresponding, organized around increasing differences, etc. (*OT* xxi)

The finitude represented by imagination becomes a critical condition of positive knowledge and effective power when specific historical techniques constrain jumbled discourses and jurisdictions into a communicable experience (*OT* 331).

Madness, mortality, delinquency, and sexuality, the topics of Foucault's historical studies, are characteristics of finite beings. In medieval and Renaissance Europe madness, for example, was sometimes viewed as a sign of God's special claim over a human being. Mortality and sexuality were among the consequences of Adam's fall. Delinquency, on the other hand, is the modern name for an ensemble of acts that once were taken to indicate human weakness or rebelliousness in the face of divine laws. Each phenomenon has been a powerful spur to the human imagination and a check on reason's pride. *History of Madness* (1961) describes the changing role of imagination in European conceptions of madness. The Renaissance regarded madness as a rebuke against reason; the early modern philosophers as a willful choice of imagination over unreason. During the French Revolution, the public was terrorized by rumors about escaped madmen. A later period expressed anxiety about women's susceptibility to influence through novels. Mortality, obviously, has always been an object of vivid speculation because of concern about the afterlife, but also a desire to outwit the fearsome processes that bring it about. *Discipline and Punish* (1975) recounts the pageantry of early modern executions and the imagery of resistance and liberation presented by broadsheets and songs about the condemned, as well as jurists', psychologists', and detective authors' fascination with hidden motives for crime. Sexuality, finally, is an object of immense imaginative concern because, as the concern for novels indicates, the acts in which it is realized are as much imaginative as physical.

Like Kant's noumena, madness, mortality, delinquency, and sexuality cannot be seen "in themselves," but only in their effects. Their most profound effects are probably manifest in the "normal" person's effort to flee or control anyone who embodies or suffers from such forces. The *objects* of these problematic concepts are inscrutable, but seem to be ways of being or interacting that render or threaten to render lawful epistemological or social experience impossible. Because they participate in discursive formations marked by a confusion between empirical and transcendental levels, I will refer to them as *problematic objects* (even though there is no object, strictly speaking, for a problematic concept). "Experience" is the field from which madness has withdrawn, in which life wages its battle against disease, lawfulness beats back disorder, and individuals are motivated by conscious interests rather than compelled by biological or unconscious forces. If one can control and systematically study the persons and behaviors in which these properly unthinkable *limits* of experience show themselves (or are believed to show themselves), one will be rewarded with great power to produce further "experience" consistent with what has gone before (see Han 2002, 125). Just as Kant posited the noumenon in order to constitute the world of appearances as a potentially law-governed whole, Foucault shows that modern societies posit madness, mortality, delinquency, and sexuality in order to constitute the world of social appearances as a potentially law-governed whole. For example: one may confine, treat, or punish the madman, but in doing so, one applies and reaffirms the laws that constitute conditions for a *sane* experience of the social world.[8]

What these historical studies propose, in short, is nothing other than the procedures for creating European knowledge of an *immanent* world as opposed to a world (imperfectly) mirroring divine *transcendence*. Modern psychiatry, medicine, criminology, and psychoanalysis—to name only

---

8. Without going into a detailed comparison with Kant, Michel Serres refers to Foucault as following the path of a "Copernican revolution of unreason" within an unsettling experience of the "immediate proximity of all points of space" in premodern European experience (Davidson 1997, 51, 40).

some of the disciplines touched by Foucault's studies—became positive sciences whenever they were able to recognize a particular social *appearance* or behavior as a *form* of the limiting conditions for cognition and action in general. Of course, beginning with an agreed-upon appearance of finitude rather than speculation on the infinite did not mean that they denied all relation to the infinite (*OT* 314). Kant's division of all possible entities into a field of "phenomenal" and a field of "nonphenomenal" appearances was an infinite judgment saying nothing about the "nonphenomenal" domain in itself. Nor did it limit the "phenomenal" domain *except on one side*, with respect to the divine understanding and those who claimed to speak in God's name. There is still infinite room for specification in the universe grasped by a finite rather than infinite (divine) understanding (*OT* 314). This is because, as Kant showed in the Transcendental Dialectic of the *Critique of Pure Reason,* reason will never be able to grasp the sum total of conditions for any appearance in a single judgment or series of judgments of the finite understanding (A508–15 / B536–43). Thus there are infinities proper to the sphere of appearances; the finite repeats itself *indefinitely* in every attempt to comprehend its own finitude. As time is the condition for psychologically exploring these judgments and appearances, historical time becomes the *infinite* continuum in which individual human knowers add to the existing store of knowledge.

The "historical" is transcendent to European reason in a different way than the *divine*, for it is an *immanent* transcendence, a continuum of historicities. In *The Order of Things,* Foucault argues that life, labor, and language were forms in which post-Kantian Europe attempted to grapple with the fact that historicities are transcendent to one another without being in any way suprahistorical. Foucault writes that they were formed from a "new fold," not within representation, but within the *finitude* Kant regarded as the first principle of possible knowledge (*OT* 341). He refers to them as "quasi-transcendentals" because they are the conditions for possible experience *within* a particular anthropological experience and a particular *episteme* of scholarly knowledges (250). But it is important to note at the outset that the "quasi-transcendentals" life, labor, and language are *not* the same kind of thing as madness, mortality, criminality, and sexuality, which I have referred to as "problematic objects." Although both can be compared to the Kantian noumenon, they represent different uses of the "problematic concept" distinguished by Kant. The problematic object is the noumenon whose exclusion from human knowledge constitutes the field of critique; the "quasi-transcendental," the *noumenon* mistakenly identified with the Ideal or unconditioned condition of all appearances by pre-critical metaphysics. Picking up the story where Kant left off, therefore, Foucault argues that the domain of knowledge unified by critique was vulnerable to further misadventures of reason—an "anthropological slumber."

Madness

Let us consider some examples that complicate this scenario, for Foucault in no way applies the Kantian schema crudely. How did psychiatry develop as a positive science amid the tumult of the revolutionary period? The asylum, Foucault suggests, formed a neutral, "closed" system in contrast to the disorder of political and economic events during the French Revolution. Insofar as the controlled conditions of the asylum permitted wardens to propose questions and laws *to* a disorderly nature, they were in a better position to govern or give advice to those governing the world outside. Despite persistent incompatibilities between medical and moral approaches to mental illness, the asylum encouraged the resolution of incongruities between scientific, political, and economic practices, in the way that one solves a geometry problem by drawing a line or projecting a plane.

The "events" traced in *History of Madness* are not events that happened to the mad (for madmen and women are too diverse to form a class) but events for those who share the forms of "sanity." These events involve the introduction and changing function of confinement, as well as private industry and the public provision of welfare assistance in Western European societies. The asylum grouped together individuals who were incapable of integrating with the newly emerging bourgeois economy, whether displaced by war or agricultural transition, isolated for religious reasons, physically handicapped, or emotionally recalcitrant. This grouping enabled the new constellation of religious, economic, and political practices arising in the late eighteenth century to produce a coherent, functionally ordered, and therefore legitimate public space.

For medieval Europeans, according to Foucault, madness was an earthly indication of the world's contingency with respect to divine transcendence or apocalyptic punishment (*HM* 14–28). For the Renaissance, this contingency was linked to the proliferation of knowledge and the imagery through which knowledge is revealed (16–25). To the sensibility of the Classical period, however, madness intimated the simple possibility of the world's failure or nonbeing. No greater power, no intuitive understanding, glimmered through the confusion of such minds. Descartes, who even in the throes of radical skepticism knows that he is not mad from the fact that he thinks at all (where the mad are imagined not to think) and because his skepticism aims at knowledge (where the mad willfully desire error) draws on the newly modern significance of the *will*, the fragile link between divine veracity and the mathematical foundations of purely human science (138–39).

Madness could only become an object of positivist medical study when such limits were accepted as *inevitable* and as the *basis* for human self-knowledge and political order. The asylum linked researchers and subjects in a common system of measures, establishing a self-evident threshold for sanity that not even an evil deceiver could threaten (cf. HM 157). It functioned as a laboratory for techniques in the control of unreason that could be applied later in the ordinary social environment. "The historical arrival of psychiatric positivism is only linked to the promotion of knowledge in a secondary manner: at its origin, it is the fixing of a particular mode of being outside madness, a certain consciousness of non-madness that becomes a concrete situation for the subject of knowledge, the solid basis from which it is possible to know madness" (HM 460). Around the walls of the asylum, in short, society could be provisionally structured; the asylum is both a part of the structure and outside it.

But the individuals designated as mad did not simply exist on the boundary of speculative or theoretical reason. They also enabled the sane to take measure of the *practical* reason or alienation manifest in the institutions and mores of their era. Confinement gave the inhuman or unimaginable, like sexuality and libertinage, a place in this world where they shed light on the relationship between humanity and its world. "The practice of confinement," Foucault writes, "demonstrates a new reaction to poverty and indigence [*la misère*], a strange, novel form of pathos [*un nouveau pathétique*], a different relationship between mankind and all that can be inhuman in his existence" (HM 55). This inhumanity affects the reasonable person as well as the pauper or madman. Decisively condemning or excluding these "monsters" allowed the political problem of arbitrary power, the juridical problem of the subject, the medical problem of the relationship between mind and body, and the economic field of idleness or poverty to find reciprocating solutions in the same social field (77). When insanity was further conceived as the *product* of this situation, it offered important diagnostic information about economic and political changes.

Far from being grounded in an ahistorical conception of human nature, from the end of the eighteenth century "madness was clearly inscribed in the temporal destiny of man, and was even the consequence and price of the fact that men, unlike animals, had history. . . . Pinel judged himself lucky to have had the opportunity to study diseases of the mind at a moment as propitious as the Revolution, as this era, more than any other, fostered the 'vehement passions' that were 'most commonly at the origins of alienation'" (*HM* 377).

It is helpful to note some distinctions in the fields from and against which various images of madness are extracted. Classical unreason (*déraison*) was a *transcendental* limit, primarily referring to the philosophical domain. "Madness" (*folie*), by contrast, was considered symptomatic of humanity's *historical* relationship with its environment. For example, Rousseau believed that human beings were mad because they were historical, not because they were animals. Samuel Tuke and Philippe Pinel, the reformers credited with founding the modern medical treatment of mental illness, were able to regard the asylum as a theater for the revelation of natural truth concerning madness because they considered both illness and health to involve the *necessarily* alienating artifice of social norms. This transformation in protopsychiatric thought was triggered by increasing public fear in the period leading up to the French Revolution that asylum inmates were spreading disorder and disease throughout "free" society (*HM* 355–57). Madness, conceived as the negative and excluded counterpart to free citizenship, provided a backdrop against which the *juridical subject* of law could be identified with the relatively new *social man* of economic liberalism. Between these two figures

> the political thought of the enlightenment postulated a fundamental unity . . . and a constant possibility of resolving any practical conflicts that might emerge. . . . The positivist medicine of the nineteenth century inherited these *Aufklärung* ideas, and took it as an established and proven fact that the alienation of legal subjects should coincide with the madness of social man in a unified pathological reality, which could be analysed in legal terms as well as perceived by the most immediate forms of social sensibility. (*HM* 128)

The legal subject and economic actor coincided, therefore, in as much as the madness of the one was identical (for all practical purposes) with the madness of the other, since the failure to adapt to economic conditions resulted in the loss of legal rights (a theme reiterated in *Discipline and Punish*). In the process, a certain understanding of economic "nature" and "necessity" was inscribed over the very real infirmities, illnesses, and rebellions of institutionalized bodies as the inescapable condition of their intelligibility. After all, to what "natural" phenomenon did madness observed in the asylum correspond, if not to traits disrupting or thought to disrupt economic and cultural life? Although "the psychopathology of the nineteenth century . . . believes that it orients itself and takes its bearings in relation to a *homo natura*, or a normal man pre-existing all experience of mental illness," Foucault warns, "Such a man is in fact an invention, and if he is to be situated, it is not in a natural space, but in a system that identifies the *socius* to [or "with," *au*] the subject of the law" (*HM* 129).

It makes no more sense, then, to speak of an "experience" of madness than to speak of one's experience as "thing-in-oneself." Of course, observers have an experience of the bodily presence, the social demand, of those who are judged mad (along with the sane) (Foucault 1984, 336), and may even question their own sanity on certain occasions; but these are appearances, and Foucault insists that different discourses identified the problem of madness with *different* appearances. But these are never experiences of madness itself, and individual gestures or statements by the mentally ill may be quite "sane." Community norms, philosophy, medicine, and political authority recognized different meanings in the presence of the mad. Philosophical "unreason" was different from medical "melancholy" or "frenzy," and neither coincided consistently with "vagrancy" or "impiousness."

The modern "myth" that medicine finally achieved an "objective and medical recognition of madness" rests on the apparent coincidence of three structures: first, "the old space of confinement, now limited and considerably reduced" (and the attendant but not entirely compatible spaces of medical knowledge and practice); second, an (ultimately internalized) gaze which identified madness, "guarded and judged it, a neutralized relation, seemingly purified of any complicity"; and finally, a space that links madness to criminality (without reducing them to each other), "authorising reasonable men to judge and divide up different kinds of madness according to the new forms of morality" (*HM* 426). The "liberation" of the mad coincided with the concealment and naturalization of the specific historical conditions from which they had been excluded, the incontrovertible laws of a market economy.

I have gone into more detail with *History of Madness* not in order to summarize Foucault's text, much less the history of European practices for treating madness, but to show how a problematic object can unify multiple practices and discourses over time and at the same time, inasmuch as it is excluded from all of them. Attention to this unifying function of the problematic object, which follows an initial division of the field of study and action, is a constant in Foucault's historical analyses of the 1970s.

Death

Birth of the Clinic (1963) takes stock of the physical and discursive spaces sutured through their relationship to *death* rather than madness. In it, Foucault examines two fundamental regimes of spatialization preceding the establishment of modern anatomo-clinical medicine, each with its own problems and limitations. Classificatory medicine (*la medicine des espèces ou des classes*) faced the difficult task of creating an unencumbered space in which medical students could be taught the schematization of diseases as natural kinds. Pathological anatomy (*anatomie pathologique*), by contrast, had to ground an enterprise of discovery whose logical and spatial limits remain open to the obscurity of the body's diseased interior. Modern medicine reached epistemological maturity when it devised a new suture between the visible and the articulable, between the "space" of sensibility, pain, and the management of bodies and the "space" of thought in which medical concepts are deployed.

Foucault positions himself, as in his previous text, on the threshold of a new discriminating practice: "In order to determine the moment at which the mutation in discourse took place, we must look beyond its thematic content or its logical modalities to the region where 'things' and 'words' have not yet been separated, and where—at the most fundamental level of language—seeing and saying are still one" (*BC* xi). During the late seventeenth and early eighteenth centuries, according to Foucault, this suture was assumed to take place in the quasi-metaphorical, quasi-physical "light" in which classificatory knowledge and the objects it ordered were rendered consonant with one another (xiii). For the tradition inherited from Plato by Descartes and Classical rationalism, light made objects clear and distinct so that they could be intelligently denoted and communicated. The objects of sight were properly intellectual essences, dispersed among contingent spatiotemporal examples. The architectural space of the clinic organized and tacitly represented this light, which underwrote a universal *mathesis* of diseases and yet, unfortunately, had to be taught to successive generations of doctors.

But gradually this light lost priority to the darkness that, in a sort of intellectual and visual chiaroscuro, gave vision its defining limit.

> At the end of the eighteenth century, however, seeing consists in leaving to experience its greatest corporal opacity; the solidity, the obscurity, the density of things closed in upon themselves, have powers of truth that they owe not to light, but to the slowness of the gaze that passes over them, around them, and gradually into them, bringing them nothing more than its own light. The residence of truth in the dark centre of things is linked, paradoxically, to this sovereign power of the empirical gaze that turns their darkness into light. (BC xiii–xiv)

In other words, medical thought at the beginning of the nineteenth century was faced with the task of linking the logical space in which the essences of diseases were conceptualized to the physical space in which a specific disease ravaged the specific patient's body. But this task was simultaneously a pragmatic and political one, first, of how to teach medicine in such a way that students could "see" what it was necessary to see in order to be capable of halting disease, and second, of how to provide for the medical care of the indigent in such a way as to preserve the health of the population. And it was not only a question of preserving the health of citizens from disease, but also from the danger that inadequately trained doctors and poorly organized medical spaces represented to public health. Foucault refers to the "system of options that reveals the way in which a group, in order to protect itself, practices exclusions, establishes the forms of assistance, and reacts to poverty and the fear of death" as a "tertiary spatialization" intersecting the logical space of nosography and the physical space of the individual body (BC 16). The clinic achieved its mythical significance for modern medicine because it permitted the solution to a number of solutions within this tertiary space: the organization of assistance to the indigent sick, the training of doctors, and the ideal role of medicine in a rational society.

Classificatory medicine and modern medicine also differed in their reference to temporality. In the former, temporality provided a common locus for the intersection of conceptual, individual, and political spaces of disease. In modern medicine, by contrast, temporality is recognized as intrinsic to disease itself. This transition corresponds to the transformation of natural history into biology discussed in *The Order of Things* (273–78, see also *BC* 89). Foucault implies that the clinic was able to furnish a stable basis for medical thought because it established a single space in which political questions and individual pathologies could be placed in a coherent and simultaneous relation to one another. The transition from one spatiality to another (from the clinic to the corpse) was accompanied by a transition from one *meaning* of temporality to another. In the eighteenth century, time was relevant for disease because it allowed epidemic events to be indexed. Later, time was seen as the form of a more robust historical process in which the pauper's or proletarian's struggle to postpone death through labor led to an accumulation of social wealth (*OT* 256–58). Thus, if the corpse provided modern medicine with a new basis for imagining the body, one may presume that it did so in part because mortality permitted modern societies to achieve a better understanding of political and economic phenomena.

These successive epistemological styles are expressed through Foucault's distinction between the *gaze* (*le regard*), which takes conditions of apprehension for granted (light and language, coordinated through the architecture of the clinic), and the *glance* (*le coup d'oeil*), which assumes responsibility for the violence it imposes on phenomena in rendering them intelligible. Where the gaze hoped to find regularity and universality in every case or symptom, the glance, by contrast, was oriented toward the particularity of a case and the unique history of a body's suffering. Unlike the gaze, moreover, the glance cannot teach. "The glance is silent . . . of the non-verbal order of *contact*, a purely ideal contact perhaps, but in fact a more *striking* contact, since it traverses more easily, and goes further beneath things" (*BC* 121–22). However, it can inquire and *discover*, and it offers medical science a new basis upon which to elaborate a theoretical discourse of disease. Ultimately, it led to a reorganization of the clinical gaze and the space into which teaching introduced novice doctors. It enabled new realms of experience to be demarcated and opened to empirical conceptualization. The glance opens a new surface to visibility—an *interior* surface whose contour is not given in advance by everyday vision but has to be shaped by the questions posed in dissection. This contoured surface, with the right technology, extends indefinitely into the architecture of the cells themselves.

The dissecting glance rendered disease intelligible because it moved through the same space of the body as disease, and outlined the progress of the disease against the same horizon that disease took as its telos; that is, the death of the sick organism. It followed the history of a disease across the body's hidden interior battlefield. "If the disease is to be analysed," Foucault writes, "it is because it is itself analysis; and ideological decomposition can only be the repetition in the doctor's consciousness of the decomposition raging in the patient's body.... Disease is an autopsy in the darkness of the body, dissection alive [*dissection sur le vif*]" (*BC* 130–31). Pathological anatomy revealed a new body, one filled with organs and capable of generating an almost infinite variety of surfaces, one for every cut of the knife. Miming death the better to outwit it, Bichat revealed a new way of conceptualizing medical thought—on the deadly model of a disease.[9]

Thus, for Foucault, pathological anatomy represents a Copernican turn in the history of medicine. Henceforth, it is on the basis of the elusive distinction between life and death that thought is capable of revealing nature by posing questions to her (*CPR* Bxiii). Life and death also structure the teacher's communication regarding objects like tissues, organs, and lesions—objects defined in terms of a recognized transcendental aesthetic of the corpse's space and the time one takes to die. "What is modified in giving place to anatomo-clinical medicine is ... the more general arrangement of knowledge that determines the reciprocal positions and the connexion between the one who must know and that which is to be known" (*BC* 137). Disease does not simply express itself or become intelligible in the bodies it contingently affects; disease is one way in which *the body suffers time*. In fact, disease is a kind of dying comparable to natural death, in which functions are disrupted and fail one another: both disease and death are "multiple, and dispersed in time: it [death] is not that absolute, privileged point at which time stops and moves back; like disease itself, it has a teeming presence that analysis may divide into time

---

9. Marie-François Xavier Bichat (1771–1802) introduced the concept of the "tissue" in medicine, explaining the spread of disease more concretely than prior physiologists, who believed disease affected self-contained organs, and explaining death in terms of the gradual dissociation of normally interdependent tissues. According to Foucault, Bichat introduced the medical *regard* and toppled the ontology of disease on which "clinical" medicine had been based.

and space; gradually, here and there, each of the knots breaks, until organic life ceases" (142–43). The indefinite divisibility of space and time shapes the modern sense for bodily materiality.

Why is death privileged as the "scene" making the visible articulable? Perhaps it is because death naturalizes and renders "self-evident" the political economy in which some sicken and others are healthy. Foucault's analyses imply that death is not a "thing-in-itself" but a problematic object enabling bodies to become sources of knowledge, labor, and power. "Medicine offers modern man the obstinate, yet reassuring face of his finitude; . . . although it ceaselessly reminds man of the limit that he bears within him, it also speaks to him of that technical world that is the armed, positive, full form of his finitude" (BC 198). Later, Foucault argued that the body was the preeminent form of intuition through which modern European thought and culture related productively to its own finitude: "The mode of being of life, and even that which determines the fact that life cannot exist without prescribing its forms for me, are given to me, fundamentally, by my body"—a body that enables people to regard the limited possibilities given to them by their historical situation as a fundamental finitude, and to perceive individual forms of practical or intellectual alienation in their knowledge and practice as open to indefinite historical improvement (OT 314). Although it purports to discover the transcendental basis for distinguishing between health and illness, the birth of modern medicine coincides with a *historicization* of poverty, illness, and death—a process that opened the social field to radical political contestation.

Embodiment is phenomenologically constitutive of meaning and experience because, as Foucault went on to argue in a series of related studies, its flourishing serves as the common—if "abstract"—presupposition of all discourses concerning government, culture, and production during the modern period. It is the "site" enabling objects of experience to form an interconnected, meaningful whole. Of course, the "body" that exists in this constitutive relationship with the idea of death (and the medical, military, and productive practices that distribute the risks of death) *is* abstract, and each individual body negotiates the vagaries of its physical ability, desires or emotional tone, and cultural situation in a different manner. As *Birth of the Clinic* states, the achievement of modern medicine was making it possible to "hold a scientifically structured discourse about an individual" (xiv), but the discourse captured the individual as much as it cured him or her. "The paradox of the norm," as Béatrice Han explains, "is that it plays individuation against individuality, as the measurement of the individual that it presupposes is effected to the detriment of the respect for the individuals themselves" (2002, 122). The development, capabilities, and life expectancy a government officially intends to foster in its population do not necessarily correspond, and are not intended to correspond, to the phenomenological situations lived by such varied beings as prostitutes, schoolgirls, telemarketers, and stay-at-home mothers.

In *Discipline and Punish*, Foucault speaks of parameters making possible a "history of the body" (*DP* 25). What the introduction to *Birth of the Clinic* argues, in effect, is that one should construct such a history by distinguishing the incompatible spatializations that have come together to form the apparent seamlessness of a historical present or period. These spatializations coexist in the architecture of the clinic, the inevitability of the corpse, and the morally reassuring image of the doctor/patient dyad. Thus the writing of history resembles the act of dissection by which the medical glance invades the body, freeing "elements that are no less real and concrete for having been isolated by *abstraction*" (*BC* 131), and Foucault eventually described genealogy as, in effect, "opening up bodies" to reveal past conflicts.

When Francois-Joseph-Victor Broussais successfully abolished the conception of disease as an entity in its own right and redefined it as a pathological relationship between an organism and its environment in 1816 (*BC* 189–91), it seemed that modern medicine had found its natural and naturally unified space, and with it the natural contours of the body. But Foucault suggests that this naturalism is deceptive: positivism may have turned finitude into an efficient epistemological ground, while bracketing the existential significance of life and death, but it also obscured the variation and instability of the spatializations in which diseases, techniques, and bodies were caught. Paradoxically, beginning from the corpse rather than the living body enabled medicine to "historicize" the frequency and forms of death but could not ensure that flourishing, rather than labor and illness, was taken as an *ethical* norm.

Delinquency

In *Discipline and Punish*, Foucault returns to the relationship between medical determinations of psychological health and legal determinations of responsibility first broached in *History of Madness*. Where *History of Madness*, however, focused on threats to public order from sexual deviance and vagrancy, *Discipline and Punish* situates the social, political, and economic offenses confronted by the postrevolutionary regimes within a broad analysis of *illegalities* (*illégalismes*). The class-identified spaces of economic practice (investment, trade, production, theft) were unified around the rights of bourgeois proprietors and merchants through the elimination and delegitimation of rights given to the nobility on the basis of blood and to the poor on the basis of Christian custom (*DP* 82–89). Cultural and juridical techniques for training bodies in relation to the *legally* problematic object of "delinquency," Foucault argues, enabled institutions and discourses to organize the sensory and discursive space of modern individuals, especially those subject to penal and workplace surveillance, into a seemingly coherent whole.

In this text, the moment of decision and discrimination is that of *revolution*, in which criminal violation and political antagonism are indistinguishable (*DP* 59–63). The internment of the criminal, like that of the insane, put the spotlight on some economic and cultural practices and thereby affirmed law's ability to cover and coordinate the social field, while allowing others to claim the law's tacit protection. "Delinquency functions as a political observatory" (281). The rationalization of punishment, Foucault argues, was only superficially motivated by concern for the "humanity" of the criminal whose mutilated body had previously served as the token of royal power (92). More precisely, it (and consequently the criminal's humanity) reflected the *limits* of a royal economy of power in which judicial positions were frequently sold in order to generate state revenue; in which the laws for classes and clergy, as well as for different regions of the country, were too varied to permit reliable trade; and in which a growing mass of impoverished and unemployed vagrants threatened the orderly practice of commerce and offered themselves as nascent objects of capitalist exploitation (76–79). These conditions predated the French Revolution and probably counted among its motivating causes. The criminalization of property offenses, like the diminution of sexual and religious offenses to the realm of "mental illness," reflected attempts to generate a new "economy of power" that would punish, through habits of reflective interiorization and work, acts of resistance to the order necessary for economic growth.

The privileges of the nobility (conditional on royal favor), the traditional rights of the peasantry, and the sheer persistence of the destitute in their struggle for existence threatened the emergent order of property and profit in a variety of ways. "The transition to an intensive agriculture exercised, over the rights to use common lands, over various tolerated practices, over small accepted illegalities, a more and more restrictive pressure" (*DP* 85). After the Revolution, and even during the period leading to its tumult, bourgeois trespass on royal privilege was increasingly accepted as a normal "use of liberty" over land and property, while that of the poor became the target for moral and penal repression: "The bourgeoisie was to reserve to itself the illegality of rights: the possibility of getting round its own regulations and its own laws, of ensuring for itself an immense sector of economic circulation by a skilful manipulation of gaps in the law—gaps that were foreseen by its silences, or opened up by *de facto* tolerance" (87). Indeed, the bourgeois "order" is only visible as such retroactively, to a sensibility that has been educated in relation to the threat or reality of poverty and imprisonment. This order legally repressed and culturally constrained those whose habits would tend in the direction of another order, such as the older peasant economy or Fourier's utopia.

Like *Birth of the Clinic*, *Discipline and Punish* describes a series of three historically successive, but overlapping schemes for organizing social and (in this case) juridical space with respect to discourses whose truth was anchored in bodily sensibility. These schemes were the *supplice* or spectacular public punishment of the ancien régime, punishment through the offender's *embodied representation* of his or her crime, and finally, *panoptic discipline*. In each of these schemes, the space of the law as written or interpreted and the space of social order which intersects imperfectly with the law were engaged with the space of the offender's bodily comportment, psychological intention, and physical pleasure or agony. In the *supplice*, for example, the pain inflicted upon the offender's body demonstrated the identity of the king's power with that of the body politic and revealed a truth previously known only to the victim and his torturers (*DP* 49, 55). In the model of punishment as "pedagogical theatre" envisioned by the Idéologues, it was not the body as directly susceptible to pain that put the masses in relation to the power of their self-governing state, but the body as *significant* in and through its suffering, the body whose subjection exhibited a *ratio* between interests and penalties to the public (101).

However, scarcely had the revolutionary period commenced than the Idéologues' plans were overthrown and imprisonment of greater or lesser duration established as the penalty for most crimes. The reason, Foucault suggests, was not only that the fledgling French republic was faced with conditions of the "great fear," in which revolutionary violence, economic illegality, poverty, and the assertion of bourgeois independence from royal control were anxiously confused. The type of surveillance found in the American reformatories and in new prisons on Bentham's "panoptic" model also exteriorized, in iron and concrete, a new type of relationship between the criminal and his or her own powers of sensibility. By substituting a duration of imprisonment for all other forms of punishment, the disciplinary model shaped the body as *individualized* in order to produce a truth regarding the necessary form of a subject of knowledge. It also provided a different "interiority" to the body than the one discovered in relation to disease and described in *Birth of the Clinic*. "It would be wrong," Foucault writes,

> to say that the soul (*l'âme*) is an illusion, or an ideological effect. On the contrary, it exists, it has a reality, it is produced permanently around, on, within the body by the functioning of a power that is exercised on those punished. . . . On this reality-reference, various concepts have been constructed and domains of analysis carved out: psyche, subjectivity, personality, consciousness, etc.; on it have been built scientific techniques and discourses, and the moral claims of humanism. (*DP* 29–30)

On the threshold of this new and fragile economy of power, the body of the criminal—predominantly the individual who lacked sufficient property to take advantage of the illegality of rights, who could only live by violating the (property) rights of others—became the experimental archetype in whose labor and anxiety were presaged the "normal" reflectiveness, lawfulness, and care for self-preservation of the bourgeois proprietor and his or her working-class imitator. The *space* and *time* of the body forced to perform labor or to engage in solitary meditation were manipulated by reformatory personnel in order to create a new *experience* of social and economic order and to bind *law* to that order instead of to the traditionally exorbitant displays of absolutist power.

The "art of spatial distributions," or architectural individuation and organization of workshops, prisons, schools, hospitals and other public spaces, is the pure form of outer intuition through which societies and individuals know themselves in terms of one another; just as the "control of activity" or exploitation of smaller and smaller temporal units in the synthesis of a gesture correspond to the increasing interiorization of the power relationship between those who observe and train and those who are observed or trained. This intensification and differentiation of the worker's, student's, or prisoner's space and time, Foucault writes, created the "natural body, the bearer of forces and the seat of duration" (*DP* 155). An increasingly fine-tuned relationship between the demands of discipline and the body's spatiotemporal responses expressed a more intimate relationship between power and the *form* of the acting, moving, reflecting social and individual body than that expressed, for instance, within the punitive economy of the *supplice*. The result was a *more effective* recombination of potentials present within existing relationships between collective and individual units of "social" space and time. For instance, if power was "articulated directly onto time," it could reorganize the speed and range of opportunities for action and reaction rather than relying upon the populace to absorb lessons exhibited in public punishment at their own rate.

The prison appears to substitute the cultivation of a morally significant interiority for the violent manipulation of the body as sign; nevertheless, Foucault points out, "a punishment like forced labour or even imprisonment—mere loss of liberty—has never functioned without a certain additional element of punishment that certainly concerns the body itself: rationing of food, sexual deprivation, corporal punishment, solitary confinement" (*DP* 15–16). There is always a "body" implicated in the act of punishment or the exercise of power; a certain form of exteriority involved with every effort to change the interior. But what that exteriority or body is, and how it registers or demonstrates power, changes from era to era. Moreover, that body is both individual and collective—which means, strictly speaking, that it is *neither*; even when we are unable to perceive the strategies that show themselves in the form of our embodied experience, bodies are always articulated according to a "political anatomy."

By substituting the spatial and temporal specification of the *soul* for the "sensibility" of the exhibited body in pain, disciplinary mechanisms make it possible to substitute the more or less "delinquent"—whether monstrous or redeemable—individual (*le déliquant*) for the essentially unknowable "offender" (*l'infracteur*) in-himself (or herself, though Foucault does not discuss the techniques for the constitution of specifically masculine and feminine delinquency).[10] "It is as a convict, as a point of

---

10. Following the analysis given in the late essay, "The Subject and Power" (in *PWR*

application for punitive mechanisms, that the offender is constituted himself as the object of possible knowledge" (*DP* 251). The convict is a physical reference point for two discourses and practices with different concerns:

> From the hands of justice, it [the penitentiary apparatus] certainly receives a convicted person; but what it must apply itself to is not, of course, the offence, nor even exactly the offender, but a rather different object, one defined by variables which at the outset at least were not taken into account in the sentence, for they were relevant only for a corrective technology. This other character, whom the penitentiary apparatus substitutes for the convicted offender, is the *delinquent*. (*DP* 251)

Just as modern knowledge takes the limits imposed by sensibility and imagination as constitutive enabling conditions, Foucault suggests that modern power increases its productivity by starting from the inevitable limits on *individual* ability as well as *state* power. The *supplice* expressed the vast extent of royal power, but also revealed the limits of its ability to affect individuals through fear and pain. The poor sometimes identified with the wretched victim of royal vengeance, and the more affluent resented its arbitrary and irrational exercise. When the death of a victim demonstrated the "exhaustion" of spectacular power, questions arose about the justice of that power. Enlightenment reformers were aware that violent public punishment carried this subversive potential and sought to establish a more constant, reliable, and productive distribution of power through punishment (*DP* 63). They hoped for power relations that would *exploit* the inevitable failings of state power and individual agency as a resource for ever more finely tuned production and flourishing. In prisons, classrooms, parade grounds, and factories, the individual's *inevitable variation* from a norm was taken as a *fault* to be overcome, and thus a motivation toward greater perfection, efficiency, and productivity.[11]

326–48), we might be able to understand the body of the offender as the "object of experience" through which prison administrators are able to get a purchase on the otherwise obscure criminal "soul" whose existence cannot be postulated except by means of a paralogism. In fact, the prison's goal would be to shape and control the degree of "spontaneity" or of "freedom" remaining to the criminal by appeal to his or her capacities for resistant self-government, which it presents as a corrupted or redeemable nature—a resistance that regularly goes awry.

11. Some criminals appeared immune to the play of interests and to the characterization of norms. The behavior of these "dangerous individuals" served as an important point for communication between the law and the psychiatric profession (*PWR* 196–98).

Foucault notes that the body of the criminal is only one in a series of normalized bodies established through the disciplinary techniques of the late eighteenth century—the soldier, student, and factory worker were individualized through a similarly "normal" division of space, time, energy, and exhaustion. "Cells," "places," and "ranks" are "mixed spaces" like heterotopias and the clinic, "real because they govern the disposition of buildings, rooms, furniture, but also ideal, because they are projected over this arrangement of characterizations, assessments, hierarchies" (*DP* 148). The carceral system built around these spaces, likewise, "combines in a single figure discourses and architectures, coercive regulations and scientific propositions, real social effects and invincible utopias, programmes for correcting delinquents and mechanisms that reinforce delinquency" (271). These are, however, the first of Foucault's heterotopias to be articulated through pleasures of control and resistance—the perverse pleasures of self-control as much as domination over others. The organization of illegalities takes place around the individual's and community's willingness to be motivated by a *negative* relationship to the limit represented by their own recalcitrance and social antagonism—to work harder, the more they wanted to work in their own way. From the finitude of *felt resistance* to norms, discipline extracts endless normative motivation.

Sexuality

The first volume of *The History of Sexuality* addresses the generative, motivating force of resistance in greater detail. Specifically, it situates this force in the context of debates over the nature of power and the importance of the state in European political theory, conservative and radical. Here Foucault is attentive to a moment of division he believes has not yet been sufficiently understood; the moment when structures of power use *one set of practices* to mask others. Power, he writes, "is tolerable only on condition that it mask a substantial part of itself"; where "secrecy is not in the nature of an abuse, it is indispensable to its operation" (*HS1* 86). In the modern era, states and some therapeutic or administrative discourses have used the fiction that power is limited to a repressive function in order to hide the operation of generative, encouraging, and self-proliferating forms of power. Unlike the division between madness and sanity or life and death, this division is hidden and generates power through invisibility. By focusing attention on the state's restriction of life, Foucault argues, we are encouraged to assist a social network whose power goes well beyond the state in fostering life and preserving it.

The various domains of social and discursive practice held together by awareness of and concern for "sexuality" are not distinguished as explicitly as in Foucault's earlier books. This is, in part, because *The History of Sexuality* is the first book in which the subject's interiority must be sutured *together with the state,* understood roughly as a matrix for coordination among at least three sets of practices. These were the practice of domestic legal or economic administration; the practice of international (colonial, economic, or military) relations; and finally, the practice of legal, therapeutic, religious, and sociological investigation and treatment. *History of Madness* gave only an indirect approach to the "experience" of madness, by way of the madness in the experience of reasonable, responsible individuals. But it concerned itself throughout with the form of sanity represented and made aesthetically present by the existence of mad persons, among whom were included the pervert, "degenerate" individual, and victim of venereal disease. The first volume of the *History of Sexuality*, by contrast, focuses on the positive self-ascription of sexuality as an ever-threatening but always attractive weakness latent in "normal" individuals.

Sexual behaviors arise along the hypothetical borderline dividing the healthy from the weak or degenerate. These behaviors are imagined to be the effects and appearances through which "sexuality," as a hidden force in itself, is known. They include direct forms of bodily desire, contact, and pleasure—but also indirect practices such as introspection, pedagogical or medical concern for the sexuality of family members, and religious, economic, or legal practices with respect to kinship. By attaching itself to these behaviors, cultivating some and discouraging others, the state pursues its domestic and international goals, while social scientific discourses claim a common point of reference in human experience—indeed, the experience that makes humans a "species" rather than a community of believers or a mere aggregate of individual beings (*STP* 77). Sexuality, a system of appearances and acts attributed to an invisible force, enabled the state to compensate for its inability to master the economic and medical life of citizens, or to completely dominate the labor and intelligence of colonial subjects abroad. It also enabled ordinary individuals to link themselves *as speakers and bodies* to the truth held by experts in an increasingly scientific society. It combined older forms of kinship and moral/religious concerns for purity, especially those which took the explicit form of prohibitions, with a concern for the life of the species.

While masquerading as the effect of a binary, repressive law, Foucault argues, modern practices of power centered on sexuality drew their energy from subjects' resistance to external norms and desire to fulfill that resistance in a communicable way. Unlike the great confinement, which assimilated forms of madness that had been regarded as aesthetically distinct during the Renaissance, the deployment of sexuality encouraged individuals to differentiate themselves from one another and to take pleasure in this differentiation at the same moment that they sought to communicate this difference in a common language, whether therapeutic or administrative. The physical places in which attention was trained on the sexually significant differences between bodies were the bourgeois home, the school, and finally the psychoanalytic office. These places were sites for confessing observations or experiences, educating about the significance of what was confessed, and curing or correcting "abnormal" experiences; but they were also sites where subjects were encouraged to identify something precious in themselves *with* sexuality as described by parents, teachers, and doctors.

Foucault proposes four primary cultural contexts for the elaboration of these differences: the care and education of children, the management of women's social and psychological health, the medical and political encouragement of high fertility rates, and the identification, demonization, or normalization of "perversions" (*HS1* 104–5). These were neither explicitly political nor economic goals. They were justified by scientific discourses whose conception of sexuality was poorly grounded, and in which research was openly tied to eugenic and colonial philosophies of race that have since been discredited (118–19, 125). From the early nineteenth century to the present day, however, agreement and progress among these domains emerged as a result of concern for the "problematic object" embodied in children, women, couples, and perverts. Once sexuality could be charged with individual disorders of thought and behavior, urban poverty, bourgeois success in accumulating capital, colonial recalcitrance, or successful domination, the individual could be integrated into the historical life of the species.

One of Foucault's concerns in this book, as in others, is to show how postulating a problematic object like "sexuality" at the heart of human speaking, labor, and biology allows phenomenal social aggregates and communities like "class," "race," and the "social" to seem more concrete and unified than they otherwise might. Another goal is to account for the *pleasure* that members of Western societies clearly take in being able to group social phenomena and then differentiate or criticize them in relation to a norm.

Foucault does not, we should note, state that sexual identity trumped all other forms of identification during the late modern period. Rather, he argues that sexuality was integral to the way in which people *explained* their identification with religion, class, or nationality to themselves and one another. In the name of sexuality, the bourgeoisie defended and extended economically successful practices against the poor and colonized (*HS1* 125–27). Likewise, in the name of sexuality, socialist critics identified the repression of and intrusion into working-class sexuality by social workers and missionaries as a legitimating arm of capitalism. It is too simple to say that the state or capitalism requires the repression of sexuality; for "sexuality," like capital, is a set of relations—relations between individuals and themselves as well as relations between individuals and the political or economic forces they confront.

Foucault does not believe that "class" and "nation" could take their modern form apart from the increasing sexualization of social mores, pedagogy, and public administration. True, inhabitants of previous ages had organized themselves in terms of caste and rank, fealty and lineage. But self-identification as bourgeois or working class, Foucault argues, took on a new meaning and complexity within systems of state surveillance and administration that proposed to increase the health of the overall population so that *markets* could operate with as little interference as possible. Nor was production the only scene in which royal administrators of the early modern period established an overall balance of social power, since they also focused on the cultivation of health—differentiated, unequal, but no less flourishing—within the national population as a whole. Individualizing the members of these populations permitted resemblances and differences to be cultivated among groups so that some individuals' and groups' failure to flourish or persecution when they flourished against the norm seemed "natural," "inevitable," or "obvious."

The social contract tradition in Western political thought, rooted in the same experience of ordered knowledge responsible for Classical analysis, attempts to limit power relations between governors and the governed in two ways: first, according to the analogy of a "body politic" in whose lawmaking and enforcing activity the powers of individual bodies must be represented; and second, according to a hermeneutic of the law, in which the original *meaning* of individual or community power is historically produced. But the fiction of the "body politic," whether democratic or autocratic, reflects and reinforces *those qualities, resemblances, and differences* among individual bodies that are held to be essential for the functioning of social order and the production of discourse that will not be a mere "thing of sand" in relation to nature. The individual's self-understanding of his or her body is a product of relations *between* individuals as well as his or her desire to escape and *act upon* those relations. Not every society believed it possible to achieve self-understanding by way of one's body or to enjoy one's body by means of self-understanding, especially a self-understanding tied to expert discourses such as medicine and economics. It is a unique (but plural) body that constitutes the particular limitation and articulation of imagination Foucault describes in *History of Sexuality*.

Wars over the origin and right of sovereignty in England and France, if not elsewhere in Europe, were often justified through interpretations of the law as a historical artifact. But they also presented the social field as torn between an indigenous and an invading or dominating race, and *dispositifs de sécurité* increased the chances that one race or another would flourish. Government for *sécurité* builds on techniques already applied in the Christian direction of conscience, such as confession. It also makes a different use of the "norm" than the disciplinary power described by *Discipline and Punish*. Discipline or "normation" intensifies movement and space by breaking them into smaller units, each of which can be brought more closely into line with a standard. It begins with an idea of what the norm *ought* to be and condemns or rejects what cannot attain the norm (*STP* 8–12, 46–48). "Normalization," the preferred tactic of security mechanisms, "inserts the phenomenon in question . . . into a series of probable events" (8), attaching itself to the population as a whole, rather than to each individual. It tries to improve the aggregate performance at the lowest cost, without expecting uniform results for each individual (58–59). Although Foucault does not specifically identify the deployment of sexuality as an element in the security apparatus, it seems that security was the broadest governmental mechanism that allowed sexuality to become a matter of state interest, fertility being a key variable in a population's health (*SMBD* 252–53; *STP* 76–77).

Although the pervert or "abnormal," with his or her associations to the delinquent, may be the paradigmatic example of what modern individuals hope to *avoid* in sexual conduct with others or themselves, readers run the danger of falling into the same habit of thinking power *repressive* rather than *productive* when they view the pervert as the repressed object of modern society, who exemplifies or supersedes the struggles of "normal" repressed individuals. "Normal" individuals are as thoroughly saturated by sexualizing relations of power as "perverts" are. Readers fall into a similar trap when they regard the state, the source of repressive power, as incapable of exercising generative power or as the sole repository of generative power in modern societies (*PWR* 372). By literally *incarnating* the problematic object of modern politics and expert discourse within themselves, sexual individuals take responsibility for resolving the impasses or internal incoherence of these institutions and extending their effectiveness. The solution, Foucault suggests in a late essay, is not "to discover what we are but to refuse what we are"—not, that is, to become disincarnate beings indifferent to pleasure, but to liberate these pleasures from the ontological requirements of "being sexual" or "historical" in the late modern manner (*PWR* 336; Foucault 1984, 359). It means to understand how these practices function in relation to a particular *episteme* or articulation of imagination that makes history the "mode of being" for empirical phenomena (*l'empiricité*), and to vary these practices as *aesthetic* forms (*OT* 293).

Kant's initial cut between the realm of appearances and that of things-in-themselves was designed to end *philosophical* controversies that might foster conflict in the political domain (or at least discredit Enlightenment in the eyes of authorities). When, in the *Critique of Judgment*, he appealed to the anthropological domain for advancement in capacities for taste and communication that would align scientific knowledge with freedom, he made unification of reason a *pragmatic* task. Foucault, by contrast, suggests that the division between communicable and incommunicable tastes and behaviors was first made in the anthropological/aesthetic sphere. By excluding the object of a "problematic concept" presumed to be embodied in certain individuals (madness, illness, lawlessness, or perversion), the remaining members of society could rest assured that they shared a common world as thinkers, political actors, and subjects of medical or juridical authority. By confining this object within marked bodies, and the bodies within specific physical institutions such as the asylum, clinic, or prison, incompatible discourses and registers of social experience could elaborate a communicable language, spatiality, and temporal rhythm.

The ability to identify such individuals presumed a specific capacity for aesthetic judgment. The judgment of madness is an aesthetic judgment, one whose criteria change over time: "What appears to us as a confused sensibility was evidently a clearly articulated perception to the mind of the classical age" (HM 54). Likewise, clinical judgment was a matter of learning how to *see* symptoms in a communicable manner. "The whole dimension of analysis," Foucault comments, "is deployed only at the level of an aesthetic. . . . The sensible *truth* is now open, not so much to the senses themselves, as to a *fine* sensibility" (BC 121). The medical gaze went hand in hand with "advice about prudence, taste, [and] skill" that troubled its practitioners' democratic faith in the self-evidence of classifications and their availability to any student of the discipline (121). Nor does Foucault presume that the development of this aesthetic was inevitable or immune to reflective reorganization: "For us, the human body defines, by natural right, the space of origin and of distribution of disease. . . . But this order of the solid, visible body is only one way—in all likelihood neither the first, nor the most fundamental—in which one spatializes disease" (3).[12] In another example, identification of the "dangerous individual" required police, psychiatrists, and judges to judge behaviors against standards of fixed residence, work hours and habits, religious and family involvement, and norms of political activity that were changing throughout the late eighteenth and nineteenth centuries, but stabilized in part by their constant comparison to the delinquent (PWR 198–99). Phrenology and physiognomy drew clear links between moral character and appearance, including racialized appearance (see AN 7:296–302 for Kant's views on judging character on the basis of physiognomy). Finally, the aesthetic by which one *senses* and *produces* sexual tension, as well as the aesthetic of *listening* in the psychoanalytic situation, were important tools by which priests, educators, parents, and doctors deployed sexuality into new social contexts over the course of the nineteenth and twentieth centuries.

Given these precursors, it is not surprising that Foucault would describe experimental practices for liberating the individual from these forms of government, power, and knowledge as "arts of existence." In *The Use of Pleasure*, Foucault defines these arts as "those intentional and voluntary actions by which men not only set themselves rules of conduct,

---

12. In *The Order of Things*, Foucault also identifies efforts to determine the conditions for human knowledge and error through neurology and empirical psychology as a kind of "transcendental aesthetic" appropriate to the anthropological era (OT 319).

but also seek to transform themselves, to change themselves in their singular being, and to make their life into an *oeuvre* that carries certain aesthetic values and meets certain stylistic criteria" (1985, 10–11). Such arts are the basis for problematizing aspects of everyday practices such as diet or sexuality (or presumably, truth-telling), giving them a moral content, and using them as standpoints from which conduct can be experimentally varied, but they do not necessarily give rise to a code of prescriptions or prohibitions (*EST* 263). Foucault surmises that in the sixteenth century, scholarly interest in ancient Stoicism reflected a resistance to confessional techniques imposed on Catholics by the Counter-reformation. In many works, Foucault himself encourages readers to resist "commentary" or "interpretation" and to substitute "monuments" for "documents" in order to resist a particular structure of power associated with those practices.

In schematization the imagination combines intuitions with determinate concepts of the understanding to produce identifiable objects of experience (or lawful experience in general, when the concepts are pure categories). The complete application of the categories to possible objects of experience generates a *world* of real, interactive, substances. Schematization (or determinate judgment, to which it is closely related) is one aspect of the power of judgment; the other, which also involves the imagination, is reflection. Insofar as their confinement pulls together a common anthropological world around the closed asylum, clinic, prison, or confessional, the objects that Foucault has described in these texts allow positive sciences and administrative practices to further refine their control over law-governed experience, natural or social. Transcendental deliberation, which sorts out the spaces, elements, and objects of experience, must take up a position *outside* the world or between worlds to subvert this schematization and allow imagination to vary the forms and empirical concepts through which we apprehend the historical world. Unlike Husserl's reflection, therefore, which *brackets* the world and varies essences in imagination but remains embedded in the same conscious stream and intends to return to the same world as before, Foucault's reflection takes place *between* "worlds" or experiences of order. Its transcendental topic is deliberately hetero-topic. This strategy is most evident in *The Archaeology of Knowledge*, which substitutes the "archive" for the totality of speakable or legible phenomena that might otherwise be treated as a "world" (*AK* 128–30).

Foucault does give historical dates for changes in perception or constitution of the world in relation to these problematic objects. His divisions have come under critical scrutiny by historians as well as philosophers. His earlier books, in particular, argue that changes in the conceptual order affect many disciplines at once, and that the presuppositions of one period are incompatible with those of the next (although these incompatibilities may not have been perceived at the level of past authors' *consciousness*). The biggest problem with his periodization is that it seems to conflict with his belief that "history" itself was not used to indicate the most inclusive being of discourses and individuals until the mid-nineteenth century. What are the Renaissance, the Great Confinement, and the Enlightenment if not "historical" periods? What were the writings and acts of people from these centuries if not "historical"?

If we remember that every heterotopia is also a heterochrony, certain misinterpretations can be avoided. Foucault does indeed argue for the historicity, that is, the actual but contingent nature, of a conceptual order in which history is primary rather than, say, representation. But history can function as a common denominator across different discourses and periods because problematic objects like madness, death, delinquency, and sexuality seem to exist *outside* the spatiotemporal forms of human intuition as metaphysical or anthropological things-in-themselves. Of course, we never have access to what lies on the other side of the boundary dividing experience from its "other," only to the boundary itself. We encounter ambiguous words and gestures in which sexuality and criminality are presumed to show themselves. Like the psychological Idea of Kant's paralogisms, sexuality, criminality, and other objects seem to unify and explain the flux of appearances because we have presumed their presence and power in organizing psychological and sociological appearances. The practice of confinement and the physical structure of the asylum are also phenomena that endure throughout the historical periods in which madness is considered a conceptual, political, or economic paradox. But their mute and stubborn positivity lends a provisional "simultaneity" or spatialization to the "experiences" of madness identified in any given age. The concurrence of overlapping historical time frames identified with "man" is what makes history thinkable as a common denominator of change across periods (*OT* 331), just as the "presence" of dangerous individuals or of a potentially illegitimate sovereign and legal code allowed the historical dimension to be opened as a potential arsenal in race wars (see *SMBD* 207–8).

Problematic objects such as madness, death, delinquency, and sexuality can persist "behind" a variety of appearances because (like the other side of the canvas depicted in *Las Meninas*) they do not "exist," per se. Rather, they mark the limits of the phenomenal world—individual and social. By examining discourses and practices that converge on paradigmatic "appearances" of the problematic object, Foucault challenges the self-evident unity of the world they define and describe. By "problematizing" this object, Foucault breaks open the world formed around it.

Is there a particular "problematic object" that makes Foucault's own thought possible? On the one hand, Foucault repeats throughout *The Order of Things* and *The Archaeology of Knowledge* that it is impossible to define one's own space of knowledge. Moreover, a consistent reference to any one form of the object might encourage Foucault to "pull it upright" as a quasi-transcendental, like life, labor, and language (*OT* 341). This is the risk entertained by scholars who read Foucault's "power" as functioning like an ontological opening in the sense of Heidegger's "being" (Dreyfus 1996). Power, like being, is bound to be thought in terms of the theological or technological forms that have been empirically available to us. In *The Order of Things*, nevertheless, Foucault does suggest that scholarly fascination with the "being of language" is a symptom that the anthropological era may be coming to an end.[13] "Bodies and pleasures" play a similar role in *The History of Sexuality*, and in "What Is Enlightenment?" Foucault suggests that the disciplines and practices of power involved in normalization are the "limits" from which contemporary problematization must begin (*HS1* 159; *EST* 315, 317). Before turning to the discursivity and materiality of language and bodies, therefore, we must turn briefly to the problem of anthropology.

The Man-Form: Empirical and Transcendental

Foucault's essay on Kant's *Anthropology* argues that anthropology is both outside critique and a necessary supplement to critique (1961, 7, 120). Anthropology is the discipline that establishes the role that empirical, historical, and changeable conditions of thought (for example, Germany's

---

13. See also "A Swimmer Between Two Words" (*AME* 173). Han suggests that figures like Gregor Mendel, who speak the truth (as we now know) without having been "in the truth" in the sense of saying things that could possibly meet the criteria for true and false statements about biology in his day, are "monsters" that only function as "a retrospective illusion induced by archaeology itself for the sake of its own demonstrations" (2002, 84).

legacy of religious conflict) should play in any transcendental philosophy; that is, any attempt to grasp the necessary formal elements of thought. Foucault agrees with Heidegger that finitude is a *defining* feature of those for whom critique is both *possible* and *necessary* because of their subjection to necessary transcendental illusions. But post-Kantian thought, Foucault suggests, forgot that critical reflection is necessarily supplemented by anthropological reflection, and that anthropology has meaning only in relation to a critical gesture that suspends particular modes of political and moral governance.

Nor does Foucault disagree with Heidegger that Kant's neglect of philosophical anthropology or, in this case, *ontology*—the *being* of the being for whom metaphysics is both possible and necessary—prevented Kant from providing a secure ground for metaphysics. Critical reflection on the sources, domain, and limits of thought does indeed require anthropological reflection on the relationship between finite thinkers and the world in which their thought becomes potentially universal; that is, communicable. But Foucault goes further to suggest that attempts to explain the finitude of this thinker, and his or her subjection to transcendental illusions, in terms of biological or historical determinants, is to mistake a *level of analysis* for a mode of *being* and thereby to capture critique within the historical forms through which we recognize a being empirically as human or inhuman.[14]

Several years later, in *The Order of Things*, Foucault argued that the modern episteme was supported by faith that the quasi-transcendental figures of "life," "labor," and "language" could serve as explanations for the various limitations humans encountered in their experience of nature and one another (*OT* 314–16). Like the British empiricists, French Idéo-

---

14. Han analyzes two arguments in Foucault's *Anthropology* essay (2002, 21). On the one hand, Foucault argues that Kant's *Anthropology* indicates the possible application of critical philosophy to the task of delimiting knowledge concerning "man"—the aspect I have stressed in making this point. With an eye toward his later formation of the "analytic of finitude" in *The Order of Things*, Han identifies this critically delimited anthropological knowledge as "fundamental." On the other hand, and more radically, Foucault suggests that the notion of *Geist* developed in Kant's *Anthropology* (as opposed to *Gemüt*) constitutes a potentially "natural" and thus questionably critical "source" of reason's ambitions, with regard to both knowledge and moral self-transformation (55). This "source," much like Heidegger's "common root," would be the unnamed and presupposed image governing Kant's own exploration of pure reason as well as the *signature, within transcendental reflection*, of reason's inability to completely grasp its own origins. For this reason, Foucault refers to it as the "originary," and *The Order of Things* argues that the search for man's origins on the basis of the "originary" characterizes most post-Kantian thought.

logues, and German metaphysicians, Kant offered an account of the representational nature of human thought and knowledge. But he grounded this nature in terms of conditions for the possibility of representation rather than in an "intellectual intuition" (242–43). Although the function of signs had changed from the Renaissance to the Classical era, representation still enabled the human being and the world to remain linked by language in a limited and therefore meaningful way, permitting the accumulation of discourses and observations that would eventually be of scientific use. Kant's approach to the problem of representation made the world an Idea rather than a given object, and revealed the pretensions of his predecessors' efforts to understand the world as an *analyzable* whole "in itself" (243). But because the "conditions for the possibility of representation" could be understood from the side of the "one represented" as well as from that of the *representing act*, Kant opened a new role for time in the Western understanding of human existence: time as an anthropological and historical horizon as well as a condition for the synthesis of representations.

Life, labor, and language are "phenomena" that make visible the historical, time-bound nature of the human capacity for representation. But from the side of the one who represents, they are also *unrepresentable* and unknowable origins, to which the sciences left behind by mathematization (such as biology, economics, and linguistics) appealed for evidence that knowledge remained an epistemologically unified endeavor. This is why Foucault refers to them as "quasi-transcendental," rather than transcendental in Kant's sense of conditioning *all* possible experience (*OT* 244–45).

> In their being, they are outside knowledge, but by that very fact they are conditions of knowledge; they correspond to Kant's discovery of a transcendental field and yet they differ from it in two essential points: they are situated with the object, and, in a way, beyond it; like the Idea in the transcendental Dialectic, they totalize phenomena and express the *a priori* coherence of empirical multiplicities; but they provide them with a foundation in the form of a being whose enigmatic reality constitutes, prior to all knowledge, the order and the connection of what it has to know; moreover, they concern the domain of *a posteriori* truths and the principles of their synthesis—and not the *a priori* synthesis of all possible experience. (*OT* 244)

The "quasi-transcendentals," in sum, are elements of empirical experience or of a scientific discourse that claims the privilege of unifying all the phenomena treated by that discourse, although they have nothing to contribute to the *a priori* or "axiomatic" discourses of philosophy or mathematics. On the other hand, Foucault suspects that any transcendental structure, even Kant's categories and forms of intuition, analyzes a historical, anthropological experience and is therefore "quasi." He views his work as an experiment or wager: "I strive to avoid any reference to this transcendental as a condition of possibility for any knowledge" but cannot guarantee success. "I try to historicize to the utmost in order to leave as little space as possible to the transcendental. I cannot exclude the possibility that one day I will have to confront an irreducible *residuum* which will be, in fact, the transcendental" (*FL* 79).

Let us briefly consider the historical conditions giving rise to three philosophical structures: first, Kant's transcendental; second, the "quasi-transcendentals" of nineteenth-century knowledge; and third, early phenomenology.

During the Classical era, natural history, the analysis of wealth, and general grammar sought to illuminate relations of identity and difference among living beings, objects of exchange, and linguistic signs by means of representational elements such as anatomical character, money, or grammatical function. These are "media" of communicable thought comparable to the media famously analyzed by Marshall McLuhan. Natural history, analysis of wealth, and general grammar engaged in this inquiry without asking how the task of organizing representation in an infinite table shaped the "message" of nature. But when scholars discovered that the relationships among organisms, economic processes, and parts of speech change at different rates and without respect for the change or continuity of human self-understanding, they could no longer avoid reflecting on the specific historicity of human beings and human self-representations. What they lacked, as in Borges's Chinese encyclopedia, was a common *site* or *rhythm* of comparison for beings whose styles of being seemed ever more incapable of simultaneous representation.

Kant showed that the possibility of representation was grounded in the finitude of human understanding, which then occupied the role of this "site." As Foucault explains: "Until the end of the eighteenth century—that is, until Kant—every reflection on man is a secondary reflection with respect to a thought that is primary, and that is, let's say, the thought of the infinite." Kant's Copernican revolution reverses the priority of the infinite over the finite: "The problem of man will be raised as a kind of cast shadow, but this will not be in terms of the infinite or the truth. Since Kant, the infinite is no longer given, there is no longer anything but finitude; and it's in that sense that the Kantian critique carried the possibility—or the peril—of an anthropology" (*AME* 257). In other words, the nineteenth century identified finitude with the biological, economic, and linguistic conditions of human existence and human ignorance of the macro processes constraining that existence, but this interpretation of finitude was not philosophically inevitable (*OT* 313).

In the style of phenomenology practiced by Husserl, early Merleau-Ponty, or Alfred Schutz, the transcendental horizon of experience conceived a way that both duplicates and grounds the range of given empirical phenomena: "All empirical knowledge, provided it concerns man, can serve as a possible philosophical field in which the foundation of knowledge, the definition of its limits, and, in the end, the truth of all truth must be discoverable" (*OT* 341). The doublet or Fold formed by the empirical and the transcendental makes covert reference to the being of the one who both knows and is known—"man"—rather than to the phenomenon of discourse itself, which modernity conceives as *expression* rather than *representation*. Until the advent of recent disciplines such as structural anthropology, psychoanalysis, and structural linguistics, the "being of language" could not be addressed during the modern period at all outside literature, an expressive rather than representative practice. However, Foucault contends that "man's being" was first identified as an indispensable "thing-in-itself" on the basis of troubles in nineteenth-century linguistic science, and (diverging from Heidegger) he predicts that the epistemologically meaningful being, "man," may soon be decomposed into another series of problems with discourse (*OT* 339). Foucault does not lament this prospect, as insight into the nature of speaking, laboring, and living "man" has made it too easy to deny the rights of humanity to those left outside.

Why is the "transcendental-empirical doublet" a problem for post-Kantian European thought? First, it misrepresents the historical record as well as the philosophical and political issues at stake in humanism. It regards humanity as a unifying thread both inside and outside history and obscures the process by which different scholarly discourses *and their objects* were constituted as historical. Here Foucault's "antihumanism" follows lines laid down by Louis Althusser, Lévi-Strauss, and Roland Barthes. Traces of Barthes's reflections on the materiality of language (especially representation) can be seen in *The Order of Things.* In *Mythologies,* Barthes argued that the concept "man" had essentially been taken over as a master signifier by bourgeois French imperialism (1972, 100–102). By identifying their own aesthetic attitudes and economic interests as the mythical referent of all language, the bourgeoisie gave a necessary, natural, and benign face to their social and economic power.

Althusser, for his part, drew upon the *Annales* school to argue that intellectual and materially productive activities were among the layers or series of historical phenomena, like natural and political events. The rates of change and observation characterizing these series could not be directly compared, so none could claim to encompass "history" as a single continuum. For Althusser, as for Foucault, the idea of "man" was a product of forces arising from many points in the historical field and giving them a provisional frame of comparison (1977, 139–40). However, Althusser also believed that insofar as historical activities could be grasped at any given point in time as a structure (an intellectual act which altered their performance, and which he, unlike the *Annales* historians, believed was inevitable), causal priority should be given to the economic domain rather than to "man" (97–99). Foucault disagreed that priority needed to be assigned at all.

The supposed identity and continuity of "man" involves a selection from a very wide and variable range of activities, objects, desires, and experiences that might be considered "human." Most of Foucault's studies demonstrate how certain practices were "excluded" or considered incompatible with humanity in such a way as to make the self-evidence of "man" seem more coherent. But even when no actual violence against the subhuman was involved, our image of the "man" in whom transcendental capacities for knowledge and action are implicit involves a limitation of imagination and a restriction of attention to morally or epistemologically neutral, but potentially enriching encounters with beings according to rhythms that take nothing from the scale of the human, for example, pleasures of the natural world or even those of the sublime. There is no transcendental that does not in some way take its shape or quality from the empirical phenomena it supposedly coordinates and reveals. It effectively limits the production or at least recognition of new events and encounters to the form of those which have been produced, recognized, or worthy of attention in the past. In other words, it materializes.

It is helpful to compare Foucault's resistance to the transcendental-empirical doublet with Gilles Deleuze's quest for a "transcendental empiricism" that would approach every empirical or sensory encounter as a condition of possibility for further encounters. This is not only true for an individual's encounter with other beings, but also with him- or herself. The identification of a transcendental *prescribes* the form of subjectivity to the Kantian understanding as much as it *describes* the way in which Kant encounters every one of his thoughts as compatible with an "I think." Foucault mistrusts hermeneutics for subjecting a given text or practice to a deeper intention or meaning that can only be further explicated, never simply escaped. Both he and Deleuze, likewise, mistrust the transcendental gesture for limiting the forms of experience that can be considered knowledge to those that enhance past or familiar forms of power and pleasure, even one's own. Thus modern morality, Foucault laments, is reduced to a single theme in both knowledge and action: the attempt to think the unthought of existing practices and discourses, rather than to vary them or even develop new forms of communication, pleasure, and production. What Foucault retains of Kant is the critical gesture, not the transcendental one.[15]

Critics claim that Foucault reproduced the doublet between conditions of possibility and empirical contents or products in his own theoretical efforts. Why should Foucault care about the "misrepresentation" of the being of language arising from its subordination to the being of man, if he believes truth is neither a matter of representation nor of human self-disclosure? Hubert Dreyfus and Paul Rabinow contend that his early (archaeological) work must aim at some kind of hermeneutic description of the conditions for producing speech acts capable of recognition as truth or knowledge (1983, xxv–xxvii, 122–25). They account for the shift to genealogy by noting that since the conditions for producing speech acts must make reference to human purposes and desires, Foucault could not continue his archaeological project without situating their description within an account of embodied power relations. Habermas does not con-

---

15. In "What Is Enlightenment?" for example, Foucault suggested that the critical project should be detached from its transcendental method (*EST* 315–16) and replaced by a genealogical analysis of the events and forms of perception that gave rise to contemporary social practices and objects of social-scientific study. He begins to one side of the common root, and affirms finitude as discursivity and materiality, rather than "man." Deleuze, again for comparison, is much more assertive about questioning even the Kantian assumption of finite understanding. The end of ontotheology, according to Deleuze, need not mean the end of a creative relationship to the idea of the infinite. Deleuze thus replaces Kant's transcendental schematism with a genetics of infinitely differentiable experience that stretches from imperceptible sensation to the affirmation of difference in an Idea of eternal return that would encompasses all ontological possibilities.

test Foucault's assessment of modern knowledge as tacitly founded on a transcendental-empirical doublet. But he asks how Foucault can claim to escape the transcendental-empirical doublet himself, since, on his reading, Foucault's own empirical descriptions of power technologies also arrogate to themselves the *transcendental* role of grounding social-scientific discourse (1990, 273–74). In other words, he believes that Foucault only escapes the post-Kantian episteme by taking Kant's place anew.

Han, finally, suggests that Foucault's goal is less to arrive at a "non-transcendental" form of critique than to arrive at a "non-anthropological" understanding of the transcendental conditions of knowledge—a variant on the transcendental that could be called a historical *a priori* (2002, 20, 36). She believes that he fails in this endeavor, although he reformulated the project several times in relation to "discursive formations," the "archive," "regimes of truth," and finally "problematizations." More important, she believes that Foucault's late writings leave tensions in his overall project unresolved. Foucault's later writings attend to the reflective abilities of a subject whose freedom consists in the amount of play allowed in her everyday practices—play which is intrinsic to the exercise of *power* as an action upon actions, and which distinguishes power from action directly upon bodies, or violence (PWR 340–42). But the subject appears to have only two options for self-constitution: either in relation to background practices that render it passive, as in genealogy, or through an intellectual process of "recognition" that threatens to reinstall the constitutive subject of Husserlian phenomenology (Han 2002, 171–72, 179–87). By distinguishing between spontaneity and determination within a single regime of truth, Han argues, Foucault threatens to put the subject of problematization back into the place so recently vacated by the transcendental-empirical doublet "man," denounced by *The Order of Things*.

Deleuze agrees with these critics to the extent that he views Foucault's work through the lens of Kant's *Critique of Pure Reason*, especially his account of the transcendental schematism. Beginning with the Classical era, in Deleuze's reading, words and things no longer functioned interchangeably as signs. The uneasy relationship between the two series of the visible and the articulable, what can be seen and what can be said, is the hidden spring of discourse and the raison d'être for modern systems of power and knowledge. These systems serve the *historical* function Kant gave in his own work to the transcendental schematism. Analogizing the distinction between the visible and the articulable to Kant's distinction between receptive (intuitive) and spontaneous (discursive) elements of experience, Deleuze suggests that the transcendentally productive imagination is one version of a *historically* productive desire to close that gap (1988, 60). One must not assume, however, that Foucault presupposes that styles of representation and their associated mental faculties are given as "receptive" and "spontaneous" *prior* to the historically situated act of discriminative reflection; in short, both Han and Deleuze make Foucault too Kantian in a way Kant himself repeatedly questioned, especially in the *Critique of Judgment*.

Now, it is possible to read Foucault as arguing that we are *not* capable of exiting this episteme, at least through an act of will or superior insight. In *The Archaeology of Knowledge*, he states that we cannot identify the rules governing our own archive (*AK* 130). It is also possible to read Foucault as a *philosopher* reflecting on the episteme involving the human sciences, therefore free in some respects with respect to their assumptions. One could regard his work as prompted by another episteme treating the same objects as the human sciences, an episteme represented by structural linguistics, anthropology, and psychoanalysis (see *FL* 80). Finally, it is possible that even if Foucault were to propose a relationship between living beings, languages, and commodities that did not pass through anthropology, readers would still interpret his work as the identification of "unthought" rules governing thought and action. Habermas is right that Foucault can only escape the anthropological sleep by standing in Kant's shoes, but he is wrong to think that Kant's distribution of the relationship between truth, the empirical domain (*l'empiricité*), and the event cannot make a non-Kantian use of thought's materiality and discursivity.

But Foucault's relation to Kant is misinterpreted if the *Critique of Pure Reason* and the *Anthropology* are taken as his sole reference points. The transcendental schematism brings seeing and saying together in acts of *determinate* judgment. Now every one of Foucault's historical studies gives ample description of past schemes for producing determinate judgments. Sometimes Foucault seems to have interpreted his own work as the task of identifying such schemes. For example, in *The Archaeology of Knowledge*, he asks if description of the archive should not hesitate to "illuminate, if only in an oblique way, that enunciative field of which it is itself a part?" (*AK* 130). Every one of these schemes is unified by the trajectory of a certain practice of *reflective* judgment that emerges from several disciplines and attempts to grasp the world through forms that communicate with the results of other disciplines. Foucault's *Archaeology* also proposes its own scheme for making determinate judgments with respect to historical materials—one that would give the role of reflective variation more prominence in historical judgment than do the schemes employed by traditional narrative history. This scheme is centered on the archive, the statement, and the discursive formation with its elements.

Aesthetic judgments bring the transcendental and empirical together in a different way than determinate judgments involving specific empirical concepts. An aesthetic judgment is an *empirical* judgment, but a *subjective* one as well. Since it is predicated of a *particular* appearance, its universality holds for the sphere of all judges (subjects), and not for all members of a class of appearances to which that particular judgment belongs. It is a singular judgment, a "judging singularity," in the sense of an event that may happen similarly across time and among persons. It cannot be grasped in a concept because it establishes a system of simulacra. By contrast to the featureless *noumenon* which contributes (by its exclusion) to the transcendental determination of all objects in experience, Foucault's problematic objects are unthinkable apart from the particular bodies in which the trained eye detects their presence.

In this version of critical philosophy, the transcendental cannot easily be unified *over and against* the empirical. As Paul Veyne stresses,

> All these practices have in common the fact that they are both empirical and transcendental: empirical and thus always surpassable, transcendental and thus constitutive as long as they are not effaced. . . . Foucault did not object to being made to say that the transcendental was historical. These *conditions of possibility* inscribe all reality within a two-horned polygon whose bizarre limits never match with the ample folds of a well-rounded rationality; these unrecognized limits are taken for reason itself and seem to be inscribed in the plenitude of some reason, essence, or function. (Davidson 1997, 228)

Nor does Foucault himself seek such unity. To the contrary, he describes the suture of discursive practices and spatializations only in order to suggest that they might have occurred differently and might in some sense still be occurring differently.

So long as we remain within the post-Kantian paradigm, Foucault suggests that the pattern of our reflective as well as determinate judgments will involve pursuing the unthought and the origin in whatever we encounter and whatever we are. For Kant, this unthought involved powers such as imagination, reason, and understanding, discovered in an act of transcendental deliberation with strong similarities to aesthetic reflection. In the *Archaeology*, Foucault proposes that "the analysis of the archive . . . involves a privileged region: at once close to us, and different from our present existence, it is the border of time that surrounds our presence, which overhangs it, and which indicates it in its otherness" (*AK* 130). Is this not precisely the post-Kantian project of pursuing an always-retreating origin or foundation behind any apparent "original"? (*OT* 333–34).

Foucault does not address this problem. He goes on to state that diagnosing the archive, or bringing diagnosis of a past archive as close as possible to today's enunciative field, "deprives us of our continuities; it dissipates that temporal identity in which we are pleased to look at ourselves when we wish to exorcise the discontinuities of history; it breaks the thread of transcendental teleologies; and where anthropological thought once questioned man's being or subjectivity, it now bursts open the other, and the outside" (*AK* 131). Although some interpretation is required, Foucault's point seems to be that the archive should not be and should not orient us toward an explanatory "unthought" behind the evidence of the present, but should disassemble the present into parallel trajectories that only meet up at rare intervals. Rather than distinguish the present from or compare the present to the past, diagnosing the archive "establishes that we are difference, that our reason is the difference of discourses, our history the difference of times, our selves the difference of masks." Here his reference to Nietzsche and especially Pierre Klossowski's interpretation of Nietzsche seems prominent; if these differences are unthought, there is no point in trying to bring them to cognitive self-presence; for the *ego cogito* is only a provisional coincidence between personae, like the conversation two people in neighboring trains might have across parallel tracks as each pulls away in its respective direction. Thus "difference, far from being the forgotten and recovered origin, is the dispersion that we are and make."

Whether Foucault's archive is an "unthought" or a "dispersion" can only be resolved by reference to reflective, not determinate judgment. Insofar as reflective judgment is *aesthetic*, it grasps these forms (including the forms of bodily and psychological self-presence) as radically nonteleological—open to risk, variation, and alternate materialization. As seen in Kant's analysis of the sublime, aesthetic reflection also allows for variation and selection of the *temporality* or rhythm through which a form makes sense or comes into focus for the viewer. Thus, although Kant did not explore this possibility, aesthetic reflection *historicizes* in the sense of setting rhythms or series of events against one another as context and content. Finally, the pleasures of aesthetic and reflective judgment are explicitly divorced from the pleasures targeted by governmental and scientific interests. Behind the body and the apparent homogeneity of history, aesthetic reflection discovers the pleasures that cannot be reduced to that body and history, discovers potentially communicable qualities, registers painful failures of communication, and experiences resistance.

In the modern period described by Foucault in *The Order of Things,* bodies are one general form of the "unthought" counterpart to subjectivity (*OT* 314, 324). Thus it is not surprising that psychoanalysis and other empirical disciplines would identify medical knowledge with the limits of individual thought and reflection, seeking to push those limits back ever further. No matter how subjectivity is structured in a particular historical moment, and no matter how well or poorly an individual human organism manages to fit into that form of subjectivity, his or her body is an immediate *image* of this otherwise intangible and invisible accomplishment. The body is an image of the ways an individual can combine with other bodies, forces, and contexts that are not immediately recognizable as forms of the "unthought"—not in order to justify or reinforce his or her present identity, but to generate new events. Its powers and weaknesses also register and represent the effects of "thought" in the visible, tangible world—a thought that is not just individual but *transindividual* and embedded in the institutions, practices, and environmental conditions whose impact on bodies is easily explicable in naturalistic terms. They indicate how far a body can survive and gain strength from encounters with the outside.

In *The Order of Things,* Foucault writes that the only imperative of modern thought is "that thought, both for itself and in the density of its workings, should be both knowledge and a modification of what it knows. . . . It cannot discover the unthought, or at least move towards it, without immediately bringing the unthought nearer to itself—or even, perhaps, without pushing it further away, and in any case without causing man's own being to undergo a change by that very fact" (*OT* 327). Although he seems ambivalent about the extent to which exploring the archive can escape this ethical impasse, therefore, Foucault is consistently scornful of those who believe the demands of thinking can be satisfied or brought to a close. Far better, he implied, to break with the modern cogito for which imagination is only a relation to the unthought of *this form of thought*. If the unthought is "not lodged in man like a shrivelled-up nature or a stratified history," but is the Other to man, "not only a brother but a twin, born, not of man, nor in man, but beside him and at the same time, in an identical newness, in an unavoidable duality," then the ethical import of our relation to the Other changes from one of establishing the "fact of our identity by the play of distinctions" to acknowledging the "dispersion we are and make" (*OT* 326; *AK* 131). Searching for the unthought is a limiting and ethically debilitating use of human imagination when it reinforces existing unities, producing alterity by accident and rejecting it in the same gesture, rather than altering who and what we are. Indeed, we might distinguish "thought" from cognition on the grounds that "thought" alters the identity of the subject or cognizing being, either individually or collectively (Foucault 1984, 388).

Materiality and Resemblance: Statements

Statements (*énoncés*) and bodies are the two problematic objects Foucault suggests may be able to lead out of the anthropological arrangement of discourses and the security paradigm of governmentality to which they correspond. Foucault's contribution to the study of discourse as a political rather than an epistemological phenomenon is as widely recognized as his effort to understand the body politically rather than scientifically or metaphysically. It is rarely acknowledged, however, that these two contributions are cut from a single cloth. By beginning our study of Foucault's work with Kant, we see that there is an intimate link between the discursivity of the understanding and the materiality of phenomenal forms, including the body. In this section, I will argue that bodies are more like statements than referents, more like events than objects. Their perceptive and affective capacities for resemblance, finally, are more important than their capacities for identity and identification.

In numerous texts, Foucault proposes that there exist economies of pleasure, pain, signification, and knowledge that are as important as those of consumable goods. The creation of scarcity in these media of exchange is as important as in the traditional economic domain (with which they intersect), for apart from God's pleasure and its linguistic expression, the potentially infinite variability of bodily states and signs would fail to indicate a common world beyond their own irruption as singularities. It is not simply that Christian Europe feared potentially unlimited conflict in the absence of such commonality. The idea of commonality is, rather, a kind of insurance against conflict, which may or may not prevent human disaster (*SMBD* 215). The world's commonality is not merely a metaphysical presumption, but a *value* to which all bodies and signs refer—in addition to whatever immediate significance they may have. In order to understand how discourse and bodies can be made sufficiently scarce to reinforce this value, it is necessary to understand in what way they are "material." This study will clarify Foucault's transition from the archaeology of discourses to the genealogy of bodily techniques, and to situate his materialism with respect to the materialism of Marxists and feminists. For economies of discourse and pleasure play as important a role in the articulation or restriction of left-wing communicability as those of traditional production and exchange.

Kant's Copernican reorientation of knowledge toward the capacities of finite knowers rather than an infinite deity imposed spatiotemporal restrictions on intuition. It grasped objects—made them objective—using judgments whose concepts could never entirely determine an object, and which could only be *thought* in time, one after the other. Spatiotemporal intuition and discursive understanding are linked, according to Kant, by the schematism of imagination. Insofar as imagination is a sensible faculty, it is affiliated with intuition. But more commonly Kant regards the schematism as the handmaid of understanding, which selects relevant intuitions for the application of concepts. The schematism "assimilates" particular intuitions which share some similarity to concepts, which grasp these diverse particulars as one. For this reason, Kant also refers on occasion to the "schematism of judgment" and calls imagination a faculty of judgment (*CJ* 5:218; *CPR* A247/B304).

In the *Critique of Pure Reason*, Kant distinguished between the transcendental schematism that rendered pure concepts of the understanding (categories) compatible with the pure intuited form of appearances (time) and empirical schemata for entities such as plates and dogs (A137/B176). Both are rules guiding judgment in its ability to draw conclusions regarding elements of experience or to present concepts in an intuitive form, but only the schematism allows one to judge that an appearance *is* an element of experience. "The concept *dog*," by contrast, "signifies a rule whereby my imagination can trace the shape of such a four-footed animal in a general way, i.e., without being limited to any single and particular shape offered to me by experience, or even to all possible images that I can exhibit *in concreto*" (A141/B180). To each empirical concept pertains a certain degree of indeterminacy in the species, breed, and size of what presents itself; thus schemata enable *resemblances* to be gathered into a presumptive *identity*; i.e., dogs.[16]

The pure forms of intuition and the discursive understanding are *relations*—relations between a subject and his or her world, foregrounding similarities between elements of the world, including the subject him- or herself as embodied knower. As discussed in Part 1, these relations are deployed differently in determinative and reflective judgment, *restricting* information about that world in characteristic ways and allowing it to be synthesized in a suitable way for human interests. The fact that one takes a geometric pile of rock "as" a church or reads a text "as" a parody allow relevant, and potentially informative, comparisons to be drawn with other examples of the concept. Likewise, the temporal order in which one approaches the rooms of an unknown building informs one's estimation of its beauty, just as the order in which a novel presents its characters shapes the kinds of expectations—and surprises—that will be emotionally significant for the reader who follows their story. No medium is capable of delivering total understanding to the reader or viewer. Print and television alter the typical patterns of intuition, association, and inference bringing embodied viewers together with particular elements or sources of information (McLuhan 1964, chap. 1). Each medium restricts the viewer's attention or allows her freedom to explore the information at a different rate, and with respect to different conjectures or points of comparison. These relations between the viewer and the medium are often forgotten as we reflect on the content that they generate—either in the world or in the viewer's mind.

In Part 1, I also used the term "materiality" for the inescapable relationship of "realization," "restriction," and mediation which ensures that every imaginative act of schematization can be transformed into *some* other act of schematization within limits which are not entirely under the subject's control. Materiality, in short, is the way any given form relates to the forms into which it may be transformed or through which it may be recontextualized. Kant's "Analytic of the Beautiful" describes a subject suddenly freed to contemplate a variety of conceptual and intuitive possibilities implicit in a given form; while "Analytic of the Sublime" describes

---

16. See Kerszberg 1997, 39–40, and Longuenesse 1998, 115–16, for discussions on the role of reflection and abstraction from similarities in the formation of empirical concepts.

a subject *torn* between measures or scales appropriate to perception, imagination, and reason. Every discourse and embodied perspective imposes certain restrictions on *what* we can see in a particular representation or environment. These restrictions are associated with the relative scale through which apprehension takes place, as well as with the subject's interests or purposes. They also establish an intimate relationship of resemblance between ourselves and the objects in which that perspective is expressed, intimate relationships which take time to create, recognize, or undo.

Appadurai's description of how communities *produce* locality by investing themselves in a place, adorning it with associations and weaving its resources into daily practice, gives a concrete example of the process by which concepts are imaginatively and actively *materialized*. Elaine Scarry and David Harvey, likewise, refer to building as an act of "realizing" the imagination. Particular objects, thoughts, and acts are required to make a site into a church or a home; this perspective restricts the kinds of thoughts one tends to have in that site (and incongruent thoughts, when they occur, are felt as anxious or blasphemous). The same site can be a home (for example, for refugees who have taken the church as their temporary residence), and such a way of *seeing* the manifold of intuition will change their anticipations of further intuition, their cognitive acts, and the ways they make those anticipations visible (building a fire, storing clothes). We forget too easily that a site (and this includes "conceptual sites" like "Foucault's work" or "the transcendental") can be interpreted and materialized in different ways. A researcher accustomed to studying cell systems may have a difficult time apprehending the level of the whole organism or the ecological system in which it lives, even though he or she knows that events on this scale affect the phenomenon he or she studies.

As I also mentioned in Part 1, however, it is easy to forget how this materialization changes the subject; in other words, how living as a refugee in a church shapes the kinds of expectations, intuitions, and thoughts of which one is capable, or living as a cell biologist or a Kantian shapes the kinds of questions one bothers to ask about any event or phenomenon. In the absence of such awareness and mobility, it is very difficult to "imagine" oneself and one's environment differently, or recognize that other imaginative frameworks are *already* present in other people's associations with those stones, spaces, and bodies. The materiality of *imagination* is far more dense and potentially conflicted than the materiality of discourses or of bodies; indeed, separating discourses and normalizing bodies is one way a governmental practice can lighten this materiality (*AK* 216).

Marx's materialism enabled him to "invert" the priority of ideal essences over historical phenomena in the writings of German philosophers and British political economists, thereby placing theoretical production alongside other forms of production. Yet Marx is all too often read as if ideas were less material than the *things* in which relations of power are sedimented and consecrated by thought. In fact, ideas reflect certain historical conditions and help shape them in return. "From the start," he comments in *The German Ideology*, "the 'spirit' is afflicted with the curse of being 'burdened' with matter, which here makes its appearance in the form of agitated layers of air, sounds, in short, of language" (Marx and Engels 1986, 50–51). Marx describes the division between theoretical production and material production as the first form of the division of labor, a division which is both enhanced and concealed by the design of machines that reapportion tasks formerly shared among craftspeople. More significant, use-values and concrete forms of labor are *really* rendered exchangeable and more uniform insofar as they are subordinated to the *abstractions* of exchange-value and labor power. Assessments of "average social labor" and the deskilling of individual tasks became possible with the emergence of large markets, the replacement of tools with machinery, and the aggregation of many workers in a single factory space. "Not only," according to *Capital*, "is the specialized work distributed among the different individuals, but the individual himself is divided up, and transformed into the automatic motor of a detail operation" (1977, 481).

Etienne Balibar has suggested that "in the critique of philosophies of history, from Marx to Foucault, there is always the question of the *nature of the material*, and therefore, of 'materialism'" (1992, 56).[17] Marx's "revolution" consisted in the discovery that profit resulted from a "gap" or subreption between the "labor power" whose value "per working day" was based on the assumed cost of the worker's survival and the actual labor-value transferred to products, which exceed that cost by the end of the day's activity. This gap—"surplus value" or "valorization"—enabled Marx to define both the materiality of labor and the origin of capital, potentially setting them free for other uses by means of his own theoreti-

---

17. See Castoriadis 1987, 13–14, for a trenchant critique of simplistic—or undialectical—interpretations of "materialism" in Marxist revolutionary theory and philosophy of history; see Žižek 1993, 135–36, for another discussion of this problem in Kantian and Hegelian terms.

cal practice. Feminists, similarly, identified "sexual difference" as an unacknowledged and uninterpreted occasion for exploitation and learned antagonism between men and women. Recognizing the relationships that have been obscured by living, material abstractions like labor power and money *as relationships* opens the possibility of constructing or liberating other sorts of relationships, as well as the concepts that would name those new relationships. One must learn to work with the materiality of these relationships if one wants to create a viable force for social transformation, a collective practice of imagination that is sufficiently "real" to motivate its participants to materialize it in productive and political relations.

In the preceding discussion of madness, sexuality, and other problematic objects, no mention was made of problematic objects in *The Order of Things*. This text is all about "man," it seems, but man and his doubles are a structure of experience, not an image of what experience forecloses. *The Order of Things* presents *language* as the real problematic object of the modern era—foreclosed but constantly evoked in the forms of nineteenth-century knowledge centered around the ideal point of "man." This is a point Foucault had already made in *Birth of the Clinic,* when comparing Kant's critique, which is motivated by the fact that "there is such a thing as knowledge," to Nietzsche's critique, which is motivated by the fact "that language exists" (*BC* xv–xvi).

However, the "being of language" changes historically. In the Renaissance, the being of language was determined by its capacity to establish and order resemblances between words, images, beings, and humans that were already written into the structure of the world by a divine author. In the Classical period, the being of language was subordinate to representation. Movement from one idea to another in the order of differences and identities, however, made use of resemblance (*OT* 68–69, 119). Condillac and Hume agree that for language to grasp things, "there must be, in the things represented, the insistent murmur of resemblance; there must be, in the representation, the perpetual possibility of imaginative recall." And yet this double requirement opens nature to a problematic proliferation of forms and combinations, a materiality that threatens the purity of representation. The reason that Kant gave the title "transcendental imagination" to the transcendental condition for the possibility of representational synthesis is that both the Leibnizian rationalist tradition and the empiricism of Locke and Hume regarded imagination as the paradigmatic seat of representations and their combination or analysis (70–71). Heidegger claims that Kant diminished the importance of imagination by asking about conditions for the possibility of *representation* and modeling these conditions on logic. Foucault, however, might say that Kant left imagination as important as ever, but foreclosed further reflection on *language* as a material force or presence (*AME* 173).

In the opening chapters of *The Order of Things* and in essays such as "Nietzsche, Freud, Marx" (in *AME*), "L'ordre du discours" (translated as "The Discourse on Language" in *AK*), and *This Is Not a Pipe* (1983), Foucault suggests that the basis for signification is a play of resemblances among forms, including the visual and auditory resemblances that are bound as signs and words in a particular *episteme*. However, as Foucault observed with respect to the Renaissance *episteme*, knowledge based exclusively on resemblance is a "thing of sand," subject to interminable shifts and associations (*OT* 30). In "L'ordre du discours," Foucault identifies commentary, the author, and the scholarly discipline as strategies for limiting the direction and apparent potential for transformation or resemblance between texts and, accordingly, between speakers (*AK* 220–25). Commentary establishes a balance between the potentially infinite linguistic combinations of which humans are capable and the limitations of a specific topic, oeuvre, or genre under interpretive consideration, ruling out a certain number of treatments in much the same way that Kuhn's paradigms focus the attention of normal science while providing researchers with a suitable range of investigative tasks (Kuhn 1996, esp. 36–38). In *Birth of the Clinic* and an interview from the 1970s, Foucault proposed that the alternative to commentary might be the identification or even constitution of structures using textual elements whose regularity or resonance may not have been recognized by the author, much as structural anthropology reconstructed myths (*BC* xvii; *FL* 21).

We can conceive of many strategies for limiting and focusing the proliferation of resemblances among forms, whether discursive, corporeal, or practical. One strategy is represented by Kant's transcendental conditions for thought, including the schematic power of imagination to group similar intuitions under a single concept. But representation, with or without transcendental conditions of possibility, is another such strategy; and so is interpretation, which may likewise have a transcendental or nontranscendental ground. For Nietzsche, the play of interpretations is as bottomless as the table of representations was infinite for the classical era (Nietzsche 1999, 82–84). In "Nietzsche, Genealogy, History," Foucault refers to interpretation as "the violent or surreptitious appropriation of a system of rules" (*AME* 378). But *The Archaeology of Knowledge* is more subtle: "To interpret is a way of reacting to enunciative poverty, and to compensate for it by a multiplication of meaning: a way of speaking on the basis of that poverty, and yet despite it" (*AK* 120).

In other words, interpretation limits the proliferation of *énoncés* and speech acts by reference to a potentially inexhaustible "unsaid," in a way that enhances the prestige of some at the expense of others.[18] It makes use of the relationship between the imagination's need to realize and to restrict, addressed in the previous section on Kant. On the other hand, representation subordinates all similar elements of a series to the one element whose form is treated as exemplary.[19] Thus, for example, any intelligible perceptual or linguistic form can be regarded as a representation and any recognizable person is imagined to represent some social milieu or historical *zeitgeist*. Historians of ideas can speak of Condillac, for example, as "representative" of the Classical era, according to features of the social environment which they believe are mirrored in his texts. Both strategies are tools in a conflict over the disposition of statements. They restrict (or fail to cultivate) the variety of texts, objects, or worlds to which "interpretive" and "representative" discourses refer.

But discourse is more than a manifold of representations formed of concepts and intuitions, or a mere vortex of interpretation and commentary on intentions. The Renaissance intuition that words and things are interchangeable signs of resemblance, for example, survives in literature, along with the force of *naming* that anchored Classical representation in sensory experience (*OT* 300). Foucault argues that during the modern era, the "being of language" was dispersed between the sciences of formaliza-

---

18. Despite his attraction to Nietzsche's account of interpretation as "a violent or surreptious appropriation," Foucault seems to regard the act of interpretation as one that necessarily ties the text and reader to an already-said content and reinforces rather than displaces the power relations within which interpretation takes place (for example, *AK* 222 and *FL* 21). By contrast, he approves of the "machinic" model by which surrealists like Raymond Roussel construct texts using procedures that involve chance and homonymy (*AME* 21–32).

Fimiani (1998, 66) argues that Foucault's goal in *The Archaeology of Knowledge* is to combat two forms of "bad infinite" to which the signifier gives rise—the bad infinite of the subject's intentions, which form the ground for an endless interpretation of the signifier, and the bad infinite of a transcendental ground which can be seen to generate alternate expressions. Makkreel (1987) criticizes Foucault's image of interpretive knowledge as limited to a "hermeneutics of suspicion." The question is whether it is possible to "interpret" a text in a way that gives it and the reader radically new *external* relations rather than weaving internal continuities between past and future.

19. In *This Is Not a Pipe* (1983) Foucault alters his terminology and defines "resemblance" as a similarity that makes reference to a model or norm, while the salient qualities of "similitudes" are less easily identified and organized by perception.

tion, the interpretive disciplines, and literature. Each of these imposed its own criteria for limiting the proliferation of words and speakers, but was subordinate to the being of "man" as finite knower. In Nietzsche's words, the anthropomorphic philosopher "considers the entire universe in connection with man . . . the entire universe as the infinitely multiplied copy of one original picture—man. His method is to treat man as the measure of all things, but in doing so he again proceeds from the error of believing that he has these things [which he intends to measure] immediately before him as mere objects. He forgets that the original perceptual metaphors are metaphors and takes them to be the things themselves" (1999, 86). Outside literature, resemblance is difficult to find in the modern *episteme*. But it remains no less essential to sound knowledge than in the Renaissance or Classical era. Resemblance returns in the *repetition* of transcendental and empirical characterizing the being of man. The anthropological *fold* is a *structure*, but it is also an *event*.

Discursive forms/events need not only be interpreted. They can also be analyzed or generated in a mechanical way, as were many Surrealist and Dadaist works. Critics have complained that poetry produced in this way lacks convincing internal depth, but Foucault's analysis aims "to seek the law of that poverty"—i.e., the relations of restriction and realization that make certain resemblances salient—"to weigh it up, and to determine its specific form" (*AK* 120). The "archaeologist" of discourse performs an *epoché* upon the "things themselves" which a perception trained "naturally" to find continuities and the activity of a constituting subject in history would tend to seize as privileged objects of reflection (47–48). She asks about the aperture of the lens used to discover these resemblances and the institutions or communities in whose discourses they were communicable. She does not *interpret* the *énoncé*, or seek the transcendental conditions for its possibility as representation, but is herself put into relation with other enunciative positions (perhaps other subjects), institutions, and historicities depending on whether a phrase strikes her *as* interpretable or a mute "monument." Thus archaeology reopens the interpretive and representative organization of words and things to consideration of the limited, but historically positive manner in which resemblances and repetitions are ordered. It shows that hierarchically organized appearances and concepts, transcendental and empirical, interpretation and interpretandum, representation and represented, can also be ranged alongside one another as "resemblances" or even "simulacra" manifesting a nonreproducible *event*. Although discourses are composed of signs, "what they do is more than use these signs to designate things. "It is this *more*," this focusing, associative, and multiplicative activity, Foucault stresses, "that renders them irreducible to the language [*langue*] and to speech"—this excess over the hermeneutics they make possible (*AK* 49).

Both archaeology and genealogy address the historical processes whereby the materiality of linguistic and corporeal forms is cultivated and restricted in the service of knowledge and social order. The *énoncé* that forms the basic object of historical and philosophical interest in *The Archaeology of Knowledge* is an event approached at an unusual range and from an unusual angle, such that the rigidity of traditional categories of analysis—and the rigid behavior of the historian who forgets his or her own contribution to the form of appearances—can be disrupted. The statement is an event "that neither the language (*langue*) nor the meaning can quite exhaust . . . unique, yet subject to repetition, transformation, and reactivation; . . . linked not only to the situations that provoke it, and to the consequences that it gives rise to, but at the same time, and in accordance with a quite different modality, to the statements that precede and follow it" (*AK* 28). In a later interview, Foucault described the historian's process of "making events" or "eventalization" (*évenementalization*) as "making visible a *singularity* at places where there is a temptation to invoke a historical constant, an immediate anthropological trait, or an obviousness that imposes itself uniformly on all," *multiplying* the causes presumed to condition this singularity, and situating the event at the crossroads of "multiple processes that constitute it" (*PWR* 226–27). Events are not simply found, but they can be produced by historical perception in ways that allow them to communicate with more or fewer events, processes, and domains of reference (228).

Because the *énoncé* is a system of *dispersion* or *relations* between similar historical contexts of speech and action, rather than a *unity* or *entity*, the archaeologist can resist his or her immediate temptation to divide historical phenomena into those which express human intentions and meanings and those which are merely functional, structural, or material. Here we should be alert to difficulties posed by the spatial connotations of the word "context," for where history is concerned, "context" means conditions for a possible repetition. Defined by its relationship with correlative contexts or repetitions, potential subjects, an associated field of other statements, and materiality, the *énoncé* indicates the limits within which a particular speech act or proposition can be meaningful for subjects or "reinscribed" or "materialized" in another place and time (*AK* 103, 108–10). For example, Aristotle's remarks on the central importance of a middle class in well-governed cities count as a different *énoncé* from Sieyes's appeal to the Third Estate, Hegel's valorization of civil society, and contemporary American invocations of the middle class in discourse about welfare and tax reform. However, the principle of the "mean" would have been recognized in a small number of Greek discourses and constituted an *énoncé* linking them culturally and philosophically. The *énoncé* is a "function" (in the mathematical sense) defining the range over which a speech act or performance can be iterated, but existing only *in* those iterations, insofar as they resemble one another and are imitable (86–87).[20]

Moreover, the *materiality* of the *énoncé* is not simply the medium, the page, or the air on or in which a phrase is inscribed or uttered—the materiality of the signifier, comparable to the materiality of phenomenal forms (*AK* 102–5). *Énoncés* are not signifiers or phrases, forms for a content the way a genus contains a species or a concept suggests exemplars. Their materiality indicates the various contexts in which associated phrases *and differently situated speakers/listeners* might have "materialized," rather than the infinite complexity or interpretable meaning of a form (genus, concept) whose scope and unity are fixed. Although an *énoncé* is unrepeatable, this is because it involves an entire group of related speech acts and speaking positions, not because it constitutes a single unrepeatable speech act or intention. For this reason, the *énoncé* can play a role in many social contexts (political, economic, theoretical, administrative)—as well as different temporal contexts—and give a subject imaginative and practical access to many contexts over time. Finally, because it refers to the anticipatory stance of listeners or archivists as well as the forms of spoken language and perceived form, the *énoncé* is at stake in struggles for power (105).

When archaeology confronts the fact that "one cannot speak of anything at any time," that "it is not easy to say something new; it is not

---

20. "Signature, Event, Context" (in Derrida 1982) takes issue with Searle's observation that parodic or improper speech acts rely on general knowledge of a "normal" performance for their idiosyncratic effects. It is because each performance can be "iterated" in improper contexts that it is even capable of a "normal" function. Derrida's iteration and Foucault's *énoncé* differ in that the *énoncé* is not a repeatable performance but the unrepeatable set of all performances that can be linked according to a certain principle of historical selection. Furthermore, where Derrida indicates the potentially infinite openness of every performance to iteration, Foucault is concerned with the historical specificity of the *limited* contexts in which a statement actually does materialize.

enough for us to open our eyes, to pay attention, or to be aware, for new objects suddenly to light up and emerge out of the ground," it is not because "unsaid" things have been repressed or prohibited by the force of formative rules; rather, they have *not been produced* (*AK* 44–45, 119). The object "does not preexist itself, held back by some obstacle at the first edges of light. It exists under the positive conditions of a complex group of relations" (45). In *The Order of Things,* Foucault stresses that the disappearance of man "does not create a deficiency; it does not constitute a lacuna that must be filled" (*OT* 342). Similarly, in the interview titled "Truth and Power," Foucault warns that "what makes power hold good, what makes it accepted, is simply the fact that it doesn't only weigh on us as a force that says no; it also traverses and produces things, it induces pleasure, forms knowledge, produces discourse" (*PWR* 120). That statements are rare or scarce, constituted by lack, does not mean that there exists a "repressed" or unsaid content existing prior to their articulation,[21] a "truth" to the relations between faculties in the individual's psychology, or a "scientific" account of material class relations that would escape the historical constraints and conditions of sociological knowledge. In other words, the *énoncé* is a form in relation to other forms, rather than in relation to a determinate content. Lack seems to be *correlative with* this (enunciative) field and to play a role in the determination of its very existence, in much the same way that Kantian imagination must restrict appearances in order to realize them as an entity identified in a concept (*AK* 110).

The rules that relate what is possible to what is actual, and which schematize actuality against the backdrop of what only later we experience and recognize as impossible for past ages, are what Foucault means by the *historical a priori*. This *a priori*, however, is not transcendental. These rules allow us to describe historical events in their *positivity* rather than as limited expressions or determinations of a *whole* whose *regulative* rather than *experiential* nature Kant would have insisted upon. Positivities are "incomplete," "fragmented" by contrast to the holism of "meanings"; they lack the "interiority of an intention" and are dispersed among

---

21. In the same way that, for Serres, madness is not an existing phenomenon identified and demarcated by reason but arises positively with the very acts of discrimination designed to ferret it out (in Davidson 1997, 42–43). On the historical priority of the symptom over the "essence" of madness, see also Foucault's interview with Chomsky in Davidson 1997, 140, and During 1992, 28. For illustrations of "rarity," see Veyne's contribution to Davidson 1997, 157–58.

potential speakers or sites; they accumulate rather than express a single teleological thought or act *(AK* 125).²²

A positivity, finally, is a "limited space of communication" *(AK* 126). This is meant not in the sense that a courtyard allows several inhabitants along the walls to gossip together, but in the sense that gossip is always an event lived differently by individual participants but only *happening* insofar as they have different pieces of information to transmit. Thus it is a logical more than a physical space, although physical space is required for its articulation. The materiality of this logical space allows the *énoncé* to operate in the seemingly distinct registers of thought and "unthinking" practice. The asylum, the clinic, and the prison are material to the extent that they enact a particular suture among economic, religious, political, and medical spaces, and relate the discourses which disclose these phenomena to the practices in which they are embedded—but with which they are never completely synchronous. The great confinement is an ongoing "event" of carceral societies, inseparable from the individual prisons, schools, hospitals, and factories in which its form has an impact on citizens. The exclusionary function of these institutions—which are themselves *enoncés*/events—demonstrates the necessary limitations that must be placed on the multiplicity of incompatible contexts for experience to take on a unified shape, to refer to itself, and to be interpreted on its own terms.

Thus the asylum, the clinic, and the prison are ways of thinking, and the materiality of these institutions reflects the finitude of thought within the world made intelligible by the exclusion of the phantasms, deformities, and illegalities which would otherwise render events too confusing and similar, in their violence and suggestibility, to be endured.

22. Foucault's use of the term "positivity" may be confusing, given the extent to which he identified with the explicitly antipositivist tradition in twentieth-century French philosophy of science and history of science (represented by Cavaillès, Bachelard, and Canguilhem). His work is also committed to undermining any identification of (ethically compelling) social or psychological norms with scientifically discoverable "normal states" in the manner of Auguste Comte, the founder of French positivist social science. For Comte, "positivist" science was the happy successor to metaphysical and mythical discourses regarding nature, including human nature: this sense of "positive," meaning "based on observation or experimentation" rather than speculation, occurs in Foucault's many references to "positive psychiatry" or "positive medicine." However, Foucault does view Comte as a genuine successor to the Kantian critical tradition, compared to spiritualist philosophers of the nineteenth century and to phenomenological or Marxist philosophies of history, which invested historical materials with intrinsic interpretive unity and spiritual significance *(AME* 437). At one point, Foucault refers to his portrait of the historical record as *constructed* from series of statements or archaeological traces as a "felicitous positivism" *(AK* 234).

This way of understanding the *énoncé* challenges some interpretations of Foucault. Dreyfus and Rabinow, as I have mentioned, read *The Archaeology of Knowledge* as an attempt to *explain* the existence of social scientific discourses in terms that borrow as little as possible from social actors' desire to find meaning in their collective world (where the alternative is to understand how the human sciences achieved their "objectivity" by regulating the production of texts and distinguishing certain "objects" and "subjects" from a background in the first place). Social practices, they contend, are the background shared by social scientific researchers and their subjects, interpreted and exhumed as the latent referent of speech, writing, and action. For Foucault, as they recognize, "Rather than being the *element* or horizon within which the discursive practices take place, . . . the nondiscursive practices are *elements* which discursive practices take up and transform" (Dreyfus and Rabinow 1983, 77). But because they conceive of *énoncés* and nondiscursive practices (institutions) as meaningful *in themselves,* the strategic backdrop of discourses and institutions is for them an even "deeper" level of meaning than that unmasked by ordinary hermeneutics (124). Genealogy, they suggest, enabled Foucault to situate his own historiographical project with respect to this meaning by analyzing the disposition of intrinsically situated *bodies.*

Although Foucault does displace and recontextualize his own project in many interviews and even at the beginning of *Archaeology* (AK 16–17), it is unclear that genealogy represents an act of belated self-situation.[23] On the one hand, Foucault's studies had always hoped to rupture the habitual schemas of historical perception that bound subjects and books, oeuvres, or ages in an unacknowledged mirroring relation. On the other hand, they identify the "situation" of mediation or between-ness linking speakers, perceivers, contexts, statements, and social or natural phenomena as "materiality" rather than "meaning." Embodiment cannot ground us in practices because we cannot know the extent of our bodies and their powers in themselves, although practices do presuppose the capacity for *invention* implicit in corporeal and linguistic materiality. In fact, I would argue that Foucault differs from Merleau-Ponty in redefining the body as a "flat"

---

23. In *Society Must Be Defended,* Foucault linked the two by saying "Archaeology is the method specific to the analysis of local discursivities, and genealogy is the tactic which, once it has described these local discursivities, brings into play the desubjugated knowledges that have been released from them" (SMBD 10–11). "What Is Enlightenment?" presents them as elements of a single critical practice (EST 315).

monument just as the archaeologist identifies an *énoncé* in place of the "document," the better to demonstrate that language and embodiment are viscerally engaging because they arise not *within* a personality or context but *between* personalities and contexts. We do better to interpret the "body" in Foucault as parallel to the *énoncé* in being an event that happens in *similar* versions. If this body is a situation, then like the heterotopia, it places Foucault *outside* and *between*, not *within* social practices.

Materiality and Resemblance: Bodies

Bodies, like linguistic and perceptual forms, are capable of resemblance to and differentiation from one another, as well as transformation and indefinite determination. But where *positivity* and *rarity* organize the materiality of discursive forms, *power* is Foucault's name for the way bodily materiality is organized or disposed, and occasions for resemblance to real and ideal appearances are cultivated and limited. The application of power ensures that some bodies (like psychiatrists or legislators) can interpret and represent others (*HS1* 93–95). Although each of us tends to identify our emotional states and pragmatic intentions with a particular interpretation of the body as well as a particular self-representation as the subject of discourse and action, these states and intentions can also be understood as variations on *bodies* and *énoncés* that are transindividual.

The power relations shaping a body indicate that in order to interact with other subjects, worlds, and institutions, it must be capable of taking on multiple forms, but within a limited range of recognizably human performances. Sexual drag and racial "passing" are examples of material embodiment that function effectively within the constraints of one role although their materiality could be worked up into another role, given the right conditions (such as exposure, or the transition between home and work or club environment). Playing one of these roles, like looking at a tissue sample through a particular kind of scientific instrument, shapes the questions one asks as well as the questions one must answer. Resistance to power, on the other hand, may be best understood as a refusal to resemble a real or imagined image, held by another participant in power relations or by a collective with which one must interact.

Foucault's understanding of power may seem very "immaterial," very "ideal" if one thinks that bodies are only affected by violence, by direct rather than indirect pressure, pain, and pleasure. Violence acts directly on the body. But power cannot and will not touch the person as "thing-in-herself." It constrains her to *appear and to perceive* in a certain way, while leaving her a margin for inventiveness and variation, indeed, for "freedom."[24] In a late interview, "The Subject and Power," he explains, "The exercise of power is not simply a relationship between 'partners,' individual or collective; it is a way in which some act on others . . . what defines a relationship of power is that it is a mode of action that does not act directly and immediately on others. Instead, it acts upon their actions: an action upon an action, on possible or actual future or present actions" (PWR 340).

Power directs the subordinate's imagination toward certain contingencies or crises and away from others. To be more precise, it *is* this directedness of imagination, rather than a force external to it. When she speaks of torture as *restricting* the victim's ability to protect his or her psychic integrity and capacity to act, for example, Scarry suggests that imagination is the first target of domination and the last point of resistance for the one who is subject to power by means of his or her embodiment (1985, 36–37). Those who *govern* hope to direct this resistance toward certain inventive ends, which are denied to the victim of violence.[25] In his last works on the history of ethics, Foucault turned his attention to the power relations, including acts of interpretation and representation, through which individuals act on themselves and resist the resemblances encouraged by others.

Leaving the "execution" of a command to the subordinate party, in however small a way, ensures his or her personal investment in the resulting action and increases the reach and efficiency of one who governs by releasing him or her from the need to "imagine all contingencies" in advance.[26] As Sandra Bartky notes, women take pride in their ability to

---

24. PWR 342; see also the lecture "Omnes et Singulatim," in PWR 324.

25. This is not to say that violent action is any less "stereotypical" or "scripted" than the action of "governing," in the sense that the one who commits a violent action *imitates* and forces his or her partner to imitate a kind of body that may or may not be physically present at that time. Violent action easily occurs in groups, I would suggest, because it is imitative but not individuated. Thus Sharon Marcus (1992) can speak of rape as a script in which bodies are constrained to enact culturally accepted postures of violent masculinity and passive femininity and "enlisted" into the service of larger social forces, while preventing rape requires male and female bodies to insist on cultivating other (probably individuating) capacities for resemblance or being willing to let the set of resemblances that define masculinity and femininity "drift," relating, one might say, through power rather than violence.

26. Amartya Sen's efforts to think about social justice beginning from a logic of situation-specific "capacities" could be rooted in a similar understanding of the relationship between power and dignity (1992).

master the details of feminine comportment even when they recognize these gestures as "responsibilities" owed to gendered social mores (1990, 69–71). Workers on a highly restrictive assembly line use their own ingenuity to maintain production when the machinery breaks, and self-employment remains an attractive ideal for many Americans even when economic conditions render contractors dependent on the whims of clients. But thereby the imagination and actions of men, managers, or clients are increased in scope and detail.

In *The Genealogy of Morals*, Nietzsche suggests that bodies and signs became interchangeable in the context of particular historical relations of commerce and punishment, though he never claims that his picture of history is anything but an explanatory *arrangement* of linguistic documents inherited from previous generations. Both Nietzsche and Marx would have agreed that "equality" among commodities as well as persons makes reference to a particular historically valorized form of comparison or "exchange-value," one that may establish some use-values, abilities, or character traits as the measure of all others (Nietzsche 1989, 70; Marx 1977, 164–69). They also understood power as the ability to reinterpret similarly situated entities as elements of a hierarchy, and to make discursive signs refer to bodily signs and symptoms as their final referent or founding context. In "Nietzsche, Genealogy, History," Foucault suggests that the human body records the traces of accidental strategies for organizing and dominating resemblances (*AME* 374–75). In other words, the body's form indicates the series of sedimented variations in scope or aspect to which human perception has been subjected in *coming to see itself as* human and distinct from an environment that it dominates in thought.

In so doing, Foucault proposes a non-Kantian understanding of the "being of language"—another way to organize the materiality and discursivity of understanding. Kant denied the identity of conceptual and linguistic representations. But neither he nor his naturalist contemporaries imagined that language was anything *other* than a representation, and assumed that the transcendental conditions of representations would also govern the use of language. At several crucial points in *The Order of Things*, however, Foucault points to Sade as a Classical author for whom, by contrast to Kant, the supposed transparency of representational language rested on an all-too-evident materiality. In its singularity and individualizing force, Sade's desire was always in conflict with the sign's generality, understood in the first instance as a name (*OT* 118–19).[27] But

---

27. In *History of Madness*, Sade is contemporaneous with the more general social projec-

because it claimed to represent what goes beyond the knowledge or experience of any individual subject, Sade regarded representation as one instrument of corporeal power among others, whether in the domain of politics or sexuality. Around the same time that Kant formulated the critical philosophy, Sade was engaged in rediscovering the material force of the language used by Classical representation—its ability to provoke resemblance and to alter bodies without being affected in turn. By repeating scenes of violence and pleasure, Sade generates anguish and desire in the reader. He uses language, including the philosophical language of truth, to act on bodies directly and to combine them in singular events, which are, in a very real sense, new bodies.

In his readings of Nietzsche and Sade, the poet and artist Pierre Klossowski explored how the world of critical philosophy might look different if Kant's finite understanding were governed by a reason driven to speculate on *events* experienced fully through repetition, rather than on *Ideas* carrying out the unifying functions of logical judgment.[28] In Klossowski's interpretation of Nietzsche, which influenced both Foucault and Deleuze, words and bodies are masks through which events, souls, and power relations (Klossowski refers to all three) endlessly *repeat and differentiate* themselves in the visible or tangible realm. For Klossowski, bodies are neither intuitive representations whose differences are submerged through recognition in a concept nor substances for a unified *apperception*. Rather, they are gestural events, repeatable gestures through which *simulacra* or souls actualize and differentiate themselves. When God declares the number of elect complete and closes the gates of Heaven, the

---

tion of imaginative contents onto the closed space of the asylum, which led to great disorder during the postrevolutionary Great Fear (*HM* 358–61). Indeed, Foucault describes the *practice* of sadism as a "conversion of the imagination" originating with the translation of confined unreason into "discourse and desire." "Sade, Sergeant of Sex," criticizes the libertine's inability to think outside the "regulated, anatomical, hierarchical" model of a disciplinary society (*AME* 226; see also *FL* 81).

28. Thanks to Brent Adkins for calling my attention to Klossowski's influence on the idea of resemblance in Foucault's thought. Foucault's essay "The Prose of Acteon" appeared in 1964 (*AME* 123–35; also see *FL* 118–19). An important point of reference is Deleuze's "Klossowski or Bodies-Language" in *Logic of Sense*, which broaches the idea of multiplicity as a "demonic" alternative to the Kantian Ideas of soul, world, and God (1990, 296–98). Klossowski gives a theoretical account of the "infinity of predicates" into which his novels dissolve presumed substances in *La Ressemblance* (1984). Beginning with "On the Mimetic Faculty" (Benjamin 1978) the Frankfurt school represents another tradition of thought about resemblance and representation, to which Horkheimer and Adorno (1995) and Taussig (1993) belong. Wiegel (1996, 39–45) refers to resemblance in comparing Benjamin's views on language and embodiment with those of Foucault.

knights in Klossowski's novel *The Baphomet* (1988) are possessed by multiple wandering spirits. Some souls, however, like St. Teresa of Avila, have voluntarily left heaven to inhabit living bodies along with their present owners and other restless spirits, thereby confusing the saintly with the diabolical. The order of knights is eventually condemned and destroyed for failing to enforce a one-to-one correspondence between bodies and the integrity of single souls. Literary images are also *simulacra* that establish relations of resemblance and influence on the reader's fantasies; language is nothing but the medium for this event of resemblance.

These literary reconceptualizations of the relation between signs, imagination, and bodily materiality make sense of certain visceral but "uncanny" phenomena in everyday life. Max Horkheimer and Theodor Adorno, like René Girard, Jean Piaget, and some psychoanalytic theorists, are attentive to the emotional resonance, excitement, and anxiety provoked in humans by instances of resemblance and imitation. They associate the formation of bourgeois white subjectivity—and embodiment—with the refusal to imitate bodies outside a certain social circle and scorn for those who are regarded as overly prone to imitative behavior (Horkheimer and Adorno 1995, 180–85). Anti-Semitism is one consequence of the European's unwillingness to resemble or imitate members of unfamiliar social groups. More recently, Michael Taussig and Saidiya Hartman have explored how aboriginal Americans and African-Americans were forced to assume the role of "mimics" for the entertainment and disgust of white explorers, conquerors, and employers (Taussig 1993; Hartman 1997). Homi Bhabha (1994) argues that the colonial subject who successfully imitates his or her political and cultural "masters" calls the uniqueness and authenticity of that cultural identity into question and causes them great unease (see also Butler 1993, 51–53). As the conflict between Kant's disdain for artistic imitators and his observation that genius must learn to emulate the *originality* of past models indicates, emulation of the master's *mastery* ceases to be a sign of subordination or flattery at a certain point and becomes, rather, an occasion for competition or tacit equality that must be repulsed with double violence.

Humans are intensely ambivalent and often attracted to forms that are imperfectly similar to themselves—as individuals and as a species. But we are also attached to systems of power that preserve the uniqueness of the "bodily matter" which enables each of us to mimic desirable others, systems of power that exclude some resemblances as "unthinkable." People who have worked to resemble members of a privileged group are often horrified at the apparent resemblance which members of the "masses" bear to one another, a horror summed up in the disparaging claim that "they" (unlike "us") all look alike. Fashion thrives on the constant creation of small differences that measure affects of admiration and disdain, differences that constitute newly discovered reasons for belonging to some community. But the most powerful small differences or similarities are those which are inscribed upon or recognized in the form of bodily anatomy, whether visible racial traits or the distinctive way of walking and speaking that marks someone's social class.

"Two broad kinds of approach to theorizing the body can be discerned in twentieth-century radical thought," writes Australian feminist Elizabeth Grosz. The first, which she associates with Nietzsche, Kafka, Foucault, and Deleuze, "conceives the body as a surface on which social law, morality, and values are inscribed"; the second, associated with psychoanalysis and phenomenology, "refers largely to the lived experience of the body, the body's internal or psychic inscription. Where the first analyzes a *social*, public body, the second takes the body-schema or imaginary anatomy as its object(s). It is not clear to me," she adds, "that these two approaches are compatible or capable of synthesis," although an earlier text, *Volatile Bodies*, proposed a model whereby they can be thought together as two "sides" of a Möbius strip (1995, 33; 1994, xii, xiii).

Inscription, which finds its clearest expression in Nietzsche's explanation of "bad conscience" as an effect of cruelty and is clearly at work in Foucault's *Discipline and Punish,* is both a real practice and a philosophical account of that practice. As a practice, external discipline and punishment encourage individuals to take their own actions, with all the muscular and conceptual or emotional aspects involved, as a "landscape" which can be scrutinized and developed in greater detail. Nietzsche's comment that the strong do not, in fact, experience "guilt" as a result of punishment reveals that those who punish have in mind some model of the kind of interiority they would like to produce in others—a model that need not be "taken to heart" by the culprit (1989, 81). Thus there are at least four elements in punishment: the bodies of punisher and punished, the imagination or "representation" animating the person or institution who punishes, and the "self-affection" or imagination that the person who suffers punishment sets to work on itself. Nor does the sufferer entirely control his or her own imagination, as discovered by theorists of monasticism. The master can say "govern yourself as I govern you," but even with the "best" will in the world, it is impossible to know the exact content of the analogy each person involved might draw between these two representations or imaginative acts; or rather, the process of coordinating this analogy takes time.

Unlike other feminists and critical theorists, who fear that the theory of inscription does away with the moral significance of emotion and interiority, Grosz's concern is that the theory of inscription tends to "decorporealize" the human body—if only by saying "the human body," rather than this or that fat body, bony body, body with diabetes, body with sweet-smelling hair or reeking of sweat, menstruating body, and so on. "Analyses of the *representation* of bodies abound," she protests, "but bodies in their material variety still wait to be thought" (1995, 31). She suspects that many feminists are anxious that attention to these details will unsettle their socially fragile claims to borrow the discourse of the philosophical or political "subject" or, to use Kantian language, reveal them to be too far outside the *sensus communis* to maintain their plausibility as communicators. "The "material" on which the punisher's imagination is inscribed with all its uncertain effects, for example, is sexed "prior" to inscription, by its capacities to ovulate or menstruate (36; see also Gatens 1996, 14–15). Nietzsche's account of the body is too "philosophical," not tangible enough to support the violence in which he claims subjectivity originates (Grosz 1995, 37–38).

Judith Butler is the contemporary philosopher most frequently accused of reducing bodies to representations used in the exercise of power. Drawing on Aristotle, Butler argues that no aspect of the body's materiality can be defined without reference to "form," and this form is preeminently given in discourse (1993, 32–36, esp. note 12). The question of whether "form" exists apart from the discourse through which we identify and communicate it is difficult, and earlier I merely touched on it by suggesting that from a Kantian perspective form can only be defined "phenomenally," that is, in relation to other forms. Butler regards materiality as an effect of power relations that create conflict between processes of formation. Language is undeniably *one* of the social practices in which form may be taken as an object or model for self-development, and the struggle between multiple forms of anticipation and invention is consciously negotiated *in language;* but visual representations and even the mimetically suggestive *presence* of other bodies are also occasions when form can be "inscribed" on an individual body. We notice male, female, or hermaphroditic anatomy based on the available images and words through which artists and doctors group resemblances and declare differences. We likewise conform to models of conduct in light of what peers and supervisors mention or fail to bring to our attention discursively.

Because Butler focuses on the application of specific norms having to do with gender, one can get the impression that every human act is *already* the product or application of some norm. I believe this is incorrect; lots of intellectual, emotional, and bodily gestures are "uninformed" in the sense that no one has bothered to notice or make them communicable *as* agreed-upon or disputed norms. What Butler does stress is the temporal nature of form: the fact that no resemblance or difference strikes the eye sufficiently to be identified (and potentially corrected or disciplined) without an appropriate interval for contemplation and repetition (1990, 140–41, 145–46; 1993, 9–10). Those identities which strike us in a flash (visual identification of anatomy, for example) are those for which words have already been developed, whereas others take time to agree upon and are more obviously matters of aesthetic apprehension. One must perceive repeatedly, and the object one perceives must repeatedly exhibit some trait before it can be distinguished and recognized. Anyone can pick up a diagnostic manual, but only after reviewing many cases and spending a certain amount of time with each patient can one say with any certainty (in the terms of that manual and its professional community), "This one has paranoid ideation," "This one is obsessional," "This one is depressed."

For Butler, "materiality" is the aspect of gestures that *resists* being formed at any given moment and requires a further inscription or repetition, rather than the unformed "remainder" or the indeterminate aspect of behavior that has not yet been theorized or identified as worthy of notice. Her main concern is how heterosexual embodiment is shaped by melancholy and resentment over social codes *forbidding* similarly sexed bodies to resonate or "communicate" in "queer" ways (1993, 63–65; 1997, 132–50). However, her insight that giving up the pleasure in certain ambiguous resemblances causes suffering and hostility can be extended to the racial and national dimensions as well. There may be ways in which white people in racist societies suffer melancholy as a result of being forbidden or never being taught social styles associated with "blackness" in the white imagination—social styles that would enable them to mix more comfortably with people outside their own racial group. (This does not, of course, justify or compensate for the privileges they enjoy as members of the majority.) Ann Laura Stoler describes the immense fear with which

Dutch settlers in the colony of Indonesia contemplated the possibility that their children, raised by indigenous "nannies," would identify with Indonesian rather than Dutch aesthetic sensibilities (1995, 146–50). Antimiscegenation laws in the American South caused untold suffering, especially insofar this ban on "feeling resemblance" was also used to justify lynching. This suggests that a considerable aspect of the body's materiality has to do with the techniques by which churches, families, and governments *individualize* or bother to concern themselves with the differences between people who would otherwise "think and feel as one" or "disagree" in unpredictable ways.

Despite Butler's reputation for upholding a purely "discursive" notion of sexual and other aspects of embodiment, therefore, it appears that she associates materiality deeply with the affective dimension, primarily melancholy. For Grosz, on the other hand, materiality clearly has more to do with the "pulpiness," "bony-ness," or perhaps the "smoothness" or tactility of the body than with its emotional tone. These do not conflict; materiality has to do with the relations between people (discipliners and "doubles," researchers and their subjects) as well as with the interaction between elements of the individualized body in its pulpiness, sadness, or sovereignty. The body's materiality also reflects the time interval over which different perspectives may identify and name the same disciplinary or subjectivizing process. A day of severe depression has a different meaning when one finds it recurring every month as part of a particular body's menstrual cycle; one day of argument with a lover can be ignored, but several weeks of tension mean that people who were provisionally one have begun to separate. In other words, materiality has as much to do with holes as with pulp, with the emptiness of the mold as with what fills it out.

Following Spinoza, Moira Gatens argues that the bodies comprising a polity are only differentiated and arranged through acts of imagination that enable some bodies to move more freely and develop more complex capacities than others. "The political body," for example, "was conceived historically as the organization of many bodies into one body which would itself enhance and intensify the powers and capacities of specifically male bodies" (Gatens 1996, 71). The quality of one's imaginative representations affects one's *being* and ability to act or be acted upon; thus sexuality, nation, and class are systems of representation that limit or enhance the intensity and efficacy of certain groups' bodily experience as well as their knowledge. "A person's capacity to affect and to be affected are [sic] not determined solely by the body he or she is but also by everything which makes up the context in which that body is acted upon and acts," she notes, adding: "When the term 'embodiment' is used in the context of Spinoza's thought it should be understood to refer not simply to an individual body but to the *total* affective context of that body" (131). Gatens has explored how early modern accounts of the state encouraged men and women to abstract from the gamut of qualitative resemblances and differences so as to form "identities" whose value can be represented, ranked, and subjected to comparison. Gatens's analysis extends beyond the production of melancholy or hostility to the affects produced around nonsexually coded similarities, as well as similarities that may result in a feeling of joy or power.

What is at stake in Grosz's conflict between these two ways of describing embodiment? The *philosophical* theory that psychic interiority results from bodily inscriptions creates an imaginative topology allowing the ambivalent, tension-filled *experience* of being both actor and patient of one's own action to be *exteriorized* and therefore lightened. Exteriorization is how Scarry argues that religious communities struggle with the compulsion to manage their overwhelming *experience* of the divinity by building it into elements of the external world, such as rituals, sacrifices, churches, and icons (1985, 170, 204–5). Projecting one's pain outward onto someone sufficiently similar to bear it, but dissimilar enough to render it less important, is a basic psychic mechanism that may account for the need to "double" man with sexual and racial inferiors who can support excessive affects of aggression, sorrow, or disgust (Gatens 1996, 35–37; Brennan 2004, 15, 119). However, the act of projection does not need to be hostile; one can also reinforce positive affects like love and joy through contagious gestures.

It is important for those following Nietzsche's analytic strategy to recall that the topology of the "master's" world or the form of social "exteriority" is not given, but had to be built through *combat*, of which punishment is merely a stylized form. The philosophical approach Grosz associates with lived experience or phenomenology is a way of *staking* out the individual's claim to a bit of exterior or social space, prior to building or combat. Because invisible things happen to me, because I have felt the demands of God or the pressures of guilt or the stirrings of desire or anger, I deserve a seat in the physical church and opportunities to use the subject positions of its theology ("God's child," "member of the people," "father," "teacher") along with the other weapons at my disposal. If inscribing is a way for those with power to create an orderly system of resemblances among particular, sweaty, heterogenous bodies using the fiction of interiority, phenomenology and psychoanalysis are ways of creating resemblances among the contents of "inner sense" and asserting their reality using an externalized fiction. If no one on the "outside" is trying to take away the phenomenological, perceptive, affective individual's rights to an *internal difference* or an *internal identification*, then it is not necessary to have a phenomenological subject to represent it socially. But because this conflict is almost always going on, it is crucial for people to be able to theoretically and corporeally "exteriorize" their conflicts as well as lay claim to "interiority."

Nietzsche describes this internalist perspective as a "slave morality" and compares it unfavorably with the inscriptive, externalist perspective (1989, 36–37). But everyone is potentially in both roles—and not always with the same people or institutions. Accounts of the lived and emotionally invested body, just like accounts of the inscribed or objectified body, are produced by people who are both "inside" and "outside" to themselves, whether or not their social roles put them in the position of disciplining or being disciplined. Individual bodies affect and are affected by each other in ways related to, but expanding upon, sheer bodily "pulpiness" (Gatens 1996, 129–30). Pulpiness, affectivity, and resistance or melancholy are qualities of the *sensus communis* that cry out for internalization as well as externalization. But in many cases people *enjoy* and find strength in their ability to be affected by certain things, so Nietzsche's blanket identification in *Genealogy of Morals* of moral superiority with the ability to affect is shortsighted. Neither account *need be* associated with the inside or outside of the human body, although this is the pattern in which historical practices of domination and government have often materialized them. Ideally, both come from the border, from the heterotopic struggle between *two and usually more than two spaces of imagination*, two ways of naming or organizing the same resemblance.

Thus I would regard the apparent conflict between inscriptive and phenomenological accounts of embodiment as an antinomy to be dissolved, rather than synthesized. I suspect that by regarding bodies in the way that archaeology regards *énoncés*—namely, as systems of resemblance—Foucauldian genealogy hints at a successful negotiation of this hinge. What resemblance is to the *énoncé, mimesis* or simulation is to the body as "transformative system" (Gil 1998). The conflict arises only when bodies are regarded as examples of a substance *that* imagines and *which* we imagine—rather than the way, as with Kant's *sensus communis*, in which otherwise singular bodies *resemble one another* and integrate recognition of those resemblances into the rest of their knowledge.

In the *Critique of Judgment,* Kant's inquiry concerns the conditions under which imagination and understanding are capable of producing empirical concepts and identifying empirical laws that determine the manifold of natural phenomena in a systematic way. Rather than similarities among intuitions (to which a concept can be assigned in a universal judgment) aesthetic judgment reveals similarities among *perceivers* and *judges* (being a subjective judgment, its universality refers to the ones who decide rather than the form about which one decides). The first introduction to the *Critique of Judgment* suggests that the concept of nature as *art* enables one to judge resemblances among natural phenomena as cognitively significant or insignificant by envisioning them as differing exhibitions of a concept realized by an artist (*CJ* 20:204′). This analogy promises harmony between the order of nature and that of the human mind.

But Kant's commitment to this analogy is tentative, as can be seen from his discussion of an *anthropological* rather than *transcendental* aspect to imagination. In the *Anthropology* Kant identifies the mimetic capacities of humans as examples of a "sensory productive faculty" (*Sinnlichen Dichtungvermögen*) of imagination. He observes that neurotic people take on the physical ailments of invalids they encounter (especially those with mental disorders) and that married people "gradually acquire a similarity in facial features" over the course of their life together (*AN* 7:179).[29] This is, we might say, Kant's nod toward the kind of phenomena that were central in Sade's attempt to reconcile desire and language through the pornographic experience. Pornography presents a scene in which some gesture or character is *so like* one's own imagined gestures or exteriorizes a "nonconceptualized" inner difference and similarity to other sexual beings that it provokes an uncanny experience of being doubled, beside oneself, or of becoming more like the Other (de-

---

29. Roger Caillois describes certain schizophrenics as losing their orientation *in* space and time to a conviction of *identity* with these pure forms of intuition, and thus a feeling of *similarity* to objects in the patient's environment. In the natural world, Caillois observes, mimicry is a form of *defense,* a way of both situating and representing oneself with respect to a predator and vanishing *into* that situation or representation. In the form of psychosis known as "legendary psychasthenia," the schizophrenia *suffers* rather than *employs* a strategy of camouflage (Caillois 1984, 28–30). See Sacks 1987 for a contemporary example of compulsive mimicry produced by a neurological condition.

picted in the sexual act) than like one (was). Pornography is one of the commonest occasions in which imagination creates unpredictable resemblances with a direct effect on bodily experience; for this reason, and not only because it could be used to slander and satirize authorities, Sade's contemporaries believed pornography had political significance.

According to Kant, the capacity to produce imaginative affinities in *thought* and *corporeality* are intimately involved with the creature's capacity for life. Kant suggests that dreaming is an exercise of imaginative *pictorial* sensory production (*imaginatio plastica*) that prevents death while sleeping. But he also associates the sensory productive faculty of *affinity* (*Verwandtschaft/affinitas*) with the play of powers in nature, especially sexual difference and erotic attraction, as well as the play of mental powers. "Affinity" refers to the "union of the manifold in virtue of its derivation from one ground" (*AN* 7:177). This faculty enables speakers to communicate by means of common perceptions and associations: "In silent thinking as well as in the sharing of thoughts, there must always be a theme on which the manifold is strung, so that the understanding can also be effective" (*AN* 7:177). But it also seems to be responsible for imaginative and physical *sympathy*, such as the mimetic reaction provoked by observing a yawn, a convulsion, or the gestures of someone wrought with emotion.[30]

Curiously, although the "sensory productive faculty" only refers to the *empirical* use of imagination rather than to its transcendental employment, Kant suggests that the union of understanding and sensibility could only be understood from an *empirical* angle as an example of affinity. We are very close, once again, to a conception of imagination as "common root" of understanding and sensibility, except that this time the common root gives rise to mental activity as well as natural, material growth. In this context, moreover, *imagination* is not a cognitive relation, but a directly material, mimetic phenomenon.

According to Kant, "the word affinity (*affinitas*) here recalls a process found in chemistry: intellectual combination [*Verstandesverbindung*] is

---

30. Imitation (*Nachahmung*) in matters of morals and thought, however, is a sign of poor character. Individuals must think for themselves—using principles, however, that are valid for everyone (*AN* 7:293). In the *Critique of Judgment*, Kant contrasts the capacity for *imitation* (necessary for scholarly and technical artistic education but insufficient for artistic creation) to the capacity for *genius* (which enables an artist to create universally communicable forms in an *original* style) (*CJ* 5:308–9). In artistic creation, the painter or poet does *imitate* rather than copy existing human works—but only in reproducing the relationship between originality and communicability that has been expressed in well-regarded models (301, 355).

analogous to an interaction of two specifically different physical substances intimately acting upon each other and striving for unity, where this *union* brings about a third entity that has properties which can only be produced by the union of two heterogenous elements" (*AN* 7:177). The relationship between sensibility and understanding, characterized as *imagination* in the first *Critique*, seems to be of this nature: "as if one had its origin in the other, or both originated from a common origin [*einem gemeinschaftlichen Stamme*]." As usual, however, he cannot explain the process of origination or differentiation. The analogy between imagination and chemical affinity also suffers from the fact that we have no justification for believing that the mind's principles are those of chemistry. We do know from the third *Critique* that Kant believed that it was legitimate to use natural forms as a heuristic for understanding the relationship between analysis and synthesis in *experience*. But here, the question is whether nature can teach us anything about the relationship between analysis and synthesis in *reflection*. How can we know nature apart from the reflectively discovered conditions for human experience?

"The play of forces in inanimate as well as in animate nature, in the soul as well as in the body," Kant writes in a footnote, "is based on the dissolution and union of the dissimilar" (*AN* 7:178). As in the *Critique of Pure Reason*, Kant declares the search for a common faculty both compelling and pointless. But the stakes of such an approach are very high, for without a justified belief that experience and nature mirror each other "in themselves," rather than for humans and humans alone (an infinite play, as Foucault would say of Renaissance scholarship), we lack assurance that critical self-knowledge will lead to truly scientific knowledge. Accordingly, perhaps, Kant *does* look away, not by citing another analogy but by *asking a question:* whether there is any purpose to the differentiation and similarities observed between the sexes.[31] This displaces the problem of *why* thought and nature should be analogized in a way that leaves it unresolved: after all, we have no reason to believe that sex can illuminate the common root any better than chemical compounds. Nor would we ordinarily believe that sexual and racial differentiation are variants on a dis-

---

31. For an illuminating discussion of the disturbances played by *Geschlecht* in Kant's effort to link the historical and transcendental domains, see Fenves 1991, 258–67. Fenves argues that *Geschlecht* gives Kant a way to ground individual participation in the progress of humanity as a whole and in the *corporeality* of communication without recourse to Platonic theories of participation, which in their most sophisticated form cannot be *communicated* discursively but only intuited intellectually.

cursive norm. Kant senses that he is on the brink of a circle or an abyss, because, having abandoned Classical reason's self-evident representational grid, he can no longer rely upon belief in a grounding *resemblance* between macrocosm and microcosm. Thus he concludes, with exemplary inconclusiveness: "In what darkness [*Dunkel*] does human reason lose itself when it tries to fathom the origin here, or even merely undertakes to make a guess at it?"

Kant finds himself thrown into confusion by the implications of "genealogy" because, like most of us, he can only imagine individuals as examples of a concept or species of a genus: substances, in short. He cannot imagine individuals as resemblances, variant actualizations of a singular differentiation. He does not know, and considers it uncritical to ask, whether a fundamental resemblance unites sensibility and understanding or appearances to things-in-themselves (*PR* 4:290). Whether the thing-in-itself, moreover, is a resemblance or an "object" is a purely speculative matter. The problematic concept clears a space for the finite employment of powers of imagination and understanding. But it also forecloses awareness of unschematized and unpredictable resemblances, allowing them to be covered with a "transcendental object = x" like a fig leaf.

What if language "materializes," i.e., offers subjects "styles" of resemblance or "ways to be similar," rather than permitting them to *recognize* forms as instances of a concept? The process of growing resemblance Kant observes between spouses would be the norm rather than an exceptional violation of individuality. I argue that there is a connection between the affinity that troubles Kant's assumption of a macrocosm/microcosm relation and the way a *singular*, preconceptual reflective apprehension resonates indefinitely among more-or-less similar critics in the *Critique of Judgment*. This would make genealogy the history of whom we have been made to resemble, to represent, and to be represented by: the history of the embedded forms taken by a *sensus communis* in empirical and anthropological conflict.

In the first volume of his *History of Sexuality*, Foucault explicitly approaches "power" from a nominalist perspective, defining it as "the name that one attributes to a complex strategical situation in a particular society" (*HS1* 93).[32] He expresses similar views about the state (*STP* 112–13, 253, 282), capitalism (*NB* 170), and, of course, "man." Not only are these

---

32. Balibar (1992), Rajchman (1985), and Flynn (2005) have addressed the importance of philosophical nominalism in Foucault's work.

"entities" not "things-in-themselves," they are *multiple* appearances, coming in many versions (and not only doubles). Unlike Deleuze, who frequently referenced the theories of "singularity" and "univocity of being" in Duns Scotus, Foucault makes no explicit reference to the medieval nominalists (though he does mention this aspect of Deleuze's thought in "Theatrum Philosophicum" (*AME* 360). Foucault's comment in the *Archaeology* that "our reason is the difference of discourses, our history the difference of times, our selves the difference of masks," suggests that he draws his insights into the mimetic capacities of language and embodiment from Klossowski (*AK* 131). But can one maintain that his reference to nominalism in *The History of Sexuality* has anything to do with the problems of resemblance that preoccupied the early texts?

Pouvoir, "power," is a name attached to a concept with multiple intuitive affiliations. The concept transcends those intuitions, subordinating differences in the singular forms of power articulated through legislatures, police, the World Bank, or religious authorities at different points in history to a conceptual identity or an original meaning. But the verb *pouvoir*—"to be able"—can also be understood as an *event* actualized in bodies and institutions, just as the *énoncé* is actualized in documents and phrases. In the essay "Theatrum Philosophicum," Foucault makes reference to Deleuze's notion of "incorporeal events" (borrowed from the Stoics) in *Logic of Sense* (*AME* 348–60; Deleuze 1990). Referring to verbs such as "to die, to live, to redden" (*mourir, vivre, rougeoyer*) as "meaning-events" exceeding the bodies they form and in which these bodies then "participate" (*AME* 350), he states: "Phantasms must be allowed to function at the limit of bodies; against bodies, because they stick to bodies and protrude from them, but also because they touch them, cut them, break them into sections, regionalize them, and multiply their surfaces; and equally, outside of bodies, because they function between bodies. . . . Phantasms do not extend organisms into the imaginary; they topologize the materiality of the body" (346–47). Neither the event nor the *énoncé* is a "form" separable from its "matter," although insofar as it is grasped by form (concept or pure intuition), it can be infinitely related to further forms. This is why there is always thought "in a novel, in jurisprudence, in law, in an administrative system, in a prison" (*AME* 267): this is also why Foucault approaches "the prison" as a singular event, "the great confinement," the "carceral archipelago," even though prisons are dispersed in place and change over time. Every confinement is a new act of power, and so is the "long-term" event produced by their repetition over time (i.e., the constant efforts at reform).[33] Note likewise that many chapter titles in *The Order of Things* can be understood as referring to events or processes even when they seem to be describing *epistemes,* famously analogized to the synchronic dimension of structural anthropology: *Representer, Parler, Classer, Échanger.* These event-objects, as I speculated with respect to Klossowski and Sade, would take the place of Kantian *Ideas* if Kant were to have taken his critique of the unity of soul, world, and God in the direction of conditions for resemblance rather than conditions for the possibility of knowing objects.

Earlier I suggested that referring knowledge to a transcendental condition of possibility, representation, interpretation, and the analysis of eventfulness were ways of organizing and controlling the proliferation of discourse. I also suggested that we think of the materiality of statements and bodies as an effect of the finitude or discursivity of the human understanding, the fact that it must both realize and restrict its application of concepts at a determinate level of detail or analogy. Interpretation establishes a hierarchy of meanings beneath all relations of resemblance. In representation, all resemblances are organized with respect to signs, the form of representation in general. Each of these strategies allows certain transformations to take place among resemblances and precludes others— they are strategies of power. Although Foucault talks most frequently about the effects of these techniques on the circulation of knowledge, the practice of interpretation also establishes relations of *personal* power or subjection in religious or psychoanalytic confession. Hume believed the self was merely a "bundle" of ideas linked by resemblance and habit (1973, 251–53). While this realization caused him great anguish (263–71), Kant used it to twist free of a metaphysical scheme in which knowledge was grounded in the (divine) infinite rather than in human finitude. Morally, however, Kant was haunted by the knowledge that he could never know his intentions in themselves or guarantee their purity (*GR* 4:407, 451). Thus a potentially infinite work of interpretation looms for the *moral* and *affective* subject within the limits set by critique and the possibilities for rational humanity opened by anthropology.

Feeling obliged to account for one's deviation from an official scheme of resemblances in ever greater detail or for one's experience of conflict between multiple imaginative spaces is exhausting: it requires actors to employ Kantian "self-affection" to neutralize qualitative variations be-

---

33. Thanks to Len Lawlor for leading me to this formulation of the problem.

tween their experiences as individuals and the communicable, socially valorized experience of other bodies. Foucault's skepticism regarding interpretation *tout court* is consistent with his resistance to techniques of confessional power that were inadvertently given new life by Kant's critique of introspective knowledge. Asking "whom we resemble" rather than about the "real" sources or meanings of action changes how bodies and discourse limit one another. When he stopped writing about imagination and phenomenology shortly before the publication of *The Order of Things*, Foucault did not stop writing about the discursivity and materiality of language—or of bodies. Rather, he talked about how discursivity and materiality could be actualized differently at different points in history.

An-aesthetic Philosophy?

I have argued that one benefit of treating *énoncés* and bodies as methodologically parallel is that it allows us to reconcile "inscriptive" with "phenomenological" or "hermeneutic" accounts of embodiment and imagination, to understand external and internal perspectives on lived embodiment as logically continuous. Another benefit is that it enables us to theorize present-day bodies as possibilities generated by a particular *dispositif* of power that may have other unrealized "doubles" or mimeses. However, few readers are likely to be convinced by these merits, for Foucault's strategy appears to neutralize the sense of visceral attraction, attachment, or trouble we associate with the idea of "meaning" rather than "matter." I would suggest that readers such as Dreyfus and Rabinow, Nancy Hartsock, and others *want* Foucault to situate his writing sociologically and psychologically because they are reluctant to abandon an emotional intensity they associate with "meaning" and conceive as accessible only through interpretation or contextualization, but not through resemblance or mimesis.

Nancy Fraser (1989, 42–43), Habermas (1990, 282–84), Peter Dews (1987), and Hartsock (1989), among others, have criticized Foucault for refusing to ground his histories of the human sciences and his political analyses in the sort of epistemologically and/or normatively universalist frameworks we find in Kant. On the one hand, they contend, Foucault makes normative suggestions about the injustice of certain social practices (past and present) through rhetorical presentation rather than argument, while denying reason's ability to establish the truth behind rhetoric. On the other hand, Foucault's attempt to historicize the subject of knowledge and ethics seems to deny the normative significance of individual emotion and judgment, including his own, thereby "aestheticizing" morality.[34] For example, Dews (1987) and David Levin (1989) have protested that Foucault's account of the human body is too abstract to explain moral evaluation. Allan Megill (1985) and Richard Wolin (1986) tend to interpret "aesthetics" as a state of desensitization rather than a sensibility that reveals worthy situations of moral concern. Far from liberating modern individuals to discover their own forms of agency, they suggest that Foucault imposes the idiosyncratic conditions of his own agency on others to whom it is poorly suited, disempowering and denying them self-defense through mutually valid reasons. For these reasons, he falls victim to the same "fanaticism" (*Schwärmerei*) or claims of privileged insight into the nature of sociopolitical "things-in-themselves" that plagued Kant's rationalist predecessors and romantic heirs.

Despite differences, these writers are united by alarm at Foucault's desire to "dispense with the constituent subject" of phenomenological experience and political action. Their reasons for defending the subject are twofold. Without a subject, on the one hand, they believe that individuals would be unable to respond to situations with "judgment," not to mention good judgment: to criticize and resist their own emotional impulses and to require others to give reasons for their own actions. On the other hand, the subject is a metaphysical support for *emotion*, in whose absence actions would lose their existential, if not their moral value. Without a subject, there would be no way to identify emotions as phenomena de-

---

34. See Racevskis 1983; Megill 1985, 203, 221, 231; Wolin 1986; Dews 1987, 169–70; Levin 1989; and McNay 1992. Some of these readings accuse Foucault of regarding all practices of domination equally because he refuses to credit "humanism" with the ability to detect "real" domination, or because he insists that power is involved in dominating as well as nondominating social practices. Lacking such moral criteria, distaste for this or that social practice can only be considered "aesthetic." According to Megill, Foucault participates in a tradition of post-Nietzschean thinkers for whom the work of art is the model for all real interaction; Racevskis suspects that the analysis of discursive or institutional forms diverts author and reader from concrete social change. For some readers, such as Wolin, Foucault's efforts to analogize ordinary social relations to strategies of war aroused Benjaminian concerns regarding the aestheticization of military machinery by futurism. After Foucault began his investigation into Greek and Roman sexual morality, based on an individualized "aesthetics" of virile existence rather than a code of permitted and forbidden acts, he was taken to task for valorizing subjects that looked on their own acts with the same amoral and narcissistic gaze previously reserved for figures from the past.

serving moral consideration and protection by others—even if the subject is supposed to ignore its emotions when judging and acting.

In *The Archaeology of Knowledge*, Foucault says that "in our time, history is that which transforms *documents* into *monuments*" (AK 7). Must a world of uninterpretable or "monumental" form be devoid of emotional intensity? In the essay "Phenomenality and Materiality in Kant" (1984), Paul de Man explores the excitement surrounding resemblances evoked by an apparently "flat" or meaningless image. Where Kant identifies beautiful forms as those provoking a play of imagination and understanding, judgments of sublimity result from an encounter with the formless, in which reason and imagination are unable to produce a stable portrait of the world's unity or a feeling of balance between the observer's powers and the immensity or power of surrounding nature. To apprehend the ocean as aesthetically sublime, Kant suggests, we must forget everything we know intellectually about its climactic or biological processes; instead, we must be able to view it "as poets do, merely in terms of what manifests itself to the eye—e.g., if we observe it while it is calm, as a clear mirror of water bounded only by the sky; or, if it is turbulent, as being like an abyss threatening to engulf everything" (CJ 5:270).

De Man refers to this image of the mirror's flatness, which contrasts with every other image of the sublime in the *Critique of Judgment*, as a "material vision" because it is "devoid of any reflexive or intellectual complication." Yet he notes that this vision is also "purely formal, devoid of any semantic depth and reducible to the formal mathematization or geometrization of pure optics" (1984, 136). The abstract landscape confronts us with the "formlessness," indeterminacy, or materiality of the forms through which we usually seek further interpretive and teleological determination. Deprived of a concrete object of resemblance, the observer's curiosity about and attraction to an unknown X are almost limitless (CJ 5:274). It forces us to acknowledge the *materiality* of our own form-giving perspective, aesthetic approach, and interest in detail as it affects and appears in the *materiality* of the natural or social forms under study. In the moment of sublime *apprehension, apatheia* and intense affective conflict converge. The subject of aesthetic sublimity recognizes her *own* supersensible vocation in the face of "monumental nature": her material specificity, that is, but also her constrained inventiveness—or freedom.

According to Foucault, as I mentioned earlier, power is "a mode of action that does not act directly and immediately on others," but "acts upon their actions . . . on possible or actual future or present actions"

(*PWR* 340). Emotions are neither wholly intellectual properties of a logical subject nor wholly bodily/neurological events, but involve bodily responses to the imaginative projection of possible actions in a social context and options for responding to, acting upon, or resisting the intended actions of others (Redding 1999). Many power relations, as mentioned in the earlier discussion of Nietzsche, play upon or reinterpret the imaginative role or *persona* that a reader is accustomed to occupy with respect to others, rather than the movements of his or her body. Foucault's examples and rhetorical practice support the idea that emotions are products of power relations, including those that bind authors and readers by means of identification with literary images. Readers are moved "indirectly" by Foucault's description of the madman ridiculed in the presence of two other would-be Louis XVI's (*HM* 498–99), the horror of Damiens's *supplice,* the numbing regulation of the Paris Maison des Jeunes, or the despair of the captured Serb partisan whose confession was torn up with instructions to "start over, and tell the truth" (*HS1* 61). They see unexpected images of themselves in such images and must decide how to interpret these affinities in the absence of a clear historical narrative. But the fact that power acts on the *imaginative* components of embodied action does not mean these imaginative components must be conceptualized as properties of an otherwise universal, standardized *subject.*

Foucault's ideas about power can be fruitfully situated in a long history of Western reflection on the connection between activity and passivity, knowledge, and the emotions. For early modern metaphysicians like Descartes, Spinoza, and Leibniz, the emotional tone of human experience indicated an individual's relative capacities for activity and being acted upon—capacities enhanced, and made more pleasurable, by accurate and more God-like knowledge. For Spinoza, all modes are unique expressions of a single divine substance. For Leibniz, all entelechies or monads are created beings distinct from God, but each expresses the totality of the universe in a qualitatively distinct manner, as coordinated by divine pre-established harmony (1998, 60–61). Spinoza thinks of "activity" as behavior governed by a mode's adequate understanding of its distinct nature and characterizes behavior motivated by an inadequate or partly imaginative understanding of one's distinctness from other modes as "passivity."[35] A Leibnizian monad, on the other hand is said to be "active" when

---

35. Spinoza's *Ethics* describes the human mind as the idea of an actually existing body, a body with unique powers and a unique physical and political relationship to other bodies (Spinoza 1982, II.11). Only God can claim a completely adequate expression of the individual's (bodily and ideational) essence (II.19). Being the idea of an actually existing body, the Spino-

its "mirror" of the universe contains more elements that explain the outcome of interactions with other monads than the monad whose perspective is more confused, therefore passive (1998, 67–68). Both Spinoza and Leibniz grant joy, fear, sadness, hate, envy, and hope important epistemological roles in discovering the true situation and active or passive interactions of the individual human subject—mode or entelechy, respectively. They view ethics as the task of cultivating knowledge regarding one's individuality, guided by emotion, in such a way as to coexist with other individuals in the *most active and joyful way possible*.[36]

But as long as the rationalists presumed human knowledge "passive" with respect to God and other "things-in-themselves," Kant thought they could only reinforce unhappy intellectual attitudes like dogmatism and skepticism, and produce fanaticism, strife, and authoritarianism in the political realm. Moreover, he suspected that such a metaphysics could easily allow some individuals to dominate others on the basis of imagination and incommunicable emotions, like the *Schwärmerisch* Neoplatonists castigated in "On a Newly Arisen Superior Tone in Philosophy" (Fenves 1993, 51). Kant's "Copernican revolution" makes human subjectivity an active legislator and creator of worldly appearances rather than a passive recipient or witness to "things-in-themselves," whose nature and coherence are fully known only by God. In his philosophy, *resistance* is a fundamental starting point for critical reflection.

For example, in the introduction to the *Critique of Pure Reason*, Kant complains that Plato "left the world of sense because it sets such narrow

---

zistic understanding is empowered to the extent that its body harmonizes with and commands other bodies with which it is in contact. Thus pleasure and love enhance our ability to know, while fear, sadness, and hatred compound our confusion and passivity with respect to a disempowering situation. Joy, for example, is both the mark of an intellectual intuition that grasps some element of the *real* structure of reality as found in God, and the method enabling one to expand this intellectual intuition into rules of conduct and desire (II.16, 19, 23, 39; III.3; IV.37).

36. Every monad, no matter how primitive, is in a constant state of "unrest" or perception; that is, appetition in correlation with other monads. No substance is indifferent to its fellows (Leibniz 1996, 165–66, 188–89). But even monads capable of reflection have difficulty distinguishing the sources of their perceptions. The more these perceptions reflect the activity of substances *other* than the monad in question, combine the effects of passions with activity immanent to the monad, or result from an interweaving of past and present presentations, the more the monad is confused and passive. In the *New Essays on Human Understanding*, Leibniz (through the voice of Theophilus) argues in a somewhat Spinozistic fashion that activity increases the monad's clarity, giving opportunities for self-observation and self-understanding, but also counsels that humans should train themselves to focus on long-term pleasures rather than allowing themselves to be further confused by the attractions of immediate pleasures (1996, 210–11). Each monad, he adds, has its own time frame for anticipating or forgetting the importance of future pleasure (202–4).

limits to our understanding," but failed to make headway in the realm of pure understanding "because he had no resting point [*Widerhalt*] against which—as a foothold, as it were—he might brace himself and apply his forces in order to set the understanding in motion" (*CPR* A5/B9). His pre-critical essays insisted that the laws of physical motion pertain to the interactions between substances rather than to their internal properties and require space and the world to be real entities over and above conceptual distinctions between substances. Even conflicting emotions, moral indecision, and contradictory propositions could be explained by reference to a common structure permitting comparison and distinction: a structure, therefore, that would stand over and against the drives, duties, or representations being compared (1992c, 2:179–84, 200). Later, of course, Kant attributed the interaction and mutual resistance of subjects and objects to a transcendental structure governing the unity of experience, but did not abandon his previous strategy of self-orientation with respect to the experience of *felt* resistance between phenomenal bodies. And in the *Critique of Practical Reason*, he proposed to differentiate rational and empirical motives for action "by this resistance [*Widerstrebung*] of a practically lawgiving reason to every meddling inclination, by a special kind of *feeling*, which, however, does not precede the lawgiving of practical reason but is instead produced only by it" (*CPrR* 5:92).[37]

Rarely does Kant find emotion epistemologically or morally informative, since it is associated with what is individual and potentially incommunicable in experience. Kant admits in the Transcendental Aesthetic that he deliberately excluded the sensuality of representations from transcendental philosophy because they were not, in his opinion, relevant to cognition (*CPR* A22/B36, cited in Schrader 1976, 154; see also Schott 1988, esp. 101–8). In Kant's ethical writings, all motives for action other than the moral law are assimilated to differing degrees of intensity in the quest for pleasure.[38] Only in the *Critique of Judgment*, where Kant considers

---

37. In fact, Kant suggests that in the absence of such conflict, a self-righteous moral agent would indulge in *Schwärmerei* (*CPrR* 5:84).

38. In the *Groundwork for the Metaphysics of Morals*, Kant stresses that the concept of happiness is *indeterminate* and corresponds to what he considers the *arbitrariness* of the inclinations that motivate us empirically (*GR* 4:399, 418). In the *Critique of Practical Reason*, Kant argues that differences among an individual's pleasures, even differences between pleasures of the body and those of social or intellectual activity, are of negligible importance, being differentiated only in degree (*CPrR* 5:23). See David-Ménard for a discussion of the psychological peculiarities inherent in Kant's reduction of all emotional tones to "intensity" in his moral philosophies (1997, 19–20).

the pleasures and pains accompanying cognitive creativity or passivity in the face of natural forms and formlessness, do more varied feelings find a transcendentally rather than merely anthropologically significant place in human self-understanding and motivation. But disinterested aesthetic pleasure results from apprehension of forms that produce a similar pleasure in everyone, rather than individuating the observer. Like pure forms of intuition and calmness in the wake of moral action, disinterested aesthetic pleasure is an "active" and informative feeling, one that appears through *resistance* to particularizing and qualitative variations in the content of experience.

Despite homogenizing most emotions and mistrusting their influence on cognition and reason, Kant's philosophy is anything but passionless. Rather, I would argue that he subordinates or interprets all emotions through what Plato would have called *thymos,* the agonistic or indignant part of the soul, which lends itself to both quarrelsomeness and the pursuit of justice. In the *Republic, thymos* is placed between the appetitive and the rational parts of the soul, and can be turned against or made the ally of either depending on the situation (Plato 1961, 439e–440e). Most important, even when it enables the soul to resist ignoble pleasure or demonstrate nobility of spirit, there is still a pleasure in *thymos.* Making resistance the foundation of transcendental idealism as well as morality gives *thymos* an invisible role in all knowledge as well as desirable action.

Foucault resembles Kant in assuming that the experience of conflict, the sentiment of *resistance,* and the pleasures and pains associated with pure form (whether this form is just being discovered or disintegrating) are more informative and empowering than the pleasures and pains of sympathy or cruelty. His choice to organize the analysis of discursive formations and administrative strategies around a history of oppressive or exclusionary institutions is motivated by resistance, not only to the practices of these institutions, but also to the *sense* and *interpretive activity* which modern strategies for producing communicable knowledge foster in members of the population who are regarded as or suspect themselves of being "abnormal."

In confessional relationships between students or delinquents and their supervisors, or psychoanalytic patients and their doctors, "the agency of domination does not reside in the one who speaks (for it is he who is constrained) but in the one who listens and says nothing" (*HS1* 62). These relationships make use of the visible surface and comportment of bodies in order to verify and establish authority. But they also "see" and train us to "see" or "feel" those aspects of our own bodies which are relevant or significant to authorities, aspects which will never be revealed (as in the case of homosexuals who remain closeted), as well as those which are designed for public performance. The asylum inmate or prisoner has to take these institutions as horizons of action and experience, but the free person potentially subject to commitment or incarceration must do so as well. They reveal and cultivate the individuality or specificity of bodies, but present these nonuniversalizable aspects of experiences as potentially "abnormal," problematic precisely as *embodied*. Such normalizing relationships "separate active force from what it can do" (Deleuze 1983, 57), decreasing what I later refer to as the "normativity" of individuals.

Because these relationships demand a great deal of attachment, identification, striving, and self-examination, and because they make use of real capacities for preference or pleasure, people are reluctant to abandon them. Unloved women, for example, may worry that life outside an abusive relationship will feel even more "dead" or "hopeless" than life with their partners, in which at least there is someone (however unworthy) to love.[39] Emotions are among the most subtle and powerful corporeal effects we are trained to see and feel. Foucault's demand for an exterior perspective on practices of social power, even if only ideal, seeks to minimize the reader's feeling of investment in relations of power that produce such unhappy forms of psychic interiority. The philosophical subject, he might argue, is one among many possible tropes organizing and making sense of an individual's subjection to unhappy relations of power. But because he does not immediately suggest a new form of interiority for all readers, hoping that they will invent their own, readers feel as if they have been denied their locus of agency, with no compensation, on rhetorical grounds rather than through "good reasons."

Although Spinoza and Leibniz would agree that emotions and the power relations producing them are highly individuated, Kant gives heuristic value only to those feelings which are universal and associated with

---

39. Feminists aware of how women have been excluded from public discourse and power on account of sex, where femaleness is associated with (undesirable) emotionality, are conflicted about the desirability of abandoning the idea of the "subject" and the philosophical or political processes that give rise to "subjectivity" because of their allegedly bad emotional effects without providing any positive emotional alternatives (Hartsock 1990; Braidotti 1991). This is one of the reasons they hope for a subject of universal rights that can be grounded in a body of the Merleau-Pontyan type, if not in a Kantian-style subject of cognition and judgment. But rights are themselves power relations that produce and suppress certain kinds of emotions regarding real or potential acts and situations of passivity.

the least qualitatively distinct forms. In "Analytic of the Beautiful," one kind of universally significant pleasure is produced by the encounter with a form for which the subject has not yet been given a determinate purpose or cognitive task. In "Analytic of the Sublime," another kind is produced by the encounter with an instance of formlessness so enormous or potentially overpowering that the subject becomes aware of an ability to resist whatever enormous and awesome forms might be imposed on her from outside. A witness to the sublime is emotionally moved or put through a whole range of apparently conflicting emotions by exposure to a rapid shift in spatial and temporal perspective, which forces her to become aware of the indeterminacy or inventive potential of her *own* behavioral and psychological forms. Thus what is most sublime, Kant concludes, is not the cathedral or the stormy ocean, but the ability to act on oneself despite external pressures, which has almost infinite scope under the right circumstances (CJ 5:264).

At the moment when resistance emerges, no particular outcome is intimated, no particular technique of power or pleasure. The time frame of inner sense is broken, and no image can be found for the *feeling* of reason's ability to transform self and world, which might replace the specific examples of moral or pragmatic action provided by authorities to every child and citizen-subject (CJ 5:259). Later, I will compare this state of mind, in which one feels ready to set norms under new conditions rather than to survive in one's existing environment, to Georges Canguilhem's notion of *normative* experience in medicine (1991, 200). The emotion of this broken encounter is so intense that Kant cannot represent it using anything but an Idea of reason. Everyone can be expected to identify with or come up with some content for a concept, but as an Idea, this pure form represents an individual's capacity for *freedom* from the expectations, the specific imaginative schemes, and the power of external parties. Kant's closest figure is the moral law, which combines the austerity of a logical universal with the judicial might of an absolute sovereign:

> We need not worry that the feeling of the sublime will lose [something] if it is exhibited in such an abstract way as this, which is wholly negative as regards the sensible. . . . It is exactly the other way round. For once the senses no longer see anything before them, while yet the unmistakeable and indelible idea of morality remains, one would sooner need to temper the momentum of an unbounded imagination so as to keep it from rising to

the level of enthusiasm, than to seek to support these ideas with images and childish devices for fear that they would otherwise be powerless. (*CJ* 5:274)

Although this complex aesthetic pleasure is communicable, to the extent of informing humans about the transcendental conditions for communicability in general, Kant adds that it may be provoked by an encounter with solitude or the idea of someone else's solitude. Strange though it seems, Kant writes, "even [the state of] *being without affects (apatheia, phlegma in significatu bono)* in a mind that vigorously pursues its immutable principles is sublime, and sublime in a far superior way, because it also has pure reason's liking on its side" (*CJ* 5:272; see Terada 2001, 83–85).[40] Thus pure form and passionlessness themselves can represent emotion if only *in negativo*, awakening an observer's capacity for preference, risk, experimentation, and affirmation.

On occasion, Foucault does offer readers images of historical scenes that moved him to certain experiences of sympathy, astonishment, indignation, puzzlement, or delight. These include his burst of laughter at the Chinese encyclopedia, his comment on the "tragic" character of *The Order of Things*, with its mournful image of man as a face traced in sand at the edge of the sea, and his confession of intense and almost incommunicable affinity for the paupers whose stories are recounted in "Lives of Infamous Men" (*PWR* 158). To these should be added his happiness at writing about authors whose own experience of writing was both impersonal and transformative, such as surrealist poet Raymond Roussel (Foucault 1986a, 182–85). Each of these cases involves the revelation of contingency in a given world or historical sense, a potential for breakdown accompanied by positive affects toward what may lie outside. Because they anticipate a new experience of the world as a whole or the failure of sense in such a whole, his best-known works have been described as employing a *sublime* aesthetic.[41] Foucault's celebration of *transgression* in the work of literary figures like Bataille, Sade, and Artaud can be understood as an exploration of normativity that challenges historically received understandings of individuation (e.g., "Preface to Transgression," *AME* 69–87).

---

40. In the *Critique of Practical Reason*, Kant considered the notion of a sensuous interest in lawfulness to be destructive of all morality, if not immediately contradictory (*CPrR* 5:38–39). However, his description of *apatheia* as a positive affect associated with pure form shows it was by no means an impossible phenomenon (see *AN* 7:254).

41. See especially Rajchman 1985, 17–22.

Unlike the "being of the law," which established a horizon for the "juridical" in seventeenth-century Europe, or the "being of man," which did the same for the "normal" in nineteenth-century Europe, Foucault seems to have identified the "being of language" cultivated in their work with the capacity to resist passive forms of interiority and to experiment with new power relations (*OT* 383–86).

However "flat" and impersonal it may appear, Foucault's work shows that formalism can coincide with great love of materiality and fury toward the power relations encouraging us to resemble or imitate certain emotional and intentional forms and to disavow others as impossible. We are more ardently attached to the uniqueness of our own materiality, whether we conceive of it as "embodiment," as "emotion," or as "freedom," than to any particular form we may assume in our own eyes and in the eyes of others. Foucault's formalism does not require readers to abandon their attachment to their own materiality, as well as the immediate pleasure or revulsion of mimetic attraction in which it is experienced. As he writes in the introduction to Deleuze and Guattari's *Anti-Oedipus*, "Do not think that one has to be sad in order to be militant, even though the thing one is fighting is abominable" (*PWR* 109). It is difficult, however, to imagine that an unknown semblance might mirror this singularity more reliably than a face or context, whether hated or beloved, with which we are already familiar.

Foucault's work may be distinguished from *Schwärmerei* insofar as it does not put a *new* sense in place of the old, but encourages readers to admit the possibility of "rising above" existing assessments of their own powers and of the potential for power implicit in their own sexual, economic, and intellectual relationships. Because he assumes that people are already in the process of acting, rather than deciding to act or in need of overwhelming persuasion, his evocation of the sublime can be analogized to a "clearing away" of heteronomous imaginative forms, a liberation of imagination from its own dead weight. In response to French historians who questioned him about the "anesthetic" effect of *Discipline and Punish*, especially on social workers in the prisons, Foucault replied that the fact that social workers may be *paralyzed* or in need of guidance on how to act does not mean that the prisoners themselves are anesthetized (*PWR* 234–37). "It's insofar as there's been an awakening to a whole series of problems that the difficulty of doing anything comes to be felt": a response which indicates Foucault's views on the relation of problematization to feeling.

But Foucault's reluctance to reflect on the epistemological significance of emotion except where the limits of experience are at stake follows Kant in painting with an intense, but very limited emotional palette. He challenges Kant's identification of the active interiority revealed by sublime experience or moral action with *legislation* (1984, 356–57), and views resistance as eminently embodied and individualizing. But in the letter, if not in spirit, Foucault follows Kant in rarely allowing the reader to witness his emotion, aggressive as well as sympathetic, except in response to abstractions—and in assuming that readers will be "universally" moved by the intense pleasures of conceptual form. It is telling that Dews, for example, expresses frustration with the abstraction of Foucault's concepts of "resistance" and "body" (1987, 163–64). In fact, Foucault's reluctance to speak about the plurality and affective differentiability of bodies until the end of his career could indicate that he, too, hoped to avoid *Schwärmerei*, even to the point of declaring "life" a historical concept anchoring the management of modern human beings in a particular cultural context.[42]

The much-vaunted "return of the subject" in Foucault's late writings on ethics is hardly a repudiation of his earlier mistrust for subjectivizing institutions and the passive interiority they tend to foster. Rather, it demonstrates a willingness to investigate alternative historical techniques for cultivating active or joyful forms of subjective interiority, such as those employed by Greek Cynics, Roman Stoics, medieval religious orders, and modernist aesthetes. As Ladelle McWhorter points out, those who find Foucault's descriptions of corporeal normalization horrifying want to retreat to a Classical (Lockean or Cartesian) conception of subjectivity, which not only escapes the debilitating individuality of the modern medical "case" but also suggests that, despite bodily differences, "mind" is something identical in each of us that can be brought to a consensus on change (1999, 161). "If it's freedom we want," she concludes, "we'll have to embrace rather than reject the developmental, normalized bodies that we have become." But since emotions are necessarily as particular as the power relations that produce them, it may be neither necessary to defend them "in general" nor most effective to mount this defense according to universal principles. In place of power relations that tend to motivate through pain (physical or psychic) or return little satisfaction in relation to the effort expended, McWhorter suggests that we will simply have to

---

42. Here is one of his greatest differences from Deleuze, who does not mind flirting or being thought to flirt with vitalism.

invent new contexts or relationships, new "assemblages" (*agencements*) in Deleuze's words, that can provoke different and more differentiated capacities for acting and being acted upon than those presently available.

The real danger of Foucault's approach is not that readers will lose faith in their own agency or indulge in aesthetic appreciation of others' passivity. Post-Kantian modernity has already *incarnated* this unhappy situation, and specific historical circumstances following the 1980s consensus on neoliberalism have prepared left-wing thinkers to read everyone, including Foucault, with melancholy eyes. The danger is rather that Foucault's readers might only cultivate the *reality* of power and agency in existing practices of the sublime—as resistance, explosive joy, or impersonal serenity. The sublime moment is necessary—in order to break down one's investment in the imaginative forms that make bodily individuation an obstacle to knowledge and dignity. This process should be liberating for all. But alongside the "active" affects arising in relation to the "being of language" and indistinguishable from *apatheia,* we must also cultivate power relations giving rise to trust, anger, humor, pride, hope, curiosity, and tenderness.

Because modern forms of subjectivity do provide pleasure and moral significance, albeit within an anxious horizon, critics of Foucault's thought are alarmed at the thought of giving them up and entering into the power relations suggested by his own texts, especially in the absence of universally binding reasons. And yet when Foucault fails to meet his mark, it is because he shares too much with Kant, rather than too little. Specifically, he expresses the pleasure of resistance using a sublime aesthetic and identifies his own agency with the capacity to create abstract forms or concepts, including the "being of language" so important in the literary essays of the early 1970s. Because he does not develop a rich account of how active and passive emotions are effects of individualizing power relations, leaving some readers with joyous and others with uncomfortable emotions that he expects to be resolved in new practices, he is judged to have acted against reason in the name of emotion and against emotion without due reason. If Foucault's project is to be carried forward by friendly readers, it will need to involve the creation of new subjective forms capable of many intense and qualitatively distinct affects, which lend themselves to moral sensitivity and whose denial can be viewed as a moral wrong.

PART 3

LOCKED IN THE MARKET

It might even be argued that the attributes of a particular political philosophy, its generosities and its failures, are most apparent in those places where it intersects with, touches or agrees not to touch, the human body—in the medical system it formally or informally sponsors that determines whose body will and whose body will not be repaired; in the guarantees it provides or refuses to provide about the quality and consistency of foods and drugs that will enter the body; in the system of laws that identify the personal acts towards another's body that the state will designate "unpolitical" (unsocial, uncivil, illegal, criminal) and that will thus occasion the direct imposition of the state on the offender's body and the separation of that unpolitical or uncivil presence from contact with the citizens.

—Elaine Scarry, *The Body in Pain*

I have argued that many modern institutions and discourses that tried to cover the social field from partial points of view managed to establish a division between true and false, identify objects and domains, and connect to other discourses by confining and studying people with socially problematic traits. Kant did not necessarily influence these institutions and discourses, but they share Kant's strategy for turning *limits* on knowledge and government into stable bases for generating positive knowledge and order. Thus the bodily forms of healthy, sane, law-abiding, desiring and desirable humanity are inextricably tied to forms of the "inhuman" who are their uncanny doubles. Both sets of bodies learned how to see, speak, affect, and be affected by each other on the basis of resonances and aversions that Kant calls pure or nonempirical feelings. The forms that a society regards as beautiful or sublime inform an investigator about the conditions for communicability and what members must not or cannot communicate, whom they must not or cannot resemble, if they wish to remain within a *sensus communis*.

Like his archaeological studies, Foucault's genealogies try to break open the world structured by such divisions to reveal the events and discontinuous strata of which it might have been composed. In Part 2, I explored comparisons between Foucault's strategy and the experience giving rise to aesthetic judgments of the sublime. In the dynamical sublime, confrontation with a formless but overwhelming phenomenon from a safe distance (in this case, the proliferation of discourses and bodies) provokes a surge of confidence in the spectator's ability to exercise moral self-legislation counteracting or equaling the external force of nature. However, I have tried to separate the feeling of power revealed by the sublime from the specific *kind* of moral action Kant believes it represents. This is because I think the feeling of power is morally significant, although Kant's identification of morality with submission to a pure law brings many problems in its wake. Here, I want to look at aspects of modern political life that claim to preserve the citizen's dignity but tend to flatten the feeling of power Kant associated with law (both moral and political). Specifically, I want to examine how modern polities make use of this feeling of power by tying it to various collective projects for managing the risk of bodily injury, poverty, and disorder. The exercise of political imagination depends on our ability to access the "pure" feelings that precede and exceed modern forms of embodiment, individuality, and citizenship—but that also resonate with an indefinite number of other people.

Foucault's lectures from the late 1970s seem to return again and again to a central point: the individual and the state are not "given" entities or substances, but ways in which similar habits or practices are bundled together for discursive convenience. The individual "is not to be conceived as a sort of elementary nucleus, a primitive atom, a multiple and inert material on which power comes to fasten or against which it happens to strike." He explains: "It is already one of the prime effects of power that certain bodies, certain gestures, certain discourses, certain desires, come to be identified and constituted as individuals" (P/K 98). Likewise, the state "is perhaps only a composite reality and a mythified abstraction" (STP 112) or one of "the terminal forms power takes" (HS1 92). But they are not spurious entities, for these concepts permit the individuals and states, once bundled or "schematized" by the imagination, to act on themselves in distinct and often quite valuable ways.

Some processes maintaining the health and productivity of modern populations only function smoothly when they are recognized as operations of a state (legitimated through law or pragmatic consensus); others only function smoothly when they are considered nonpolitical. Power works "only on condition that it mask a substantial part of itself" behind humanizing activities like medicine or education that are thought to be natural or traditional (*HS1* 86). The productivity of knowledge often benefits from masked power as well, as when researchers believe they are pursuing a question for the sake of pure science rather than making a contribution to military or industrial technology. Likewise, individuals claim power only over those activities that provide a good anchor for self-government, such as sexuality and criminality, though these may also be of interest to larger organizations. The individual takes credit for his or her labor, for example, and not for "labor power," which is organized and maintained by the employer. Foucault's stresses this nominalist point in order to put Marxist concern with production into a continuum of power-producing activities, and to deflect left-wing attention from the question of whether the state should be regarded as first among enemies or a necessary building block for a new society. His goal appears to be breaking what psychoanalysts would call the political activist's imaginary identification with the state and with the individual, often regarded as enemy of the collective.[1]

Imaginary identification is a process through which a person takes stock of her skills and relations as a totality by projecting them onto an image, much as the subjects in the famous frontispiece to Hobbes *Leviathan* understand their collective power (or equal powerlessness) through the image of the sovereign. The salient parts of this image can only be discerned in language, much as the salient parts of Kant's objects of experience can only be identified through concepts of a discursive understanding. (This also means that some aspects of the image can be ignored, and undesirable aspects of experience blotted out insofar as they are never imaged, or never identified as such.) This image may be visual but also has a tactile and auditory component; it is tacitly referenced, for example, when one "feels at home" in a room that is intended for use by people in one's social group. But this image has an ontological impact on the person who identifies with it. The *being* of the Hobbesian subject, and the quality of his remaining powers, changes when some of those powers are transferred to the sovereign, whose body gains proportionately in importance and ability. Identification with the subject of any discourse—such as scientific knowledge, law, or religion—alters someone's being. It also alters

---

1. Some of the crucial writings on imaginary identification are "The Mirror Stage as Formative of the Function of the I" (Lacan 1977, 1–7) and "Ideology and Ideological State Apparatuses" (Althusser 1971). Karlis Racevskis's (1983) reading of Foucault's early work emphasizes his attempts to undo such identifications.

the objects of the discourse that now exist for these subjects. According to Foucault, the "individual" and the "state" are images and objects that need not be taken for granted and need not be contested in a life-or-death struggle, because they are malleable imaginative forms through which members of a complex society differentiate and act on themselves as well as one another.

In his 1979–80 lecture course, which began several months after Margaret Thatcher became prime minister of Britain and a year before the U.S. election that brought Ronald Reagan to power, Foucault remarked on a growing "phobie d'état" on both sides of the Atlantic, among right- as well as left-wing politicians. He traced this phobia to the nuclear race and the Nazi or Stalinist legacies in Europe. But he also warned his audience repeatedly against believing that the state is a single "thing" worthy of fear or hatred. The state, he claimed, is "the correlative of a certain manner of governing," and it is more important to focus on the pros and cons of the manner of governing than the state itself (*NB* 7, 79). As with "power," he encouraged them to be "nominalist" about the state (*P/K* 122–23; *NB* 4–7; *STP* 112). This warning is in line with earlier claims that power should be considered a relationship (not a possession) that requires subordinates to maintain some portion of their own freedom and to enhance their capacities (*PWR* 120). Thus the state is not the source or repository of power that should rightfully be returned to the people. We only say a state *exists* where an effective practice of governing organizes institutions and discourses, producing power relations between people that may be frustrating or maddening but would also be difficult to unweave without a great loss of order and efficiency.

Many of the ways people understand the source and power of their own individuality, such as intelligence, property, and sexuality, were developed in tandem with security-based and liberal governmental practices that reached their apogee in the postwar welfare state. Of course, these personal traits may have profoundly impersonal correlates in a community's biology, geography, and political structure. But what matters for political imagination is the way they enter into people's sense of their own being and ability to act on that being, to entertain and survive risks. The social movements that interest Foucault are the ones that challenge the right of expert discourses and state or private institutions to shape their members' being in relation to an ever more finely tuned idea of individuality and an ever more comprehensive social vision.

According to Foucault, the characteristic power relations of modern states, found in activities such as education, medicine, warfare, and economics, can be traced to techniques of pastoral Christian guidance in the European Middle Ages and Renaissance. Monarchs applied these techniques at the level of the population as well as to individuals. Although first developed by despots, the bourgeoisie used the moral prestige of law and natural rights to claim this form of power as their own at the end of the eighteenth century. In the twentieth century, movements across the political spectrum called upon law once again to resist this form of power. The waning of premodern community structures and conflicts arising from cultural pluralism also led to a revalorization of law and kinship as bulwarks against anomic individualism.

In the 1970s and 1980s, however, "phobie d'état" led to a massive reorganization of normalizing power rather than to its dissolution. This is the phenomenon known as "neoliberalism," which made common cause between economic libertarians and cultural conservatives in Britain and the United States, and exerts constant pressure on the political culture of democracies in Europe, Eurasia, South America, and Africa. It combines aggressive national and international legislation for the defense of business autonomy with reduced legislative and financial attention to social services and labor. To exit the impasse between law and normalization represented by this reorganization of normalizing power, Foucault suggests that we must rethink individualization in relation to a collective that is *neither* the state nor a premodern community. "The political, ethical, social, philosophical problem of our days is not to try to liberate the individual from the state, and from the state's institutions, but to liberate us both from the state and from the type of individualization linked to the state" (PWR 336).

If Foucault is right that modern states are more thoroughly legitimated by the normalizing power that emanates from civil society (in collaboration with state institutions) than by that which emanates from the law, then the appropriate question for those who want to resist normalizing power is not how law can be used against the state, but how power can be exercised in relation to norms. To answer this question, we will first examine the path by which state sovereignty shifted from the juridical domain to the practice of security in the early modern period. Markets and public spaces are domains in which the modernizing state created "freedoms" for individuals that would implicate them in the collective project of security. The state made use of bodily materiality and discursivity, as well as the features of imagination and discourse addressed in earlier sections, to give these freedoms a productive shape. Bodily differences eventually associated with the biological notion of "race" helped articulate the feeling of danger surrounding these freedoms. Having explored how risk imaginatively shapes the bodies of modern citizens, we will then discuss the concept of *normativity* that might allow them to detach from particular historical forms of the state or individual without losing agency or imagination.

From *Raison d'État* to *Phobie d'État*

Most of Foucault's writing was done during a period when the North Atlantic had a relative consensus on the virtues of the welfare state and a mixed economy (that is, an economy in which some industries are nationalized or government takes a role in setting prices and wages). While trying to recover from the Great Depression and then after the Second World War, Western Europe and the United States abandoned laissez-faire liberal or authoritarian governments and replaced them with welfare states bolstering "ordinary" people's rights through social entitlements, labor protections, and consumer-oriented regulation (Yergin and Stanislaw 2002, chap. 1). One reason for this historic change was fear of the kinds of racial and class-based social unrest that had characterized the prewar era and had led to fascism. Another was the need to convince groups in the North Atlantic and the newly decolonizing world to avoid the temptations of alliance with the Soviet Union. Western Europe and the United States used a distinctive moral justification for their shift in government policy. In the newly decolonizing world, similar results were achieved by a combination of moral appeals, direct aid, and violent subversion against Soviet-aligned or potentially Soviet-aligned states. But the kind of state promoted at home and abroad by the United States was Keynesian. In other words, it prioritized full employment through government investment, funded by loans from the IMF and IBRD (International Bank of Reconstruction and Development) if necessary.

The welfare state's administrative "way of life" was contested in France, the United States, and many other countries around the world during the 1960s. Their confidence buoyed by rising economic expectations, young people and social movements became more critical of social inequalities. They were especially attuned to the suppression of minority ethnic traditions and colonial or neocolonial enterprises in Africa and Asia, necessary for first-world affluence but now justified in the name of the Western standoff with communism. British, American, and French counterculures had distinctive indigenous roots. But those who had grown up with postwar prosperity all thought of the Vietnam War as the imaginative site for an experience of "heterotopic" citizenship.[2]

In the United States, civil rights activists insisted that the national imaginary must include and respond to the experience of African-Americans, not just whites. They were deeply troubled that the apparent unity of American experience was purchased at the cost of divisions and violence in geographically distant places like Vietnam. France, on the other hand, had been deeply torn by the decolonization of Algeria, which until 1962 had been a department of the French government, not just a colony. Arabs were nonpersons subject to vastly lower wages, loss of land, and official or vigilante violence within their "own" country, Algerian France—just as African-Americans had been repressed in the South during the worst years of Jim Crow. In mainland France, police violence against the youth uprising of 1968 resurrected the uncanny sense of living in two versions of France during the Algerian War; one benign and humanistic, the other complicit with terror (Ross 2002, 40–48). In other European countries, such as the Netherlands and Germany, antiwar activism and cultural experimentation were ways of grappling with the everyday legacy of Nazism, which had been surrounded by silence during the recovery period of the 1950s.

The ostensible objective of these movements was to bring the welfare state's increasingly deadly militarism, whether expressed in the nuclear arms race or in proxy wars like Vietnam, under more democratic control. But another, more dramatic goal was to change the patterns of social hierarchy and surveillance through which women, racial and sexual minorities, the poor, and political dissidents were belittled or controlled in workplaces, schools, hospitals, and psychiatric offices. Although these hierarchies were produced and reproduced in civil society, employers and the state tacitly enforced them by declaring them outside the scope of law enforcement or public discussion.[3] By arguing that such conditions interfered with the full use of female or minority citizenship, activists hoped the state could be persuaded to enact protections and entitlements. They sought to enshrine their changes in legislation and to bring regula-

    2. See Gitlin 1989, 184–85, 262–82; Ross 2002, 80–81; and von Dirke 1997, 35.
    3. For example, many states prioritized stranger assault of any kind while ignoring spousal abuse or rape (until the 1970s). In the American South, the state enforced white racism and backed extrajudicial white violence.

tory power to bear against potentially discriminatory employers. This project was crystallized and expanded in the United States through the 1964 Civil Rights Act, which banned discrimination in public accommodations, as well as the 1965 National Voting Rights Act, passed as a result of intense multiracial organizing on behalf of African-American voters in the South. Both acts helped focus new minority political agendas at the national and municipal levels during the late 1960s and 1970s, and offered a model for feminist legal reforms throughout the latter decade. Making use of these protective regulations required plaintiffs to engage in class-action suits that were often time-consuming, expensive, and humiliating, but which succeeded in improving access to government services, political clout, and relations with police for both women and minorities.

The project of *expanding* the existing notion of citizenship using the state's juridical, legislative, and regulatory functions coexisted with and was often confused with a second project to redefine citizenship from the ground up to include African, Arab, Asian, female, working-class, and other minority experiences as legitimate and inevitable ways to be fully American, British, or French. American citizenship, for example, had been coded as male and founded on the right to exploit Native American and slave or immigrant labor since well before the founding of the United States, but sexuality and cultural esteem were also important elements of the *aesthetic* conditions for American communicability. Redefining citizenship from the ground up, giving citizens powers and freedoms that most had never held or thought desirable (for example, the right to be homosexual) would have entailed a radical redefinition of the state. Indeed, it would have detached the state from various ethnic and sexual understandings of the American "nation."[4]

The controversy and enthusiasm generated by these social movements can only be appreciated if one recognizes their *transnational* character. Some members of these movements were Marxists and allied themselves in some way with international socialism, although less frequently with the Soviet Union; others were motivated by religious or liberal values. Many identified with anticolonial movements and developing nations;

---

4. Foucault helps to explain the ordinariness as well as the radicality of this second project when he claims that the state is not a preexisting entity in which individuals participate or fail to participate as citizens. The state is not a container to be filled; it is *composed* by practices of governing that emerge as much from civil society as from the juridical or administrative apparatus. Although the state's activities had been justified since the nineteenth century by the security and prosperity of a *nation*, states have been around longer than nations and were not always constituted for a nation.

their image of the future reflected the official discourse of nationalism in those countries. These "imagined communities" (to borrow Benedict Anderson's phrase) were not utopias, because they were formed from the actual behaviors and fantasies of people who wanted to stress their similarities to people in other parts of the world or their own countries whom they had been taught to regard as alien. Even when these imagined communities were religious (like many later forms of transnational identification during the 1980s and 1990s), they indicate the existence of struggle over the shape of *this* world, not the invocation of another (theological) world. During the 1960s and 1970s, these movements had remarkable success in changing the ways younger generations understood their individuality and their relationship to the state.

According to Foucault, these movements were distinctive because they tried to give ordinary people control over the conditions of subjectification, rather than to end domination or exploitation (though he acknowledges continuities between these three) (*PWR* 331–32). These movements were also unprecedented in targeting practices in civil society, such as racial discrimination or relations between the sexes, rather than practices in the state or economy. Indeed, the political culture of struggles against normalizing power differed in key respects from the tactics used by more traditional movements against domination and exploitation, such as the antiwar and labor movements (Gitlin 1989, 212–21). To prevent the emergence of equally normalizing forms of individuality within the radical sphere, Foucault tended to recommend that these struggles over subjectification remain situated and not attempt to form an overarching "imaginary" for themselves (*EST* 316). This approach is exemplified by his reluctance to treat sexuality as a source for identity or a general object of liberation, affirming, instead, the plurality of "bodies and pleasures" (*HS1* 157).

Although the counterculture shared many values with political radicals, they were usually mistrustful of efforts to work within or exercise media power over the formal political structure. Many participants in both groups admired urban minorities, arts, and marginal status vis-à-vis the law and "straight" economy. Public recognition of affiliations between these groups, however imaginative and utopian they might have been (many working-class blacks were quite opposed to the counterculture), was followed by a backlash by the "silent majority" during the early 1970s when crime rose sharply in the very inner city neighborhoods that were supposed to be receiving the benefits of civil rights and antipoverty legislation. A series of crises in the larger economy led the political mainstream to undertake a major reassessment of the welfare state. These crises were facilitated by the deregulation of currency markets in 1971, the energy crisis of 1973, and a sharp increase in U.S. interest rates, whose effects were felt around the world. In Britain, unemployment and inflation created an unworkable financial situation requiring assistance from the World Bank. In their wake, protections for organized labor were slashed and national industries were privatized (Harvey 2005, 57–63). Faced with similar difficulties in the United States, Reagan rolled back environmental and labor regulation of industries and employers (Singer 1999, 161, 194). These are key elements of "neoliberalism," which Michael Peters summarizes in this way:

> Economic liberalization or rationalization characterized by the abolition of subsidies and tariffs, floating the exchange rate, the freeing up of controls on foreign investment; the restructuring of the state sector, including corporatization and privatization of state trading departments and other assets, "downsizing," "contracting out," the attack on unions, and abolition of wage bargaining in favor of employment contracts; and finally, the dismantling of the welfare state through commercialization, "contracting out," "targeting" of services, and individual "responsibilization" for health, welfare, and education. (18–19)[5]

The authors of neoliberalism believed that national industries are inefficient and do not return sufficient profit for governments to justify the public expense, while regulation inhibits business and requires disproportionate tax dollars to enforce. Because most Americans recognized the benefits of New Deal protection for labor and the elderly, some of the most crucial elements of the welfare state remained intact for several decades. But they believed that protection was unwarranted in the cultural domain (see, for example, Kristol 2004). Conservatives excoriated the

---

5. It may be worth distinguishing several related terms. "Globalization" refers to the effect of new communication technology, free trade agreements, regional trade blocs, the development of export processing zones, and highly mobile currency trading to produce goods using components and workers from several parts of the world. Neoliberalism is the *domestic* art of government that creates conditions for globalization. It encourages and justifies the "post-fordist" business practices described by Harvey (1990). Neoconservatism is the *cultural* and *political* ideology that renders neoliberalism compatible with traditional forms of community, at least in the United States.

lower- and working-class urban minority communities whose real or imagined tolerance of sexual freedom and drug use was adopted by members of the white counterculture. For the New Left and the counterculture, the protections of the welfare state were minimal conditions for political and cultural experimentation. For libertarians and monetarists, on the other hand, the protections of the welfare state were obstacles to financial and commercial experimentation. As is well known, the libertarians formed a coalition with social and religious conservatives in both Britain and the United States, initiating several decades of deregulation over financial and industrial practice and reorienting government activism from domestic social policy toward favorable conditions of international trade.

This coalition's electoral success was due to its apparent defense of traditional communities and families against the expansion of countercultural, New Left, and minority groups with increasingly transnational affiliations. Although the religious right in the United States could have been considered a "counterculture" in its own right prior to the 1980s (Armstrong 2001, 201), it spoke for an aesthetic sensibility that felt disoriented and embattled by social changes, especially in women's rights, race relations, and the nascent globalization of the U.S. economy. However, the libertarians' embrace of open currency markets, opposition to protective tariffs, and repudiation of Keynesian goals like full employment were sure to create enormous dislocations in communities and increase stress on families. Despite their stated desire for less government interference with business, the internationalization of currency markets and industries required tax incentives, protection for financial institutions, tolerance of emerging tax havens and offshore production zones, and the formulation and enforcement of new trade laws. As Harvey has argued, this vision of free trade is quite as utopian and activist as anything the left ever conceived (2000, 176–77). Nonetheless, the New Right package was promoted to skeptics as a lesson in common sense and realism, as opposed to the social engineering or constructivism of American Democrats or the New Left. After the fall of the Berlin Wall, socialist ideas were no longer associated with treason, but fell into the category of *Schwärmerei*.

Foucault describes the eighteenth-century process of Enlightenment as one in which forms of knowledge were consolidated and brought into a stable, that is, communicating, group of scholarly disciplines or sciences. During this process, local technical knowledges and histories were subsumed into discourses treating conflict of any kind as contingent, empirical, and temporary. The primary goal of genealogy, by contrast to these pacifying and unifying discourses, is to recover the memory of forgotten struggles (*SMBD* 8; *P/K* 82–84). One way of understanding what happened during the 1960s and 1970s was that many histories of struggle that were little known in the North Atlantic gained a wider audience: the histories of African slaves and their descendents, of colonized peoples, of labor and regional resistance to government, of women in all classes. People whose competence and sense of reality were rooted in those struggles or whose disqualification as speakers had facilitated the creation of scholarly disciplines (for example, those who were subjects of social scientific research but not researchers of any sort in their own right) began to elaborate alternate accounts of the past and made themselves heard, to a certain extent, by academics and the general public.

These alternate histories were embedded in the techniques used to socialize and dominate people as *bodies*. According to Foucault, the hierarchical practices employed in schools, hospitals, prisons, and homes had been coordinated through the course of the eighteenth and nineteenth centuries by a style of government aimed at fostering the *security* and *health* of populations rather than their conformity to law. This style of government is roughly contemporary with the rise of European absolutism and subsequent liberal regimes. But the legitimacy of this style of government and of those who administered the state had been fiercely contested since the birth of absolutism. Challenges to royal despotism were framed in terms of a war between races who represented a biological threat to one another's existence and way of life. This biological and historical way of imagining social antagonism persisted into the revolutionary and liberal period, when economic and military activity, including colonialism, could be justified on behalf of the security and health of the "people" rather than the regime. But who are the "people"? During the 1960s, citizens on whose behalf wars and ecologically destructive economic programs were supposedly waged reflected on their internal diversity and transnational affiliations and wrote new genealogies of struggle against the state racism of liberal societies.

In the *Genealogy of Morals* (1989), Nietzsche observes that any knowledge-producing enterprise, even genealogy, must be oriented by an image of the future as well as one of past origins. Otherwise it remains merely reactive, the tool of a subjugated people who want to gain power without changing the form or fact of domination. Such images of the future, I would argue, rearticulate the *sensus communis* through which each spectator finds him- or herself resembling and capable of communicating with his or her fellows in a potentially universal way. The New Right mobilized its adherents around a utopian image of small-town and suburban white America from the 1950s in which the real social conflicts of that time were solved or ignored (see, for example, Coontz 1992, esp. chap. 2). The New Left drew on a variety of images for its possible future, including images of armed struggle, artistic movements, sexual and affectionate practices, self-help communities, and agricultural or educational development programs in the postcolonial world. Like the New Right, it took inspiration from idealized moments in the past, such as American anarchist communes or the Wobblies. However unjustly their adherents had been repressed in the past, most of these movements recognized the importance of communicating via an image of the future and uncovering images of how *past activists imagined a better future* (Brown 1995, 72–75).

For many scholars, the welfare state served as a point of imaginative access to those past struggles and their potential futures, not just symbolically but in terms of the real opportunities for leisure, education, and cultural experimentation that social protections made possible for workers and the middle class. Attacks on the welfare state (hardly the most imaginative of institutions) ended up blocking these intellectual processes of retrieval and projection. The neoliberal consensus allowed genealogical investigation of past conflicts to continue within the university, but it brought an abrupt end to the process of uncovering forgotten images of the *future* that had accompanied past conflicts. These images were not locked in the heads of scholars but accompanied the everyday playfulness and invention of people struggling to negotiate their subordinate status within power relations.

Those who entered the public by way of community organizations and labor, antiracist, or peace movements rather than the university hardly fared better. Although most American and European leftists were aware of serious injustices within the Soviet Union, "actually existing socialism" provided another *image* for the future of those alternate pasts. The collapse of the USSR was used as further proof for the impotence of any project challenging the utopian economization of everyday life (e.g., Singer 1999, 49–55). Because these were struggles over the extent to which working-class and minority bodies would actualize certain events rather than others, *signify* the ability to act on one's own actions, or serve as objects of others' normalizing judgment, I think it is not going too far to say that many felt this shift in public imagination as a kind of bodily depression.

Migration of Sovereignty

How is it possible for history to be so embedded in the body that the end of the welfare state could provoke depression? In this section, I want to examine Foucault's account of how bodies feel or assume the stress of governmental technologies such as *raison d'état* or liberalism. Although ideas and practices involving "freedom" have affected a larger and larger mass of the world's population since the European early modern period, they imposed new inequalities in the process. Democratization and even the welfare state's social protections made space for freedom and defended it in legal terms, but the cultivation of freedom was guided by the overall goal of reducing risks to the state and its population. The ideal space of the *market*, an object of knowledge as well as an element in power relations, played a crucial role in building an association between sovereignty and *security* that had previously been grounded in *law*. Law and the public sphere, which have often been regarded as the self-evident tools of democratization, encouraged patterns of similarity or hierarchy in keeping with state goals of security and health, affecting the *pure aesthetic feeling* or *communicability* of those bodies.

The European social contract tradition, rooted in Christian theology, thinks of freedom as a natural capacity and right of human beings. Kant, more cautious, says that nothing in natural science precludes the possibility of freedom. However, he obliges humans to practice or *create* freedom by selecting actions according to a test of possible universality or impartiality, an obligation they regularly fail to meet. Foucault gives another genealogy of modern freedom, naming it as an element in governmental technologies that took markets as their ultimate laboratory for social coordination and transformation. Like Kant, Foucault does not believe freedom can be "known" in the sense of natural or social science, but unlike Kant, Foucault believes that genuine freedom can be experienced without structuring one's actions according to the model of scientific or juridical laws.

Raison d'état was an approach to government associated with Machiavelli in the late Renaissance and early modern period.[6] Where medieval

---

6. A detailed discussion is found in *Sécurité, Territoire, Population* (STP), one lecture of which has been translated as "Governmentality" (PWR 201–22). Gordon provides a thorough

kings had governed by waging war and enforcing law over subjects oriented toward the Christian afterlife, the goal of *raison d'état* was to make a "state" out of subjects and their activities and to preserve the state as long as possible against natural processes of decay (including war, but also internal strife). Theorists mined the writings of Roman historians for an understanding of the state as a substantial political entity with its own rationality. Practitioners discovered that techniques of pastoral government, which individualized church members on the basis of particular virtues and sins rather than regarding them as part of an indifferent mass subject to prohibitions or commands, could effectively contribute to *raison d'état*. These techniques coexisted with and often borrowed authority from the juridical structure, usually concentrated in the person of the king.

Raison d'état viewed secular splendor and power as rational goals enabling a state to participate in a regional balance of power. Its most important mutations are *discipline, sécurité,* and *liberalism.* As discussed in *Discipline and Punish,* practices of *discipline* grasp bodies according to an ever more finely tuned spatiality and temporality. They individuate bodies, break down single gestures into multiple components, establish an ideal performance for each component, and apply pressure to bring each part to perform at its norm. By contrast, the *dispositif de securité* discussed in Foucault's 1978–79 lecture course, *Sécurité, Territoire, Population,* grasps bodies and events as masses, using statistical techniques (*STP* 21–23). It tries to discover regularities in the behavior of a "population," such as birth- and deathrates or the frequency of shortages, that would be invisible to a naked eye, no matter how attentive (*STP* 24, 76–77, 107–8). These statistical traits characterize a population's internal physiology in the same way that rules of grammatical change indicate the inner historical development of languages, or the operation of internal systems indicates an animal's place in the evolutionary order. Eventually, these traits were identified with the "laws" of political economy in the liberal practice of governmentality that followed *raison d'état* (*STP* 109–10).

Although these practices of government did appear at specific historical moments and places, Foucault believed earlier practices coexisted with and often reinforced later ones (*STP* 7–10). For example, children and criminals were regarded as agents of positive or negative social health from the

---

summary of this course and its successor, *Naissance de la biopolitique* (Burchell et. al. 1991, 1–51). See also "The Political Technology of Individuals" (in *PWR* 403–17).

sécurité standpoint in the nineteenth century, but these risks were managed or corrected in schools and prisons using tactics from the arsenal of discipline. In the twentieth century, political strategies associated with early *raison d'état*, such as deterrence, coexisted with the security mechanisms of the postwar welfare state. Rather than regard these forms of governmentality as following one after the other, therefore, we should look at distinct practices that are available to many regimes and often combined. A subject or citizen may be touched by one or more of these restraining, ordering, or motivating strategies, each of which has an affective impact and may provoke reflection or strategic action.

Early and later techniques of *raison d'état* differ in the scale and the level of detail at which they assume human behavior can be observed and affected. Early "pastoral" techniques provoked rebellions in both the religious and secular domains because they were too intrusive. *Dispositifs de sécurité*, by contrast, dealt with all phenomena at the aggregate level. Many theorists of *raison d'état* believed that the state *could* generate exhaustive knowledge of economic and social activity within the territory it governed, as exemplified by François Quesnay's effort to create a "Tableau Économique" describing the economic behavior of the realm (*NB* 288–89). But later theorists believed that the state should not interfere excessively with trade or family life on the basis of this information.

Early and later techniques also differed in their method and scope of application. *Raison d'état* identified a whole domain of social practices to be ordered and optimized at the national level. Seventeenth- and early eighteenth-century "police" thus monitored and took charge of road maintenance, efficient markets, food supply, the expulsion or confinement of vagrants, and public health, in addition to any crime-prevention activities (*STP* 326–36; see also *PWR* 317–23). Having discovered what tended to happen with the aid of statistics, the police helped it happen more regularly and productively, employing price controls, setting import/export quotas, and authorizing guilds to control the number of tradesmen.[7] Ef-

---

7. The *Polizeistaat* developed with greater speed and complexity in Germany than in France and England, although in the context of principalities rather than a single national administration (*PWR* 138–39). Germany took the lead in theorizing the activity of "police," because the Prussian state drew upon the universities for personnel at the turn of the eighteenth century (Wolff taught on *Polizeiwissenschaft* at the University of Halle); whereas in France it emerged unsystematically from the everyday practice of administrators concentrated around the king (*STP* 325–26). For an extensive discussion of the transition from the Prussian *Polizeistaat* to the *Rechtstaat*, including the role of Wolff, Kant, and Smith, see Caygill 1989, chap. 3.

fective policing was necessary to enhance the prosperity and splendor of the state, allowing its king to maintain a standing army and the respect of other nations (*STP* 321).

The object of security-based *raison d'état* is "population," a statistical being affected by and capable of affecting its environment (*STP* 79–80). Foucault's study of "population" adds an important supplement to his comments about the being of man in *The Order of Things*.[8] In that text, Foucault explained the epistemic shift away from representation and toward "man" with reference to specific discoveries that introduced a historical dimension into the study of life, labor, and language. He also gave a capital role to Kant's unification of the *conditions* for representation by means of a transcendental subject. But these explanations give disproportionate weight to the discursive dimension, akin to the sins committed by historians of ideas when they prioritize the history of philosophy over other levels of eventfulness (see, for example, his criticism of Cassirer: *DE* 1, 548). In the 1978–79 course, by contrast, "population" emerges as the mediating concept between Classical and modern epistemes. Population is "the operator of transformation which turned natural history into biology, analysis of wealth into political economy, general grammar into historical philology; the operator which made all these systems and groups of knowledges tumble toward sciences of life, of labor and production, and of language" (*STP* 80). Although nineteenth-century "man" may be a transcendental-empirical doublet at the intellectual level, at the administrative and moral level, "man" is the way subjects of a governmental technique experience their implication in the mass of a living population (81; see also *SMBD* 190).

The transition from discipline to *sécurité* (and from there to liberalism) follows a self-critique or self-limitation comparable to Kant's restriction of finite understanding to application in the realm of appearances. This self-limitation crystallized when a certain set of activities falling within the study of population and the management of "police" were identified as topics of true or false statements related by an internal logic, rather than by mere statistical observation. These are the activities we think of as "economic" in the narrow sense: production, trade, and pricing. Just as it gave a pragmatic rather than theoretical account of "man," *Sécurité*,

---

8. On ideas about population from the seventeenth century to the twentieth-century welfare state, see "Governmentality" (*PWR* 215–18) and "The Birth of Social Medicine" (*PWR* 134–56).

Territoire, Population offers pragmatic reasons for the birth of political economy. Delimiting this specific set of activities as an object of secure but limited knowledge enabled government to become a genuine expertise, a science like other nascent sciences of the day, rather than a mere "art." Identified in some sense with the totality of national forces as "things-in-themselves," yet standing apart from them, the state became a regulative Idea governing the expansion and intelligibility of the market as a "phenomenal field."

This internal shift within *raison d'état* can be compared to Kant's "Copernican revolution," for it allowed the exercise of sovereignty to be justified in terms of the laws governing *human* economic knowledge, rather than through divine right or knowledge of divine law (*NB* 13–15; see also Burchell et al. 1991, 15–16). Political economy therefore shifted the grounds of legitimate sovereignty away from law, except insofar as law might enter into the preservation of production and trade. The process had already begun with the expansion of state involvement in society through police. Foucault notes with some irony that although the police had great discretion to command subjects on economic and social matters and sometimes acted in a disciplinary manner, most of their activity took the form of issuing regulations (*STP* 348). But he insists that these regulations were expressions of utility, not sovereignty—so it would be improper to confuse them with "laws" in the sense of commands or prohibitions. The gradual resettling of sovereignty in the social and epistemological rather than the juridical domain was cemented by liberalism.

When Foucault refers to the state as a regulative Idea of political rationality, he means that it is a "principle of intelligibility for a reality already given, for an already established institutional ensemble" (*STP* 294) but also the objective or situation to be *produced* through the activities of this ensemble: "One only understands what the State is in order to make it exist better in reality" (295). Like an Idea of Kantian reason, it makes no sense to talk about the state as an entity "in itself." The "state" is composed of existing security mechanisms insofar as they resemble one another and present themselves as the originating ground of administrative activity. But in its "regulative use," the state makes it possible to discover laws within the manifold of human social experience and to encourage certain behaviors, certain patterns of affecting and being affected by others, or changes in the physical landscape. One might object that this analogy is weak, for the *noumenon* was the object of a merely problematic concept, while the limit of state reason is *phenomenally embodied* in the physical, biological individuals whose desires and activities it reveals and coordinates. But insofar as the *noumenon* is hypothesized only in relation to real phenomena, there is a special sense in which it, too, is embodied in the scales and levels through which phenomena are knowable and pliable, or opaque and resistant (i.e., their materiality). For example, in authorizing "autonomous" economic actors to create and enforce regulations, such as factory owners who provided housing and education for their workers or maintained armed security forces, the state avoided public "liability" for oppression, while establishing an ever finer social order whose channels were compatible with and protected from legislative change (Gordon, in Burchell et al. 26–27).

Early students of trade and production recognized that the categories of political economy were artificial forms by which society's activities could be treated as a "nature" in the sense of lawfully ordered appearances (*NB* 17–18). However, these categories were now internal to a problematic of human reason rather than one emerging from religious custom or relying on pure military strength. Foucault compares the invention of the "market" as a liberal schema to the invention of many other schemas in which power relations were backed with a structured discourse and vice versa: madness, which was not an idea in doctors' heads but a juridical category that became an object of knowledge within the special context of the asylum; criminality, which was identified as an object of knowledge in institutions that existed to punish specific actions; and sexuality, which first became an object of discourse in specific power relations like confession, direction of conscience, and medicine (*NB* 35–36). "The stake in all these efforts regarding madness, sickness, delinquency, sexuality, and those I am discussing now—is to show how the linkage between a series of practices and a regime of truth forms a *dispositif* of knowledge/power that effectively marks something that does not exist into the real, and submits it legitimately to the distinction between true and false" (*NB* 22). Delimiting the market "in the real" enabled the state to extend the range of activities over which it could be said to have "true" knowledge and to anchor the criteria for this truth in the nation's economic vitality.

At the level of sovereignty, liberalism entailed two changes in *raison d'état* that are significant for Foucault's comparison to Kant and for modern efforts to exercise power with respect to laws and biopolitical norms. The first is that it begins from a limited sphere of empirical knowledge, rather than a metaphysics of labor and land. Early liberals argued that exhaustive knowledge regarding the domain of people's choices and inclinations was a chimera. This was an important shift from the physiocratic belief that the sovereign should gather exhaustive knowledge but use it sparingly (*NB* 63). Now, it seemed that to know society and its economic phenomena as a lawful whole "in itself" would require an administrator to transcend the bounds of possible experience. To claim knowledge of the hidden motives and subtle transactions resulting in aggregate economic phenomena was exorbitant, *Schwärmerisch,* to intervene in the individual actions of firms and buyers sheer folly. Foucault even hazards that "naturalism" is as good a word as "liberalism" for this new governmental rationality.

The second difference is that *raison d'état* enhances the health, productivity, and military strength of a nation in order to pose it as a presence within a balanced European order, while liberalism takes responsibility for *indefinite expansion* (*NB* 53–57). Foucault refers to such flourishing as biopower (*HS1* 140–41). We have seen how limiting the application of understanding's concepts to the phenomenal realm permitted Kant to anticipate an indefinite growth of knowledge regarding nature. Likewise, Adam Smith (whom Foucault reads as Kant's contemporary) predicted that a mutual and potentially infinite enrichment of nations would follow from a scientifically justified refusal to meddle in details of economic activity (*NB* 58–60). Although he believes rational self-legislation by increasingly constitutional governments *ought* to abolish inter-European wars and the expensive practice of standing armies, Kant suggests that *natural* processes of competition and accumulation will inexorably accomplish the same goals (*CJ* 5:430–34; *KPW* 47–51; 90):

> Civil freedom can no longer be so easily infringed without disadvantage to all trades and industries, and especially to commerce, in the event of which the state's power in its external relations will also decline. But this freedom is gradually increasing. If the citizen is deterred from seeking his personal welfare in any way he chooses which is consistent with the freedom of others, the vitality of business in general and hence also the strength of the whole are held in check. (*KPW* 50)

Kant also acknowledges that "it is hard to develop skill in the human species except by means of inequality among people" (*CJ* 5:432). Although increasing inequality creates misery for the poor and violence against the wealthy, he believes this state of tension is purposive insofar as it leads to the development of human dispositions. Cultural development of specifically human capacities is an "imperfect duty" enjoined by respect for human rationality (e.g., GR 4:423, 430; MM 6:390–91).[9] But the point is that even Kant, whose respect for law prohibited him from considering revolutionary action in the name of abolishing bellicose monarchies, believed that the modern state had two sources of legitimacy: (1) "juridical" sovereignty, and (2) natural processes organized into a phenomenal field by political economy. The market is an artificial space of self-limited state reason in which concepts and intuitive strategies are embodied. This artificial space, according to Kant, should enable liberal/constitutional/imperial governments to act upon the entire world in a naturally understandable as well as a morally ennobling way (NB 59).

In sum, the *market* is the final addition to Foucault's series of ideal spaces, within which a controlled domain of knowledge was posited by contrast to some avatar of the "unthinkable," such as madness, death, criminality, or sexuality. Like the asylum, the clinic, the corpse, the prison, and the confessional, "markets" are artificial spaces over which knowledge and power can be exercised in ways that permit transformation and organization of the surrounding world. As artificial and ideal, they also provide a common point of reference for disciplines and governmental or civil society practices that might otherwise conflict. Marx thought of the market as a site of contradictions, and Foucault is more likely to regard it as a compromise formation between incongruous discourses. But both Marx and Foucault agree that the market shaped and reshaped the bodies of modern individuals who are both "free" and part of a "population," object of a political economy and a biopolitics. Just as control over the asylum permitted various forms of administration to be coordinated in postrevolutionary France around an image of man as *animal rationale*, so too control (albeit a limited, structured control) over markets as competitive spaces enabled man to be individuated as *homo economicus*.

Risk Technology

Foucault's contribution to reflection on liberal discourse is his attention to the importance of *risk*. The concept of "risk" entered administrative discourse in connection with social phenomena whose frequency adminis-

---

9. A long series of authors from Hegel to Lenin and Fanon have explained the role of the colonial world in providing materials, labor, and markets for this expansion. Foucault's definitions of imperialism and colonialism are unclear in this section.

trators of the eighteenth-century *Polizeistaat* would have preferred to reduce (*STP* 63–66).[10] Riots, epidemics, and shortages were statistically regular phenomena interfering with regular improvement of the territory's average strength and productivity. But members of the population to whom the state has left freedoms can still be enlisted in the fight against such risks by means of an appeal to their *interests*. "Interests" and "risks," like the "market," are political "phenomena" rather than things-in-themselves (*NB* 46–47). Like the objects of astronomy and biology, they are also elements of a "nature" that can be rationally understood according to laws under the regulative guidance of pure reason. But as I have mentioned, interests and risks also correspond to *relationships* between the state and individuals that cannot be known-in-themselves. These relationships are what Kant and other liberals tried to get at with the term "freedom" (*liberté*).

From the side of the one who governs, freedoms (especially property) are what reason and utility dictate should be left alone; while from the side of the public, freedoms (especially property) are what morally belong to the individual and may be legitimately protected from state interference (*NB* 45).[11] However, the meaning and exercise of freedom differ from one governmental practice to another and do not exist outside or prior to state reason. Freedom is not a universal "which would present a progressive accomplishment, or qualitative variations, or more or less serious setbacks and concealments over time"; freedom is an actual relationship between a government and the governed, an object of mutual demands (*NB* 64). In Part 2, I identified it with the margin of inventiveness exercised by individuals who are subject to constraining power relations, but remain able to imaginatively vary the outcome of their actions. For gov-

10. See also Ewald 1993, which condenses some of the philosophical themes of *L'état-providence* (1986), his massive history of French laws relating to insurance and their political impact.

11. In her study of leading federalists, Nedelsky (1990) shows that by treating the right to property as paradigmatic of all individual liberties to be protected against the tyranny of majorities, and by structuring the legislative process so that those without property would face greater obstacles in gaining elected office, the Founders struck a balance between political equality and the desire for long-term economic stability. Morris and Madison treated private property as the *limit* of the state's power to act against individuals, but they did this for reasons of principle, as the limit beyond which majorities *should* not have power, whereas Smith and the Physiocrats regarded it as the limit beyond which states *could* not act without disrupting and harming the very process of prosperity which was their goal. Of course, the most striking case of property as the limit of state power was white ownership of black bodies. Holland refers to the Dred Scott decision—in which citizenship could not be extended to a black man, even a free one—as the limit of federalist "techniques of government" (2001, 96).

ernmental rationality, freedom is an *unknown* in whose exercise the government is forced to recognize the materialization of its own limits. Interests and risks are the positive and negative poles by which governmental reason apprehends its subjects' freedom.

Recall that liberalism developed within the practices of *raison d'etat* when the economy was identified as the portion of social activity most accessible to state knowledge, as well to the expression of police power. *Security* in the liberal art of government still entailed preventing famine, disease, and riots. But with time, security began to imply security against the risks of civil freedom. Freedom, in turn, was suspected to lie at the root of famine, disease, and riots—as well as the useful phenomena of trade and production. Thus the unpleasant statistical phenomena observed by earlier administrators were interpreted as *phenomenal manifestations* of the risks associated with freedom. These are the circumstances under which "risk" was transformed from an empirical statistical content into a kind of relation between individuals and groups; or, as François Ewald explains, a "category of the understanding" and not an intuition:

> In insurance the term [risk] designates neither an event nor a general kind of event occurring in reality (the unfortunate kind), but a specific mode of treatment of certain events capable of happening to a group of individuals—or, more exactly, to values or capitals possessed or represented by a collectivity of individuals: that is to say, a population. Nothing is a risk in itself; there is no risk in reality. But on the other hand, anything *can* be a risk; it all depends on how one analyzes the danger, considers the event. (Ewald, in Burchell et al. 1991, 199)

The possibility of translating statistical, population-wide risks into the risks of liberty and vice versa allows individuals to be held morally responsible for the "good use of liberty"; i.e., for augmenting or minimizing social, medical, and economic phenomena originally identified as morally neutral objects of governmental reason. Kant, for example, identified a "good use of liberty" as one that has a structure identical with the structure of knowledge. Its law could be "a universal law of nature" (*GR* 4:421; Foucault 1984, 372). Individuals whose liberty seems not to be motivated by interest or regulated by self-legislation present a special threat to this regime, as shown in the case of the "dangerous individuals" of nineteenth-century psychiatry (*PWR* 196–98).

The ideology presenting human beings as naturally driven to improve the general good through the pursuit of individualistic desires arose at the point when too much government posed a greater risk than insufficiently detailed attention. However, liberalism was no less a practice of governmentality than its predecessor, *raison d'état,* and as the nineteenth century advanced, liberals developed statistical discourses enabling individuals to be ranked in relation to overall patterns of normalized or deviant behavior. Jacques Quetelet's concept of the "average man," in particular, came under attack by Marx for giving scientific legitimacy to the exploitation of differently situated workers according to the most profitable assumptions about their common needs (Marx 1977, 440).[12] Despite implicating each worker in the condition of the population as a whole, methods characterizing the outlier as "deviant" placed social burdens of avoiding risk—of resembling a dying or dangerous body—on the shoulders of persons who were increasingly aware of their individuality and potential conflicts with group belonging (Burchell et al. 1991, 45). Thus regulatory ideals and dangers became *possible lives.* For example, health and military preparedness, economic competitiveness, and educational achievement are *normative* ideals corresponding to the adequate management of life, labor, and language at the level of the population. But individuals are responsible for proving their fitness, patriotism, economic worth, or intelligence to themselves and their communities.

The invention of risk or the application of a risk technology to events and relationships satisfies two goals of the security-oriented state. On the one hand, the individual's anxieties or feelings of exposure to danger can be tied to his or her sense of belonging to a mass with statistical laws. This diffuses social antagonism and makes each responsible, in some way, for the behavior of the whole. Colin Gordon points out that the concept of social risk allows responses to social problems to be presented as matters of *insurance:* "creative simultaneously of social justice and social solidarity." While revolution exceeds the coverage provisions of almost all insurance plans, the existence of various forms of private and social insurance programs at the macroscopic level deflects attention from class and other antagonisms; functioning ultimately as *insurance against revolution* (Burchell et al. 1991, 40–41). By allowing the individual to situate him- or herself securely with respect to potential disasters, it contributes to his or her net power.

---

12. Kant discusses the relationship between the Ideal of human beauty and the "average man" identified through statistical comparison in *Critique of Judgment,* 5:233–36.

On the other hand, identifying and even creating threats to public health, safety, and prosperity—external or internal—gives the public, in all its cultural and bodily heterogeneity, an investment in the singular identity of a population with natural laws, products, and losses. This prevents it from fragmenting or reorganizing in ways that might foster the flourishing of different bodies. Sometimes the threat is external: immigrants, foreign economic forces, or hostile nations appear as foes requiring a patriotic discipline and consensus-reinforcing social action from members of a population. Sometimes it is internal: crime, homosexuality, falling student test scores, drug use, and teenage pregnancy are presented as "risks" to which any individual or community may fall victim if not supervised and examined on a continual basis. One of the leading forms of risk is the population's potential to cultivate transnational affinities and resemblances—a danger imaginatively embodied in Jews, communists, and radical Islam at various points in American history. Thus technologies of risk have historically contributed to projects of "state racism," whereby the state uses regulation, economic power, and military means to enforce the dominance of one racialized group as if it were under attack by another (although Foucault does not suggest that risk technologies must *necessarily* serve racist ends) (*SMBD* 254–56). Perception, conceptualization, and imagination of risk continually unify the nation and its biological forces against other communities and races, sifting them from an originally indeterminate flow of peoples and populations (Bartolovich 2000, 16–17).

Since modern power concerned itself with the preservation and intensification of a population's biological life, states have become willing to entertain increasing risks and to inflict far greater death and devastation on one another and on their own citizens than those caused by nature and rioting in early modern Europe. "Outside the Western world," Foucault notes, "famine exists, on a greater scale than ever; and the biological risks confronting the species are perhaps greater, and certainly more serious, than before the birth of microbiology" (*HS1* 143). These famines are thoroughly historical, insofar as they result from patterns of economic distribution and political conflict that differentiate the meaning and quality of first-world lives from those of the rapidly industrializing world. Of course, "one can always argue that life and health are things beyond price. But the practice of life, health, and accident insurance constantly attests that everything can have a price, that all of us have a price and that this price is not the same for all" (Ewald, in Burchell et al. 1991, 204). It is at this level, where misery in the newly industrializing world is all too frequently cited as the result of uncontrolled population growth, that we can see the races and their life-chances being *defined* through the differential deployment of sexuality around the globe.

On the other hand, the fear of being considered "abnormal" or of presenting a genuine threat to their fellow insurers constrains the freedom of those who have the option of varying their responses to the environment—or of varying the environment itself (Butler 1993, 100–110; see also Adair 2003, 33). The suffering of those who are active targets of social hostility or allowed to subsist on the margins of society warns anyone who might feel *capable* of acting on their own actions or the actions of others not to challenge prevailing practices. This is the most intimate way a risk technology acts on the political imagination of citizens, presenting possible futures in which the body is at stake. The example of poverty in the former colonial world is often used to remind Western citizens of the benefits of their existing governmental structure (even when they are the same structures responsible for the poverty). Because they recognized the role of these "problematic objects" in stabilizing social conflict, activists in the transnational movements to reformulate citizenship I described earlier approached the problem of collective risk management from the perspective of domestic and foreign groups who embodied risk in the imagination of wealthy and/or majority citizens.

In sum, law *stabilizes* the population's relation to risk, rather than providing a clear zone of normative authority, and extends a similar promise of stability to individual citizens. It also involves citizens in a complex double game. Law protects the space of freedom opened by liberal government and mediates between the individual's fear of risk and his or her hope of power and pleasure. But law invests bodies with freedom and risk unequally. While the impartial structure of law gives everyone in a given class the right to collective protection against risk, understood as an anonymous and statistical phenomenon, the parallel structure of the *norm* in security-oriented states requires the individual to assume responsibility for "not posing a risk" to his or her fellows by falling into the category of the abnormal.

Public Space

The foregoing history helps us understand how the basis for state sovereignty could "migrate" from its juridical activities to its risk-prevention activities, while continuing to enforce the law and to recognize law as a medium of individual freedom. It also helps us understand how tension can build between activists who want to use law as a weapon against state racism and citizens who, without being consciously committed to state racism, nevertheless believe that law should protect them from the idea of social disorder historically associated with the bodies of poor or minority citizens.[13] The tension between "juridical freedom" and "normalized freedom" is reproduced in the tension between the American public sphere and the counterpublics or mass movements that the bourgeois "public" regards as endangering the population or incapable of serious dialogue about the nation's collective future.

Generally speaking, the public sphere encompasses all the physical and imaginative spaces in which people confront one another as strangers around a general "topic" or "event" in such a way that afterward they may relate differently to themselves and one another. For Hannah Arendt, who regards the public as an element of all true politics, what is essential is that individuals gather either as actors, witnesses to action, or spectators of a *world*, an object of common perception (1958, 50–58). Analogizing political judgment to aesthetic judgment, Arendt writes that in politics "it is as though taste decides not only how the world is to look, but also who belongs together in it" (1968, 223). How do they know who belongs together, prior to all exchange of reasons? Because the objects or events in front of them produce a state of mind in which they feel capable of communicating—indefinitely, if not universally. Sometimes those objects and events concern a shared architectural or cultural world; sometimes they concern humans and human actions. A public space may resolve itself around a leader who introduces a new chain of actions, but I would argue that this can also happen if a "dangerous individual" introduces division into the community or offers the possibility of unifying an already-divided populace against him or her.

Ideally, citizens and collectives use public spheres to problematize the compromises or incongruities between the discourses and practices they are ordinarily forced to live at the level of the body. They experiment with the perceptual and temporal scales through which shared events seem communicable, in the same way that the Kantian critic becomes aware of the potential plurality and necessary limitation of forms that

---

13. In the late-twentieth-century American context, the classic article on social disorder and public space is "Broken Windows" by James Q. Wilson and George L. Kelling (2004), which originally appeared in 1982.

promise cognitive purposiveness. Although problematization has cognitive aspects, my interpretation of the public sphere stresses the importance of the concrete sensory dimension. Bodies symptomatize a community's sense for who may communicate and whose view of an event or aesthetic form is worth taking seriously. These kinds of symptoms shape citizens' convictions that the variations on individuality and the state they imagine as a result of their encounter with events and forms can truly lead to *more* sense—that is to say, exhibit greater purposiveness—than political life currently provides.

According to Foucault, the "public" was a phenomenon considered by seventeenth- and eighteenth-century theorists of the security state in tandem with "population." Population, as mentioned earlier, is an ideal way of grasping the living resources of a country as a whole in order to protect and optimize their performance. The *public* is "population taken from the side of its opinions, its ways of acting, its comportments, habits, fears, prejudices, demands; it is that on which one has a hold by education, by campaigns, by convictions" (*STP* 77, 45). Although the freedoms associated with a public pose risks for government, other risks arise if these freedoms are suppressed; at the very least, circulation of ideas allows a government to monitor its potential enemies. Like disturbances around the scaffold, riots are occasions where the "population" acts as a "subject" rather than being determined by outside controls; it is not surprising, therefore, that riots were among the first statistical phenomena to which security mechanisms were applied (see *DP* 59–63 and Roberts 2003, 76–79). However, the elite response to riots also reveals that publics can easily regard one another as manipulable "populations." Populations are what, in the colonial situation, the public confirms its own rationality by "observing" or "propagandizing" rather than "talking with."

Many theorists of the public sphere begin their discussion with Kant's "What Is Enlightenment?" which gave the bourgeois public sphere a unique role in the modernization of European states (*KPW* 54–60). In this essay, Kant defends a limited public role for the free press and reflects on the extent of intellectual autonomy in Prussia in comparison to preceding eras. Criticism as a philosophical doctrine and method corresponds to a political practice—exchanging reasons without engaging in direct action. Borrowing the words of an autocrat, Kant proclaims, "Argue as much as you like and about whatever you like, but obey!" and even claims that too much civil freedom raises the stakes of intellectual discussion so high that the masses would rather remain unenlightened than risk being the subject of radical experiments (*KPW* 59). In another essay, he suggests that preventing humans from publishing or communicating "removes their freedom of *thought*" (247).

But Kant never imagines we can or will think with *all* the others. The participant in his public is a *"man of learning* addressing the entire *reading public,"* the bearer of a certain class and gender identity, someone equipped with elite means of persuasion (*KPW* 55). Kant assumes, likewise, that the right to legislation in a constitutional government where publicity functions as a political *resource* as well as a *check* upon illicit use of state power will be limited to men with property, for a wage worker lacks the necessary experiences of autonomy to offer a constructive opinion and "allows someone else to make use of him" (78). This limits the number of things, obviously, that can be discussed, and the number of people who can participate; who shall correct the property owner's written assumptions concerning the ideas of workers or women? Slaves are not mentioned, for they *are* property.

The bourgeois public sphere is not, therefore, "incidentally" bourgeois insofar as it involves speakers and writers who happen to be bourgeois. Like sexuality, publicity is one of the ways in which a certain segment of educated merchant society *constitutes* itself as a distinct class. While Kant stresses the pursuit and acceptance of reasoned criticisms of one's views, other contemporaries stressed the importance of mannerisms identifying speakers as "respectable" (Roberts 2003, 76–80; Warner 2002, 104–5, 110–12). But the constitution of a bourgeois public sphere says nothing about speakers left outside or defined as "unrespectable" in new ways, who may constitute a "plebian" sphere or "counterpublic" on certain occasions.[14] Nor should the focus on discourse mislead us into thinking that bodies are unimportant in the distinction of whose speech will count as public. As we will see, embodiment, media, and geography are ways in which states and publics attempt to materialize or direct the imagination of participants, acting on their freedom and encouraging them to relate to themselves in communicable ways.

Although the idea of the "individual" is often regarded as abstract, forms of individuality are not given apart from the social codes that make some bodies more credible as speakers than others. People only *discover*

---

14. The most important works in this discussion are Habermas's *Structural Transformation of the Public Sphere* (1999) and *Between Facts and Norms* (1996). A great deal of discussion has responded to Habermas's conception of the public sphere. See Negt and Kluge 1993, Calhoun 1992, Crossley and Roberts 2004, Hill and Montag 2000, and Warner 2002.

or begin to reflect upon their own individuality when they fail to communicate or to share their parents' and peers' reactions to collective phenomena; that is, recognize that their ability to communicate is limited and that they must actively seek a public in which they *do* communicate. Freud (1961, 26) writes that the psychic ego is formed concomitantly with the body image, and that the individual's intellectual grasp of his or her own body space is shaped by past injuries, practical successes, and illnesses. It also bears traces of all the verbal encounters in which it has borrowed from or conceded energy to others. The other's bodily presence is the first sign that sounds and symbols in its vicinity are worthy of mutual respect, sincerity, and orientation toward agreement by rational criteria— Habermas's conditions for communication (see Gil 1998, 107–21, and Gardiner 2004). But human bodies are not equal in their ability to signify this worth, and are constantly competing with other bodies to be "more" individuated and "more" worthy of consideration. Thus their bodies represent the *capacity* to perceive aesthetically but are also *what* one perceives and judges.

In short, the body often marks the limit of what can be made communicable for a "patrician" or "bourgeois" public sphere that prides itself on evaluating events from an abstract, but also a competitive standpoint. Women and the poor were excluded from the bourgeois public sphere, which consolidated the association of masculinity with citizenship, publicity, and universality. On the other hand, women, workers, and the poor are never fully offstage even when they are in the spaces men and owners consider "private," because their duty is to "appear" helpful or deferential to those who have power over their livelihood. A similar dynamic held for populations subject to European domination, in the sphere of religion as well as politics; public discourse gave rise to a notion of European-ness as always more successfully Christian than African-ness or Asian-ness. Public spaces could only play a legitimating role in European state formation after modern national character had been defined through two processes: overseas colonization and the assimilation of regional subnationalities and languages into large administrative units (Bartolovich 2000).

Foucault described something like a "plebeian" public sphere forming in the gossip, songs, and speeches around the scaffold in the Renaissance and early modern era. During this time, according to Habermas, the only "public" consisted of royal pageantry and terrifying displays for subjects who were by definition spectators, not "participants" in anything—at least not from the court's point of view (1999, 5–12). When a pauper or criminal who was the target of royal punishment became the focal point for rebellion and solidarity, however, he or she facilitated the airing of conflicting opinions regarding justice, oppression, or the relative responsibility of the king or God for collective evils. In London, the final statements of criminals hanged at Tyburn were witnessed year-round by a large public and published widely afterward (Roberts 2003, 76–77). Although populations can behave in erratic and violent ways, John Michael Roberts suggests that eighteenth-century references to the "mob" (from the Latin, *mobilis vulgus*) are more likely to express aesthetic disdain for the class, not the behavior, of assembled people (74–75).

Under certain circumstances, however, elites do recognize intentions in the speech of the poor people rather than the dangerous symptoms of an animal mass or population (Farge 1995, 3). Arlette Farge's studies of prison records from the reign of Louis XV show that ordinary Parisians were vocal in their judgments of the king and assiduous in their efforts to sift rumor and error from the probable truth of stories circulating around them. Without exactly forming a consciously "imagined community," political gossip expressed a distinct social point of view, as well as values and criteria for persuasion. A royal edict condemning Jansenism, a populist form of Catholicism promoting, among other things, people's access to the Bible in French, provoked immense controversy and led to the circulation of clandestine newsletters in which the actions of poor and working Parisians were reported and discussed. "The *Nouvelles* accepted as 'reason' things which others despised or affected to ignore: the ensemble of cultural reactions to some divulging of information or religious questions; ordinary people's ability to learn; the importance of reading; the dignity of the servant, equal to that of his master; the possibility of influencing the event that one has lived through, known, and seen" (49). Although officially it regarded the people's chatter as beneath serious concern, the *ancien régime* employed spies to monitor the opinions of common people, feared its supposed passions, and incarcerated subjects for spreading news or engaging in seditious talk.

African slaves in the United States engaged in political reflection and resistance, especially using the idiom of religion. The Underground Railroad and uprisings against whites testify to the presence of such counterpublics, whose voice was extended by northern African-Americans and abolitionists. On the other hand, slaves sold at public auction form neither a public *nor* a counterpublic (Bartolovich 2000, 20). Controlling someone's bodily expression of emotion and judgment can discredit their right to invest that body with an original meaning, as shown by the antebellum American practice of making slaves dance and "entertain" themselves while traveling to and from slave markets (Hartman 1997, 32–36). The chain gang and auction were exercises of power designed, as far as possible, to make a subject people's words into mere "symptoms" even for themselves, in order to extract the maximum energy from whatever freedom remained in their laboring, chained bodies. Anyone who must do extra work to distinguish the expression of reasons from mere "behavior" is likely to face discrimination in a public space and have less faith in the efficacy of his or her participation.

In the United States labor unions and political parties have played an important role as public spaces for discussion, the formation of group interests, and consensus on national identity. But regulatory restrictions on *labor* (and state permissiveness toward corporations that seek labor in the international market) have damaged its function as a public sphere for workers who were excluded from the "bourgeois" forums of the media, academia, and the law (Aronowitz 2000, 94). These mediating structures were especially important at the local level where citizens learn to take an interest in politics and to gain confidence in forming positions and actions or reactions. As the welfare state retreats, churches have returned to the fore as sites for the provision of social services and "public" debate and information sharing.[15] At the same time, the business lobby or political action committee with a single constituency has maintained its power. Unlike labor unions and parties which represent people in their *whole lives,* these organizations need not take more than one or two strategic imperatives into consideration at any moment or imagine how competing imperatives will be coordinated by the voter or citizen who lives with the resulting laws. The increasingly televisual nature of national campaigns and the government's protection of corporations' right to charge enormous sums for campaign airtime have forced parties to focus on the national level and to compete with entertainment in its own medium rather than to provide a different mode of public life altogether (Patterson 2003).

Oskar Negt and Hans Kluge (1993) suggest that "non-bourgeois" public spheres contain all the speakers and authors whose words are regarded as somehow at or "beyond" the edge of bourgeois communicability. They

---

15. See Warner 2002, 83–85, on American sermons and revivals as important occasions for rhetorical creation of a public. See Ehrenreich 2004 on churches as alternative sites for social welfare.

argue that the imaginative fragmentation or conflict I have called heterotopic is experienced primarily by workers and racially or geographically marginalized groups whose fundamental life experiences are not central for the bourgeois public sphere. When abstraction and an ability to distance oneself from particular examples or situations is a condition for respectable discourse, working-class speakers who present their opinions and reasons in dialect or performatively are often regarded as unable to speak "normatively" but only to provide "examples" for the elites.

The "proletarian public sphere" may not exist as an actual imagined community, but it is an ideal point from which the universal pretensions of the bourgeois sphere and its damaging effects on people's normativity can be challenged. Negt and Kluge observe that bodies do not only serve as markers for class belonging, but that in combat situations, the massing of physical bodies gives a sense of reality to working-class ideas or demands that the bourgeois public sphere would like to regard as "misinformed" or "fantastic." However, their account of the proletarian public sphere, at once problematic object and "regulative ideal" for a revolutionary society, does not really explain the existence of counterpublics or the way dialects and cultural practices break up the fertility of nonbourgeois discourse; even in Germany, Jews troubled the presumed unity of the Enlightenment public sphere and Turks constitute a distinct subsphere of the contemporary proletarian public. Multilingualism poses a serious problem for any state in the postcolonial world that wants to use the public sphere as a site for legitimation, since bureacracies often use European languages known only by the middle or upper class, who thereby reinforce their cultural power.

But the "material" and "discursive" aspects of public understanding are not just instantiated in bodies and languages. They are also evident in the sensory context of public interaction or the differentiation of public from private acts. Although the American public sphere is often analyzed in relation to the idea of "free speech" or First Amendment rights, public reason does not simply require freedom from "censorship" or illegitimate state restrictions on the flow of information. It also requires that citizens have access to unexpected encounters and points of view and be exposed to persons and places where such encounters might take place (Sunstein 2001, 30–33). A person or community is only "cosmopolitan" when it contributes to the potentially infinite circulation of ideas and experiences; as Michael Warner argues, a public is only formed where people encounter one another as strangers (2002, 72–76). This means that the public must not be closed off from the "world," though it will inevitably focus attention on certain events or topics and ignore others. But the relationship to strangers is geographically and linguistically embodied; in Kantian terms, citizens must have public *intuitions* as well as *concepts* if they are not to suffer from "imaginative blindness."

What happens on Wall Street or in Lincoln Center is considered more likely to be noteworthy than what happens in the Bronx or Iowa City, just as what happens in the United States or France has a more significant global effect than what happens in Cameroon, and the kinds of bodies that are allowed in certain geographical areas or discouraged from entering *marks those spaces and events* as more or less universally communicable. Immigration and travel laws, therefore, shape the extent to which concepts can be matched with intuitions and communicated more or less widely. In Cass Sunstein's view, the danger of Internet technology is not only that it will reduce speakers' access to a potential audience as consumers "tune out" unwanted or unexpected information, but also that it will prevent citizens from encountering the kind of situations and ideas that shape the lives of fellow citizens but might not be discovered "on purpose." Americans from privileged walks of life are far more likely to be familiar with the *ideas* expressed by those from other social groups, as presented in journalism, than they are with the *look and feel* of the situations that gave rise to those ideas—far more likely to know what positions urban blacks have taken on school vouchers, for example, than to have actually walked through a poverty-stricken school.

One reason why it is impossible to simply liberate the "proletarian public sphere" in the manner suggested by Kluge and Negt is that *attention* forms an intrinsic limitation on the scope of communicability. "Attention-directing" is an exercise of power that acts on the actions and other bodily materializations of the public. Parties and unions are not just contexts for the sharing of information; they are also contexts for the identification of events, even events that concern only the members of that mediating public. (Ah, this is what we're living through together; this is what you'll never understand the *way* we do; or, this is what you'd have to *do* to see things the way we see them.) Joan Tronto (1993) uses the phrase "privileged irresponsibility" to describe people who can ignore or overcome certain practical problems in life, certain members of the population, or the practical problems of those other members. One of the most difficult tasks of any insurgent social movement is to grasp the attention of those in power, because this attention is itself a measure of power.

The speeds and "slownesses" of biological and informational rhythms are also material aspects of public spaces. They do not simply transmit or verify propositional claims regarding the world, they *direct citizens' attention* toward this or that element of the world. For example, William Greider gives a horrifying portrait of how slowly clean-air legislation is changing the actual practice of power plants in the United States because of litigation and because environmental agencies and energy lobbies often trade top personnel (Greider 1992, chap. 5). The speed at which one receives news, like food and medicine when one is hungry or ill, makes the difference between being able to use it or not use it. It takes time to remove environmental toxins from the soil, and time for occupational stresses to disable a worker. This process is only capable of being presented as an "event" to concerned citizens from a certain vantage point in time and community involvement or distance (Warner 2002, 96–98).

Advertising structures the public sphere by subsidizing the distribution of its most widely received print and broadcast media. Advertising also constitutes a sphere in its own right, because it shapes one's sense of the other kinds of people with whom the public is shared, even if only by subtly appealing to a range of fantasies rather than to actual lives. Michael Warner hypothesizes that advertising-driven and consumer identities are important because they allow individuals to be *embodied* while still participating in a public. Through products, we "trade in" our own bodies for improved—that is, sexier, smarter, and wealthier—models (2002, 176). By wearing or using products with the proper logos, we both communicate and *become* more communicable. Thus in some respects the triumph of the market reflects a displaced frustration with available forms of public communication rather than consumer fascination or the self-evident efficiency or moral rightness of liberal governmentality.

One obvious problem with advertising is that the primary criterion for communicable events and attitudes is whether they will make money for a small group of manufacturers. The public function of advertising (as opposed to its commercial function) arises because these private interests want everyone to know about the same product. Yet every advertiser wants the bystander to look at and remember *his or her ad* rather than the others. Advertisers place ads where they believe target audiences will best appreciate them (Cartier watches in the *New York Times*, malt liquor ads on urban billboards—and internet advertising tailored to the viewer's most recently visited sites of interest). The competition for attention by thousands of advertising images a day, posted on public transit, television, Internet borders, school bulletin boards, and print media creates its own kind of stress on the viewer who is interpellated, minute by minute, as athlete, executive, seductress, or homeowner (see Warner 2002, 86–89).

A second problem is that advertising only addresses those who are likely consumers, and forces people to identify their public personae with their consumer personae, much as Kant's public sphere only addressed people insofar as they were scholars. But those who have less money or whose tastes are difficult to package as a product disappear from advertising and the cultural or news media supported by advertising. The unemployed, rural, or disabled are rarely called upon as public witnesses or participants—except, perhaps, by personal injury lawyers. When advertising substitutes for most other common but unexpected experiences, then those who are not represented as worthy of *advertising to* lose moral importance in the eyes of the ad-watching public. They remain unimagined, indifferently and ominously similar in their deficiencies. And when advertising unifies viewers around a presumed set of fears—whether baldness, bankruptcy, teenage drug use, or bathroom mildew—then individuals are encouraged to express their normativity through consumable cultural traits that are explicitly *denied* to others. In the process, the bathroom mildew of the unenlightened casts a discrediting light on their lives and expressiveness, justifying decreased funding for affordable and safe public housing.

Regulation of the media and of public health—including environmental and work safety, as well as access to medical care—are among the most significant factors shaping people's ability to "come into the public" as living beings at all. In both technological media and health, the "media" of power, the "media" of financial gain (the worker's bodily labor power), and the "media" of communication are conjoined. Communication requires a body, whether that involves the gestures of a conversationalist or time on a radio station or access to a web server. Those who have access to the latest technology, like those who have adequate time and energy for absorbing and responding to information, are significantly advantaged compared to those who must rely on older or debased forms of media. In most of the United States and large parts of the world, a technological gap in information access is replacing the problem of political censorship (though both may work together to paralyze populations) (McLaughlin, in Crossley and Roberts 2004, 171–72). Likewise, AIDS and other diseases that disproportionately affect the poor and those in impoverished regions consume all the resources and personal energy that could otherwise go into economic innovation and the generation of communicative political power. Citizens with health problems become, not surprisingly, less able to participate in everyday civil society as well as the public sphere.

Security-based and liberal governmental technologies found it prudent to leave room for the exercise of economic freedom and discursive or expressive freedom. But the fact that the public is never free from the perception of risk that emanates from ordinary disciplinary and cultural institutions like education and medicine ensure that the public space will structure itself hierarchically as the expression of one (sentient) population that must protect itself against threats associated with the public *or* private presence of dangerous individuals. In this way, a thriving public sphere can be perfectly compatible with practices of state racism and can even bring new opportunities for conflict to the attention of the state.

The Normal and the Normative

Foucault has been criticized for acquiescing to the historical shift from political legitimation through law to political legitimation through security or disciplinary practices. Jürgen Habermas, like fellow Kantian John Rawls, believes that law has not *yet* lost its power to legitimate or criticize state activity.[16] But he fears that this power, which prevents citizens from being mere means to one another's ends or mere clients of state services, is damaged by the increasing complexity of government and civil society (Habermas 1996, 375). Moreover, as the state relates to citizens in a more "normalizing" way, citizens begin to relate in instrumental rather than communicative ways, destroying the emotional and moral fabric of the "lifeworld."[17] Most critical theorists agree that the process is only has-

---

16. Habermas refers to law as a "transmission belt" by which structures of mutual recognition from everyday interaction are able to bind the anonymous structures of administration and economics (1996, 76). The law's main function is to enforce norms citizens *must* obey, regardless of private morality, and to resolve social confusion arising from religious or cultural pluralism (122). The other function of law is to ensure that once it has become unhooked from sacred traditional power, the state's power reflects the norms implicit in *communicative* power rather than becoming merely arbitrary or despotic (145–50).

17. Habermas's notion of the "lifeworld" combines Husserl's notion with Wittgenstein's "form of life" and refers to the relatively unreflected background of linguistically mediated, recognition-based social practices (1987, 117–19; 1996, 21). He contrasts the lifeworld to "systems" or "autonomous organizations connected by delinguistified media" such as money or hierarchical power (1996, 154).

tened when intellectuals turn a blind eye to the (potential) normative content of law or refuse to identify power-free norms by which their political arguments may be judged.

Although Habermas's comments are often taken up as representing one "side" in a purely philosophical dispute with Foucault over the importance of political norms or the validity of ethical relativism, some historical (and national) context is useful. During the 1970s, Foucault's ideas were cited with great approval by a small group of political thinkers referred to in the French media as *nouveau philosophes* (Dews 1979). Some of them, like André Glucksmann and Bernard Henry-Lévy, were former participants in the anarchist/Maoist action groups involved in May '68. They shared a basic romanticism about the working class and marginal populations and a fierce opposition to the Soviet political system, which they viewed as a natural culmination of Marx's ideas as well as an overvalorization of control in Western rationalism. Although their criticism of "reason" and domination brought them closer to Horkheimer and Adorno (1995) or even Heidegger, Foucault's support for Soviet dissidents and his well-known distance from Marxism made him a real or imagined fellow traveler. Other thinkers associated with the *nouveau philosophes*, such as Luc Ferry and Alain Renaut, were equally fierce *critics* of May '68 and attempted to ground the new (but short-lived) intellectual movement's defense of human rights in Kantian rationalism, espousing a position closer to that of Habermas and Rawls (Ferry and Renaut 1990; Renaut 1993).

Habermas recognized this movement as an attempt to shift media and university attention away from the liberatory projects of 1960s radicalism, but his mistrust of the welfare state was shared with a broad swath of leftists whom Foucault tried to engage in his lectures from the 1970s, when the French left was torn between those who wanted to pursue electoral legitimacy (and gained it at the start of the 1980s, with Mitterand's election), and those who insisted on political and cultural struggle apart from and against the state. One might argue that this mistrust for the state is the point around which capitalist hegemony was eventually reconstituted by the middle of the 1980s, although Ferry and Renaut and cultural conservatives on both right and left believe the responsibility lies with the individualism and consumerism fostered by the countercultures. Habermas's reasons for defending the law against the norm, therefore, are not *just* based on principle but were also part of a theoretical constellation whose elements were gradually disentangled and appropriated by different (non-Habermasian) political strategies in different countries as the twentieth century came to a close.

Habermas and Rawls, of course, do not draw their critical norms from the law but criticize law in light of an "ideal speech situation" or "original position."[18] Habermas is particularly troubled by the fact that Foucault's only alternative norms come from the body, especially (he claims in *Philosophical Discourse of Modernity*) the body in pain. Habermas argues that Foucauldian "resistance can draw its motivation, if not its justification, only from the signals of body language, from that nonverbalizable language of the body on which pain has been inflicted, which refuses to be sublated into discourse" (Habermas 1990, 285–86). If our only critical standpoint is an empirical understanding of the body, then not only will empirical needs and experts' knowledge about those needs trump the individual's right to participate in the definition of needs—individuals will also be caught in systems of power that define them as clients, permanent unequals lacking the authority to participate in processes of collective self-definition (Fraser 1989, 154; Habermas 1996, 79).

Having seen how Foucault explains the appearance of *freedom* within a governmental practice that did *not* assume it as a natural given but as an important safety valve within a security technology, we must ask whether is it only possible for citizens to exercise power over themselves and their collective form of life through law, or whether this is possible within the realm of normalizing power itself? George Canguilhem's distinction between normal, pathological, and normative states of human health points toward a moral standpoint divorced from law. In *The Normal and the Pathological* (1991), Canguilhem pointed out that what is "normal" for a given individual or population depends on the dangers and resources it confronts. Medicine does not even begin to identify structures and functions in the human organism unless a patient's negative evaluation of his or her condition spurs investigation. A pathological condition

---

18. The ideal speech situation is a regulative concept, implicit in the rules for everyday communication among equals who trust each other and share common cultural references (Habermas 1996, 322–24). This norm reflects the mutual expectations accepted as legitimate by humans who share a relatively stable, egalitarian form of life, and he believes the public can use legal institutions to limit the encroachment of economic and technical rationality on this form of life. The original position is a reflective state in which Rawlsian citizens assess the relative importance of rights, liberties, and other structural features of an ideal government while suspending knowledge of the actual social advantages and disadvantages they would experience if it were put into practice (Rawls 1971, 12, 18–22, 137–39). Both are critical standards for the evaluation of existing states.

is destructively abnormal, preventing the organism from carrying out its usual functions. But normativity implies that the organism would be able to perform adequately or create new standards of performance under new conditions; indeed, it could choose between states of "normalcy."[19] He or she can distinguish, in other words, between *qualitatively* better or worse experiences of freedom, and not simply between freedom and coercion.

For example, many patients return to normal functioning after recovering from an illness or disability, or even remain within the "average" abilities of their population. Norms are statistical creatures, and under certain circumstances disease itself can be "normal," as when the recipient of a transplant must take medicine regularly to suppress his or her immune responses. But former patients often lose their ability to confront new dangers and experiences; the same transplant patient must always have access to the medicine, must avoid certain aggravating activities, and so on.[20] "Disease is still a norm of life but it is an inferior norm in the sense that it tolerates no deviation from the conditions in which it is valid, incapable as it is of changing itself into another norm. The sick living being is normalized in well-defined conditions of existence and has lost his normative capacity, the capacity to establish other norms in other conditions" (Canguilhem 1991, 183). An overweight person may be "normal" in the sense of having a stable body weight that is often found in his or her population, but "pathological" in that there are situations he or she would not think of entering because they would be too physically taxing or too socially unpleasant. The overweight person's normalcy, like his or her pathology, has a historical dimension: our living spaces reduce the occasions on which being overweight is physically traumatic but multiply those in which it constitutes a social impairment. The loss of normativity is a moral loss, though there can be no absolute upper standard for individual human potentials.

Statistically speaking, as David Harvey notes, the life expectancy in

---

19. For Habermas, by contrast, the word "normative" implies offering a standard for judgment that can be intersubjectively validated. Habermas uses the term "cryptonormative" to describe Foucault's expectation that readers will draw normative conclusions from his empirical descriptions of systems for social control (1990, 276, 282).

20. Foucault refers to these ideas in his introduction to Canguilhem 1991 (also found in *PWR*) and in an interview regarding the logic of social security in contemporary democracies (*PWR* 373–79). Although poverty is an important indicator of vulnerability to illness, depression, accident, and crime, the gap between wealthy and poor members of a society also seems linked to higher morbidity in better-off members of the population (Wilkinson 1996). Thus social relations and even political or economic structures must be considered aspects of "normal" human biology.

poor neighborhoods of Baltimore during the 1990s was "comparable to many of the poorer countries in the world," standing at 63 years for men and 73.2 years for women (2000, 136–38; see also Sen 1992, 114). Within the United States, this is an "abnormal" state of affairs. But it becomes less so in comparison to those facing extraordinary risks of war, famine, and disease in all parts of the world. For some early demographers, "normal" referred to the "average" form of a certain trait throughout a population; for others, "normal" referred to the "most frequent" traits found in that group of people. Yet the average strength of a man, considering a sample including both office workers and bodybuilders, or the amount "most men" in that population can bench press, says nothing about what the athlete can achieve under optimal circumstances.

> Everything happens as if a society had "the mortality that suits it," the number of the dead and their distribution into different age groups expressing the importance which the society does or does not give to the protraction of life (Halbwachs 53, 94–97). In short, the techniques of collective hygiene which tend to prolong human life, or the habits of negligence which result in shortening it, depending on the value attached to life in a given society, are in the end a value judgment expressed in the abstract number which is the average human life span. The average life span is not the biologically normal, but in a sense the socially normative, life span. (Canguilhem 1991, 161)

Thus the norm is an external standard for a given organism, although that organism may have enabled researchers to discover the norm by self-professed suffering. Normativity, on the other hand, is a unique qualitative state in each organism that cannot be measured in advance because it relates that organism to unforeseen events. A person's normativity is threatened when what happens to be statistically normal in one situation or for one individual is expected to be normal in all situations and for all. A more serious danger is that what is *normal* in the sense of permitting basic functioning will be substituted for what is *normative;* that is, the conditions under which it is possible to risk and develop one's capacities. Because what is downright pathological can be statistically "normal," and because neither is defined outside a given environment (biological and social), the most important goal of medicine in Canguilhem's view is not the restoration of normalcy but of *normativity*.

Because the normative individual is capable of risking him- or herself in new situations and has reason to believe that he or she will be able to invent responses without losing prior functions, normativity is similar to the experience of power generated by the sublime. According to Kant, although the sublime is not a moral experience, it reveals our capacity to adhere to the moral law even when confronted with overwhelming natural obstacles (*CJ* 5:269). Since what is sublime is the witness's state of mind, not the phenomenon being witnessed, political phenomena may provoke a feeling of sublimity in addition to those of nature. Kant refers to war as a sublime phenomenon because it provokes a feeling of strength in fearless warriors (263). The absolute monarchs who formed early modern states governed on the basis of overwhelming splendor or demonstrations of power, such as Damiens's public execution, which could be described as sublime by a spectator who regards the sovereign's vengeance as a natural spectacle and believes him- or herself to be safe. This analogy suggests that certain conditions enable us to *see through* the monarch and assume responsibility for what is sublime in ourselves: "a superiority over nature that is the basis of a self-preservation quite different in kind from the one that can be assailed and endangered by nature outside us" (261).[21] Although security is not the same as normativity, security allows people to *exercise* normativity—through law or otherwise. When protected against the risks of nonhuman, political, or psychological nature, citizens feel equal to one another and to the tasks of national life. The question Canguilhem raises is whether security structures *do* make citizens normative and *which* subjects are able to engage in creative risk, rather than struggle to be normal.

Both mathematical and dynamical sublimes acknowledge the existence of incommensurability between measures of the human, the world, and reason itself. First, the spectator's ability to confront varying scales of natural complexity does not originate in his or her body *apart from nature or society:* diving gear is necessary to appreciate the sublimity of the deep sea, telescopes to physically bridge the gap between earthly and

---

21. Perhaps Kant's image of the thinker confronting the "starry sky" can be compared to the modern consumer facing down his or her environment's toxicity like "canaries in a coalmine." Each image, after all, emphasizes the individual body's limits in the face of differing degrees of risk and kinds of indefinite measure, although it fails to capture the qualitative uniqueness of each embodied spectator's experience of the sublime. Insofar as the canary's health is an indicator of human strengths and weaknesses, it also conveys the relational nature of embodied risk. See Gibbons on the human body as measure, both part of and distinct from nature, in Kant's "Analytic of the Sublime" (1994, 145–46).

interstellar frames of reference, vaccinations to travel in foreign lands—or rather, in many cases, to live at home. Second, the body itself incorporates many measures, not only physiological measures such as blood pressure or T-cell count, but also the administrative assessments of the individual health care a municipality can afford to provide, the just and expected cost and quality of housing for a given amount of labor, the "competitiveness" of a nation vis-à-vis its neighbors, the methods of toxic waste disposal available and appealing to those who govern a society. Third, there is no *single* body type whose experience of conflict with natural immensity and power serves as the exemplary type for the moral law: there are different thresholds at which children, the elderly, people with AIDS, and war veterans can be "disinterested" in the violence they experience and regard it as a thrilling test and proof of inner strength.

The *Critique of Pure Reason* insists that bodies cannot be known in themselves. In the sublime, we see that the body is always known in relation to *internal* measures of power as well as the scope of the natural world and society. According to Canguilhem, pathology can be judged and corrected in relation to the norm of prior (painless) living or acting, but teaches us little about what a patient or citizen might achieve if placed in new circumstances or allowed to create them. Feelings of pure aesthetic pleasure and pain, however, indicate an aspect of the finite human being that goes beyond the empirically known body and enables people to define *desires* as well as needs, to be pure "practical" reasoners as well as clients. Habermas is wrong to believe that Foucault's critical standard is the body in pain, for this body is pathological. Normativity does not reside in the relief of pain or satisfaction of needs, but in a bodily and psychological state that is sufficiently secure to pursue desires and inventive capacities.[22] The image of the human body expresses *both* a certain measure of physical safety and an inner power that can only be measured by the risk associated with "massive mountains climbing skyward, deep gorges with raging streams in them," and so on. This feeling of combined pleasure and pain informs us of *the extent to which our feeling of moral power is communicable;* that is to say, how far or how close we can come to each other without violence or degradation.

22. See "The Risks of Security," an interview from 1983, for Foucault's views on the relationship between dependency and normalizing power in the modern welfare state. "There is indeed a positive demand: a demand for a security that opens the way to richer, more numerous, more diverse, and more flexible relationships with ourselves and others, all the while assuring each of us real autonomy" (*PWR* 366).

Here it is important to recall that the "aesthetic" does not just refer to the look of things. It also indicates our ways of *seeing, feeling,* or *exerting* and *suffering power.* The aesthetic is not just a relationship between the forms of human sensibility and the forms of given natural or artistic objects; it is also a relationship between the forms of sensibility that *differentiate* human individuals from one another or bring them together in common experiences or expectations. In Kantian terms, one might say that no two individuals have exactly the same forms of outer and inner sense, space and time, or exactly the same schematic ability to bring concepts and intuitions into harmony. But only purposeless forms—such as nature, art, and perhaps historical events—are capable of revealing these subtle differences, which strategic action and communicative action have every reason to overlook because they create unmasterable complexity. It is not, then, the "normal" and isolated body of medicalized and commodifying discourses which should serve as a moral standpoint for political criticism. But neither is it the *pure universal* before which one is obliged to mistrust one's *individualizing normativity* along with one's empirical inclinations. Rather, this standpoint must be found in states of pure aesthetic pleasure in which citizens to come together *as* a public and differentiate themselves *as embodied* in the act of communicating.[23] As we have seen, "Nietzsche, Genealogy, History," *Discipline and Punish,* and *The History of Sexuality Vol. I* make it clear that the body cannot be considered as a natural phenomenon "in itself" but *only as the expression* of evaluations that represent different scales of social perception and scientific, social scientific, or cultural analysis and investment.

Foucault's concern at the time of writing *Discipline and Punish* was the migration of penal practice into the individualizing, normalizing work of case management. Although he expressed concern about the application of law according to norms drawn from the social sciences, Foucault did not believe that law should ground sovereignty as if individual bodily situa-

---

23. Levin (1989) argues for a form of bodily normativity associated with the *reversible flesh* described in Merleau-Ponty's *Visible and the Invisible,* a capacity for intersubjectivity built into the organism and developed through education. My reservation regarding this and some other Merleau-Pontyan approaches to political philosophy (such as Rogozinski 1996) is that they underestimate bodies' aggressive attachment to and desire to intensify their own individuality, with all the potential problems for consensus and power negotiation this entails. Although there are great differences between *Phenomenology of Perception* and *The Visible and the Invisible,* moreover, the idea of a single flesh presumes a norm to which all parts of that body conform, a norm whose existence I cannot conceive of outside history and struggle.

tions and histories were all the same or did not matter.²⁴ Besides, it is far too late to invest law with this power, or to invest law with this power *without first dismantling normalizing practices* in civil society. Bureaucratization did not begin in the late nineteenth century, much less with the post–World War II welfare state; it developed from a governmental technique that was adopted by European governments in the seventeenth and eighteenth centuries. "If one wants to look for a non-disciplinary form of power, or rather, to struggle against disciplines and disciplinary power, it is not towards the ancient right of sovereignty that one should turn, but towards the possibility of a *new form of right*, one which must indeed be anti-disciplinarian, but at the same time liberated from the principle of sovereignty" (P/K 108; my italics). Normativity is not directly threatened by regulatory law and its potential for bureaucratization, since these have the genuine potential to provide citizens with security against collective dangers like disease, old age, and economic instability. However, it *is* dimmed when citizens are persuaded that something incommunicable about their experience, in its freedom and individualizing sensibility, poses a risk to themselves and others. To go behind or beyond both the law *and* the norm, it is necessary to interrogate the practices through which people individualize themselves and one another, and believe themselves to be mirrored in or threatened by the state.

In *The Order of Things*, Foucault identified life, labor, and language as "quasi-transcendental" forces giving rise to the phenomenal objects of biological, political economic, and linguistic knowledge, and corresponding to fundamental powers within man as "transcendental-empirical doublet." Life, labor, and language are what needed, therefore, to be protected against corruption and risk by the security *elements* of nineteenth-century governmental practice. I would like to propose that sexuality, property, and intelligence are "quasi-transcendentals" that give a form to contemporary neoliberal practices of governance and are believed to reside within each citizen as a potential power and source of risk. These are the qualities and potentials that should make it possible for individuals to generate "human capital" and compete even in the absence of social security mechanisms, according to the neoliberal art of government.

24. Foucault distinguishes the power relations of normalization from those Hans Kelsen referred to as "grounding" norms of legal systems, though he does not explore this comparison in detail (*STP* 58). Canguilhem's concept of normativity is derived from the medical sciences, not law; moreover, it is an individualizing relationship to norms rather than a common essence underlying them.

Relations of property, intelligence, and sexuality limit the scope of communicable experience and allow these limits to be *interiorized* by the individuals they govern. On the one hand, as Arendt has indicated, property is a way of relating to others and to the physical environment that confirms one's personhood and right to enter a public or political space; modern Western elites have often interpreted propertylessness, collective ownership, or nomadic tendencies as signs of imperfect humanity, individuality, or agency (Arendt 1958, 61; Stoler 1995, 128; Daniel 2000, 69, 76). On the other hand, since most individuals have no "property" other than their ability to labor and what space or possessions they have by virtue of continued employment, property can also be understood in the Marxist sense as a series of complex power relationships between bodies who command and those who must exert themselves in order to possess the external trappings of individuality. Techniques for ordering space and for subordinating individuals on the basis of their relationship to "property" (for example, through cadastral maps) were developed in early modern Europe, perfected in European overseas colonies, and reinforced against the newly industrialized working class throughout the course of the nineteenth century (Scott 1998, 36–44). Strictly spatial or material meanings of property overlap with the meanings and opportunities for mobility and self-direction available in a given social milieu.

Social equality in liberal societies, such as it exists, is not entirely a function of property but of the *opportunity* to develop property relations in which one's ability to resist changing conditions and to engage in active "pursuit of happiness" is sustained. Actual property, equality, or happiness is something to be *earned* through work and risk, and frequently it is viewed as a function not of power relations but of innate "intelligence" or "talent."[25] Yet intelligence and talent can only be demonstrated in the context of power relations. Compulsory education was an important part, as Foucault himself explains, of strategies for inculcating obedience, attention to detail, and a sense of responsibility for deviations from the social and economic ideal in children of the poor. It was an effective way to give state and social workers influence within the private sphere. Their goal was to improve public health and to ensure that the state would have a ready supply of students whose labor and attentiveness could be turned into obedience and inventiveness by bureaucratic structures.[26]

25. This view can be traced to Adam Smith and Locke, for whom the right to a portion of social goods resulted from the opportunity and willingness to transform those goods through labor.

26. See Brunschwig 1974 and Delanda 1997, 227–47.

Characterized as "genius" by Kant, intelligence is a touchstone of communicability and creativity, involving an ability to imagine and anticipate the reactions of others. Significantly, Kant's disdain for Africans and some Native Americans is couched in terms of their "apathy" and lack of "talent" (1965, 110–11). Nevertheless, tests and evaluations of obedience, creativity, problem-solving skills, and so on, imposed by teachers penalize those who fail to communicate, inscribing perceptions of ethnic, class, or gender difference associated with students' "appearance" or "performance" as a problem against which they must permanently struggle. While appearing to praise imagination, they also instill the conviction that *what there exists* to be communicated and imagined is far more common and unified than the experience of recalcitrant students would suggest. More than limiting imagination through exalting understanding (as Horkheimer and Adorno, for example, might argue), such systems of education make the student's body or physical presence a sign of gaps, potentials, or fundamental incapacities.

The intelligence tests developed by Alfred Binet and R. M. Yerkes in the French and American contexts respectively defy Kant's association of intelligence with originality in judgment. Rather, they associate it with the ability to anticipate similar interpretations of common phenomena and to solve abstract problems in ways that reflect the habits and expectations of other middle-class students (Gould 1981, esp. chap. 5; Young 1990, 206–10). But these psychologists do follow the German idealist conception of genius in treating intelligence as if it were a single *natural* power within the human mind to be developed through culture. Such an assumption verges on *Schwärmerei* in its speculation about fundamental powers. It also inscribes creative and self-legislative activities that enable humans to transcend nature in the natural-scientific classification of humans as a species with sexual and racial divisions (see Mills 1997, 59, and Eze 1995, 227).

If we ordinarily judge intelligence according to a multiplicity of technical and abstract skills, empathetic abilities, willingness to persist in tasks for their own sake, and capacity to imagine new solutions to power situations, how can we understand it as a single natural "thing," any more than the property to which it supposedly gives access? Elite standards of intelligence often overlook the directive aspects of women's labor and give those who can engage in abstraction or memorization opportunities to direct or design others' activity, even when these abilities are not necessarily relevant to the activities being directed (Young 1990, 202–5). Habermas conceives of these pragmatic skills as "non-communicative" knowledge of the lifeworld; Foucault referred to what ordinary people knew about their lives apart from or in resistance to power as "subjugated knowledge" (Habermas 1987, 135; *P/K* 81–83; *SMBD* 6–8). According to Marx, the division between manual and mental labor, which has become one of manual and managerial or design labor within the modern corporation, is itself a form of property; the manager disposes of the labor power of others just as he or she disposes of the products resulting from that labor (Marx and Engels 1986, 53).

In *Discipline and Punish*, Foucault argued that the Enlightenment transition from punishment by *supplice* to correction and discipline was a more efficient way of implicating nonbourgeois bodies in social projects, making them share responsibility with the wealthy for reducing disorder and producing without complaint to meet expanding social needs. What I have referred to as the collective "risks of freedom" were thereby made a special problem for the poor to solve. So too, in the American context free labor was considered to be a more efficient structure for the use of impoverished black bodies and their creative, inventive energies than the direct domination of slavery. Just as the disciplined individual is responsible for his or her deviation from the norm, so too the freedman and laborer are responsible for their poverty, their "defective" relation to property by contrast to the bourgeois owner or sharecropping landlord. As Hartman observes,

> Indebtedness was central to the creation of a memory of the past in which white benefactors, courageous soldiers, and virtuous mothers sacrificed themselves for the enslaved. . . . The emancipated were introduced to the circuits of exchange through the figurative deployment of debt, which obliged them to both enter coercive contractual relations and faithfully remunerate the treasure expended on their behalf. Furthermore, debt literally sanctioned bondage and propelled the freed toward indentured servitude by the selling off of future labor. (1997, 131)

After emancipation, African-Americans were encouraged to express recognition for the "responsibilities" of freedom by laws, extralegal violence, and social service discourses aimed at producing a docile, geographically rooted, and "respectable," that is, sexually restrained and hard-working, class of agricultural laborers.[27] Rather than grant social equality, therefore, emancipation enabled power relations to seize black bodies in a new way that produced blackness as an "unworthy" or "incompetent" relationship to property and to penalize African-Americans who were economically and socially assertive, often through lynching or the destruction of goods. Sexuality, lack of education, and poverty were the terms in which their bodies were positioned as the raw material for solving the economic contradictions of post–Civil War society. These are sane bodies made to appear insane in light of relations of power that do not stand over them but seem to be hidden *within* them, and whose "enunciative poverty" they experienced as personal debt.

Foucault contends that the "implantation" of sexuality as a capacity and a danger at the heart of each individual built on older practices of kinship as well as religious discourses concerning sin, grace, and the flesh. In the context of the early modern state, sexuality proved to be a point at which charitable organizations, educators, public medicine, and legal institutions could combine and justify their efforts to improve the health, productive capacity, population size, and political malleability of their populations (Stoler 1995, 190). As an object of science, sexuality enables individuals to become subjects and objects of discourse in ways that reinforce their economic, emotional, and political obligations; the association of sexualities with greater or lesser deviation from the model of animal coupling gives them a basis in nature. Sexuality also integrates individuals into economic and social roles and differentiates among them by serving as the point of identification and anxiety addressed by advertising and consumption (Salecl 2004, esp. chap. 3). Like property and intelligence, finally, sexuality is both that which individuals "have" by virtue of their embodiment, and that which they never have "sufficiently" or "properly." In this way, sexuality always presents an occasion for individual or collective danger, at the same time that it enables individuals to experience a kind of equality to the natural forces around them. It refers to the margin of invention which the individual must assume to preserve the stability of a power relationship in which not only intimacy or sexual pleasure are at stake, but also the freedom from violence, want, or shame.

Life, labor, and language are empirical conditions for a certain experi-

---

27. See Holland 2001, 63–65, for a discussion of similar techniques, including appeals to religious interiority and public shame, for socializing and disciplining the white body to "feel American" during the period of the American Revolution.

ence of "man" as subject of desire. But researchers and the lay public are never confronted with anything but individual speech acts, living beings, workers, or exchanges. Life, labor, and language only *become* empirical conditions, therefore, when treated as ideal objects in the context of an expert discourse that feeds back into the direct government of individuals—and often overrides what those individuals "know" about their own needs, workplaces, or mutual understandings. Life, labor, and language are objects of the sciences that formed the nineteenth-century epistemological trihedron, whose relation to man as "figure of finitude" remained to be worked out in the human sciences. But they also have psychological correlates, by which individuals orient themselves with respect to the expert discourses that may justify their conditions of employment, medical care, and participation in public or private discourse. These concepts were once linked by colonial domination and now have global reach by virtue of migration and the internationalization of market economies.

Conflating morality with post-Kantian philosophical anthropology means that one must endlessly weigh one's preferences and experiences against the "unthought" to which an expert always has better access than oneself—and strive to bring the conflicts *between* these discourses into line before one can claim to be fully human in a normative sense. In modern disciplinary societies, the responsibility for any perceived dissonance among discourses and practices is given over to the ignorance, moral weakness, biological defect, or perverse cussedness of individuals—specifically those individuals who are regarded as being inferior in class or race (*DP* 193). Sexuality, property, and intelligence are simply names for our abilities to negotiate these conflicts in daily life. No one knows exactly how sexy they are, how much they are worth financially, or how smart they are. These are abilities that always prove themselves in situation-specific ways. None of them refers to a single "object" but to a body's degree of latitude in creative response vis-à-vis kinship networks, opportunities for intimacy, access to education and employment, and responsibility and rights to medical care and social insurance. Nor could they possibly lead differently positioned bodies to actualize their possibilities in identical ways, for the sexuality of racialized bodies, male working-class bodies, mothers, executives, and poor children is inevitably different and *intended* to be different. Like property and education, sexuality is as much a system for differentiating and evaluating bodies as it is for recognizing their existence.

However, one cannot simply dispense with one's investment in these abilities. They signify an *individuating*, evaluative desire, one that persists even in the absence of recognizable grounds or legitimate expectations. In American culture, being an "individual" (exceeding or deviating from the human norm in some positive manner, exhibiting special capacities or developing unusual skills) is valorized as a necessary condition for escaping from authoritarian power relations and assuming a dominant role in the governance of oneself and others (Sennett and Cobb 1972). The Enlightenment ideal of personal autonomy and universal human dignity can be found in democratic justifications for the state as well as in demands for equality of economic opportunity unfettered by the formal restrictions of caste and class. Yet the systems for education, health, and economic activity legitimated by the European Enlightenment simply gave rise to new ideals and hierarchies of merit that justified superior and subordinate roles in all areas of social life.

Every individual who hopes to exercise the full Enlightenment ideal of self-governance as an equal must prove him- or herself not just equal to but "above" his peers, thereby tying his or her deepest aspirations to "simultaneous individualization and totalization of modern power structures" (Sennett and Cobb 1972, 64–65; PWR 336). Persistent hierarchies of intelligence, discipline, bodily ability and economic prowess confirmed elite beliefs that it is useless to rethink these capacities or reorganize the relations of power in which they are exercised. Those who were found to be merely "average" or "normal" on the community bell curve could not complain about their eventual station in life or the way stations are conceived and exercised, because it was their own fault for not doing better. They were both more willing to take blame and to place blame—and to assume scarcity as a horizon for making sense of their own life choices. They "resembled," to borrow terminology from our previous section, without inventing new images or categories of resemblance in relation to which they could be considered "original."

Above all, striving to be an individual within systems of examination and confession that operate by treating all humans as instances of a "norm" generates a tremendous amount of labor and activity from every member of the community while leaving roles and rewards, especially emotional rewards, relatively limited. The social value of the "individual," Richard Sennett and Jonathan Cobb might say, is to be "normative," to be recognized as a capable judge of what is and can be "normal" for oneself and others. The American valorization of professional roles reflects a vague popular awareness that professionals not only have some choice over hours and projects, but more important, that they have some say, among peers, as to the norms by which their work will be judged (Young 1990, 213). Ultimately, "normativity" in American culture means having demonstrated the intelligence, health, and economic power to have a say in how intelligence, health, and economic power themselves shall be defined and what difference they shall make in people's lives. In relation to this ideal, almost all Americans and the vast majority of the world's peoples are "pathological."

Law fails to uphold the moral values of dignity, equality, and freedom if it does not emerge from, and reinforce, a situation in which citizens are free to risk, alter, and expand their capacities rather than struggling to meet a norm. This situation, which Canguilhem calls normativity, is represented in Kant's aesthetic judgment concerning the sublime. Intelligence, property, and sexuality are ways in which modern individuals relate to themselves as potentially normative beings, although they often struggle to perform these qualities or abilities in merely "normal" ways as determined by experts. Although like law itself, intelligence, property, and sexuality are composed of multiple bodily and discursive practices and are not "things-in-themselves," these axes of interaction between individuals and the state cannot be dismantled without substituting other forms for the experience and exercise of normativity.

Heterochronic Interval

As I mentioned earlier, Kant warns readers against the temptation to engage in a genealogical analysis of the state's real origins (see *MM* 6:339–40). First, this is because he believes the resulting conflict may destroy the efficacy of law, in whose priority he firmly believes, in addition to security, whose benefits he never disparages. Second, the results of historical inquiry into the state's origins could not be made public to, or accepted by, all parties in the constitution. Those who believed one version rather than another would necessarily be acting unlawfully, as a mob (*sich dazu rottirte*) or as a conspiracy (*MM* 6:340; *KPW* 126). Third, even were an existing state to be proven the offspring of war, nations are responsible for finding common ground that can lead to peace and legal resolution of hostilities (*MM* 6:346–47). The idea of an original contract, therefore, is only a regulative ideal; otherwise "we would first have to prove from history that some nation, whose rights and obligations have been passed down to us, did in fact perform such an act . . . before we could regard ourselves as bound by a pre-existing civil constitution" ("Theory and Practice," in *KPW* 79–82).

The excruciating violence of civil wars in the past century suggests that Kant's desire to treat the original contract as a regulative and not a constitutive Idea is wise. It prevents the state from going beyond the bounds of either security or law and attempting to determine the essence of a population as "thing-in-itself." In this way the population remains free for progressive redefinition as a "public" over the course of Enlightenment (*KPW* 57–58). Refusing to enquire into origins, moreover, prevents those who *have* been oppressed from chaining themselves to that history and exhausting their resources in expressions of *ressentiment* (*pace* Brown 1995, 71–75). But this refusal also tacitly holds those who claim to be oppressed responsible for overcoming the tone-deafness, the resistance to communication, of those who have conquered them or constructed their public space and state around elite codes. In a situation of social conflict, Kant's approach hands the state and public space over to those who *need not* invoke the rights of race, sex, or history because they already have the power.

One can read *Society Must Be Defended*, Foucault's lecture course from the Collège de France in 1975–76, as a direct response to this challenge from Kant. The first and most important aim of this course is to describe the historical moment at which a common or unquestioned "experience" was explicitly broken into a field of conflicting claims regarding the historical origins and meaning of that experience. Foucault's examples are drawn from the period immediately preceding the English Revolution, when radical Protestants appealed to Anglo-Saxon traditions and laws and opposed them to the traditions and laws of the conquering Normans, from whom the king claimed descent (*SMBD* 49–51, 99–110). Likewise, French nobles who resisted Louis XIV's absolutist unification of sovereignty and the king's own historians drew on conflicting views on whether the indigenous Celts or the invading Franks were the true heirs of Roman imperial civilization (115–37). As more events, documents, and customs were invoked as evidence for the legitimacy or illegitimacy of the ruling or ascendent power, a whole field of material was "historicized" (167–71). The creation of genealogical depth in this field corresponded, in some respects, to the creation of historical depth in *The Order of Things*, following the discovery of internal rates of change in Indo-European languages, natural structure, and trade and production.

The second goal of Foucault's lecture course was to show how the constitution of subnational positions with contested historical antecedents (in Britain and France) prefigured the "biologization" of the political field in nineteenth- and twentieth-century racisms. This is one of the crucial moments at which the human body came to represent not a lineage, but a kind of *historical experience* that might be ascendent or threatened, therefore tinged with the affects of hope or fear.[28] The existence of "dangerous individuals" indicated to the law-abiding citizen that he or she was caught in a heterotopic imaginary, torn between hope and fear. The illegalisms, sexual status, health, or economic unproductivity of dangerous individuals suggested that a particular imagined community had internal weaknesses that actually increased the strength of potential adversaries (*PWR* 178). During the Revolution, rumors spread that escaped asylum inmates were responsible for terrifying acts; during the mid-eighteenth century, and especially after the cholera epidemic of 1832, the poor came to be considered an urban health hazard (*PWR* 152–54). The most important example of this dynamic is probably the European fear of being "contaminated" by contact with colonized peoples or of "degenerating" through mismanaged sexuality.

Foucault's third goal was to demonstrate that many adversarial interpretations of the past or present participate unwittingly in the same "genealogizing" process as those which produced the devastating state racisms of the Nazi and colonial powers. The first volume of *The History of Sexuality* argued that the bourgeoisie did not regard their own sexuality as the same kind of object or worthy of the same kind of concern as they regarded the biological potential of the working class (or, for that matter, the colonized). For the bourgeoisie, sexuality was a focal point for practices to strengthen the next generation's economic and educational capacities (*HS1* 122–27). In the case of the working class, sexuality was a focal point for practices aimed at establishing conformity to a certain model of domestic order (121–22). *Society Must be Defended* argues more explicitly that the division of the historical field into dominant and subordinate economic class experiences follows the same logic as its division into racial or cultural experiences (79–83). States whose legitimacy rests on protect-

---

28. In response to an audience question, Foucault explains that he is not trying to "trace the history of what it might have meant, in the West, to have an awareness of belonging to a race" or of the rituals used to "exclude, disqualify, or physically destroy a race" (*SMBD* 87–88). However, in later interviews he said that all of his work had been an effort to understand "the genesis . . . of a system of thought as the matter of possible experiences" (1984, 336). The discourse of race war, I would argue, produced such an experience (despite its nonsystematic character).

ing the security of a specific class or fostering its health and productivity are just as vulnerable to racist *Schwärmerei* as states whose legitimacy rests on the security and vitality of a specific race or lineage.

In both of these lecture courses, one can see that Foucault tacitly reads Marx as a fellow genealogist, one who creates a certain kind of historical knowledge by dividing the field of an experience whose supposedly universal communicability is a ground for the modern state (see *FL* 140). In Marx's case, the class division is revealed within the experience of the workplace and commodity exchange, in which workers become signs of lesser value than the products they produce. In Foucault's case, the biopolitical division is revealed within the practices for dividing sane from insane, healthy from sick, indigent from productive, and normal from perverse, degenerate, or racially inferior. Each of these divisions could delegitimize the state that maintains them, but it *will not* unless a different form of knowledge and different practices of power emerge to take their place.

Marx refers to this form of knowledge and power as "class consciousness," but what Foucault describes is something more like a *sensibility* or aesthetic apprehension making some topics of communication and some potential interlocutors seem more reliable than others. Until a new form of sensibility is created, citizens will live out the coexistence between imaginative "forms of life" associated with different "races" at the level of bodily anxiety and/or empowerment. The Third Estate eventually managed to occupy an antagonistic "national" position within the race war of *ancien régime* France while creating a new meaning *for* the state as universal (*SMBD* 220–24, 236).[29] This is the position leftists have been trying to capture ever since. However, Foucault suggests that in order to do avoid waging a battle against owners as representatives of an embodied "experience" on behalf of an equally embodied working-class or plebeian "experience," (a strategy that would easily lend itself to state racism), the left must alter the *way* it orients itself toward social-scientific knowledge (*SMBD* 261–63; *PWR* 296–97). In his 1979–80 course, Foucault warned the French left that it could only win the legitimacy to govern if it developed its own governmental rationality—one that was nonliberal (*NB* 93–95). Neither the Socialists, who took charge of French politics just a few

---

29. *Society Must Be Defended* situates the question of the present with respect to the conflict in racialized historical perspectives at the time of the French Revolution: which race or class, "in the present," contains the forces that make the "nation" and therefore deserves recognition through the state's universality? (*SMBD* 236–37).

years before Foucault's death and ended up governing as liberals, nor Foucault himself, managed to solve this dilemma.

Foucault proposes to read Kantian modernity as an *attitude* toward historical time rather than a determinate period. Although he believes Enlightenment will affect all humankind, Kant is unclear whether this will be true in the sense of spreading globally or in the sense of altering the meaning of "humanity" for those whom it touches (*EST* 306). Kant's reflection on the increasing tendency of individuals to make up their own minds on scholarly matters suggests that this attitude is "aesthetic" inasmuch as it alters the contours of communicability and individuality at a specific moment in time (*KPW* 58). Enlightenment will change the *being* of the beings who deliberate over the difference in the present; it will change the form of their individuality as well as their collectivity. A practice that has only a geographical spread, such as freedom of conscience, falls short of Kant's goal if it fails to change the imaginative and practical meaning of that geography. Moreover, Kant anticipates maturity only along a *via negativa,* as the search for a "way out," rather than the completion of a collective or individual process whose *telos* has already been established.[30] One generation cannot bind the reason of the next because no particular form can be envisioned for the "adult" form of humanity.

According to Foucault, Kant "is not seeking to understand the present on the basis of a totality or of a future achievement. He is looking for a difference: What difference does today introduce with respect to yesterday?" (*EST* 305).[31] This difference is something for which by definition we have no existing concepts and cannot bring under a determining judgment of the schematism. In this sense, modernity involves a particular exercise of imagination, one that does not necessarily give rise to a particular goal or fantasy, but which reveals the kinds of perceptions and actions that are possible or impossible *at a certain moment, in media res.* "For the attitude of modernity, the high value of the present is indissociable from a desperate eagerness to imagine it, to imagine it otherwise than it is, and to transform it not by destroying it but by grasping it in what it is" (311). Habermas is right, therefore, to suggest that one of the most important functions of the public sphere is the identification of problems

---

30. For an extended study of maturity and enlightenment in Kant and Foucault, see Owen 1994.

31. His later interviews described the ability to see multiple lines of fracture rather than a single stabilizing contradiction in present-day social relations as "problematization" and as "thought" (see *FL* 296 and Foucault 1984, 334–35, 388–90).

for collective consideration (1996, 359–61). But Foucault warns about the temptation to interpret the experience of criticism in a way that casts the dangerous individual, fanatic, or *Schwärmer* as its inevitable shadow. "One must refuse everything that might present itself in the form of a simplistic and authoritarian alternative: you either accept the Enlightenment and remain within the tradition of its rationalism . . . or else you criticize the Enlightenment and then try to escape from its principles of rationality" (*EST* 312–13). Like humanism, and like the state, the Enlightenment involves a number of rationalities and practices and has been claimed by warring parties.

Subjects of Enlightenment must take the risk of being obedient or disobedient about the wrong things, or failing to communicate the distinction to which they have committed themselves in the manifold of the present. However, the witness must also endure the Enlightenment, biopower, or neoliberalism a little to know what in them is worth rejecting—by contrast to something *better* and worthy of affirmation. This is the positive side of Kant's injunction to private obedience. Unlike Kant, Foucault does not believe all possible aspects of experience must be conceived in a lawlike way. But it is true that disobedience might *destroy the present* before what is "eternal" in it—that is, solid or lasting enough to be imagined and communicatively heroized—could be grasped as such. Although perception of any event, including the new style of conduct called Enlightenment, will provoke aesthetic disagreement among observers, this disagreement cannot be confused with the opposition between obedient and disobedient, or critic and fanatic. It takes *time* for an event to become perceptible to many people, much less to gain one or more debatable meanings. It takes time, likewise, for the aesthetic/historical observer to decide whether this difference is worth heroizing or allowing to affect one's resemblances. The luxury of *having time to judge* is an example of the kind of security that makes us aware of our normativity.

On the other hand, Kantian Enlightenment is a process of communication among finite rational beings. The indefinite unity of the public is structured by lacunae and choices of topic or scale. Even when this limitation does not take the form of conflict or dissent, it is inevitably present. Just as no single concept or appearance can present a complete intuition or object as it would be "in itself," so too no public can present a coherent opinion without *some* limitation and exclusion; human attention can only be split in so many directions at once. But just because a discursive discussion cannot be "intuitive," or a material "communication situation" remain "ideal," that does not mean that relations of individuation and communication cannot be better balanced. It is possible to "include" people in a public in ways that deny them the power to alter the terms of inclusion, but there are also forms of direct action that can counter such exclusions, *showing* rather than describing how their world works. Likewise, there are forms of peace that allow poverty, lack of health care, and overwork to produce many of the same effects as war.

The temptation to ignore divisions in the present or divisions in the social and historical field by forbidding genealogical inquiry is powerful. For a common example, consider whites in the contemporary United States who believe the effects of racial segregation or of conflict with the Soviets in the postcolonial world "finished" because they no longer feel invested in these struggles. But so too is the temptation to resolve these divisions by gathering the present into a single "unbreakable" event on whose contours everyone can agree, and to produce such an event if none present themselves. The United States and European liberals quickly claimed the fall of the Soviet regimes and the "Iron Curtain" in 1989 as the triumph of a single economic system. Although people rejoiced internationally at the end of a tyrannical political state, like the jubilant spectators of Kant's French Revolution, nonetheless the meaning and the historical flows in which Arabs, Central Americans, and West Germans situated this event varied (see, for example, Mernissi 2002, 3–5). The fall of the Berlin Wall should have opened room for democratic socialist movements in Eastern Europe and blossoming of axes for political contestation and organization.

The U.S. government's compulsive efforts to give a single meaning to the horrific attack on New York City in September 2001 and to organize public sentiment against "fanatics" who are also racially differentiated from the majority betrays a similar impatience. Fixing the scope and meaning of an event too quickly works against the discovery of new affects and resemblances in the geopolitical arena. Compulsive repetition of the phrase "everything changed" shows, among other things, how badly Americans *wanted* an event that would organize the materiality of bodies and discourse into a coherent whole. The disappearance of the Soviet Union, after all, did not end nuclear proliferation or reduce the risks ordinary individuals entertained on a daily basis because of capitalism. To the contrary, it seems to have left them all the more exposed, as the "triumph" of capitalism justified withdrawal of the welfare state.

Of course, this is neither to downplay the immense suffering of the families affected by 9/11 nor to say that in some way Americans were "asking for" a terrorist assault. Nor does it excuse the apocalyptic use of religion to conceal the multiplicity of cultural trajectories and strategic interests within diasporic Islam. It is to say that the way in which this trauma has been identified as foundational for an international public space indicates a hidden desire on the part of Americans *and* their adversaries for a totalizing event that could unify nations without forcing them to confront the history of divisions according to class, gender, and slavery or empire. It is difficult to heroize a "present" whose difference is more complex than a single act of past or future violence. Although citizens of the world did gather in sympathetic solidarity with the victims of New York, many people have used the attack on the Twin Towers to justify forgetfulness about history, without being willing to recognize their cosmopolitan resemblance to economically precarious and conflict-ridden citizens in poorer nations. Thereby, of course, they try to prevent anything from *really* changing at all.

Crisis in Liberalism

Foucault suggests that economic crises, whether in capitalist or socialist states, may indicate a crisis in *liberalism* as a regime of power/knowledge rather than a crisis in the market, the paradigmatic locus of liberal knowledge and power (*NB* 70–71). More important, he adds that since the market is a structural element in a certain art of government, aspects of the market can survive potentially radical changes in composition if the art of government or its epistemology changes (170). Neoliberalism marks a change in the way governments understand and try to create markets, rather than a "late" outgrowth of contradictions in a fundamentally unified and historically unique phenomenon called capitalism.

The political legitimacy of socialist states depended just as much on economic knowledge and management as that of liberal states—although the kind of economic knowledge Marxists value is different from the kind of knowledge generated by classical or neoclassical economics. In fact, since their official philosophies hold that the state is an expression of economic forces rather than a framework for guiding and monitoring them, they could be said to rely more heavily on the "security" model of governmentality than liberal (i.e., capitalist) states do. But Foucault did not theorize the process by which neoliberalism took hold in the North Atlantic, partly because he did not live to see the supposed "confirmation" of its truth, the fall of the Berlin Wall and the end of bureaucratic socialism in Russia and Eastern Europe.

During the 1980s and 1990s, as I have mentioned, Britain and the United States began to withdraw social protections for labor, to privatize and deregulate industries or public utilities, and to abandon the goal of full employment. One possible reason for the shift away from Keynesian liberalism is the contradiction between the ostensibly egalitarian *juridical* basis for the state's power to protect its population and the actual *security* basis for this power. Activists expected that the law, as the ostensible source of state legitimacy, would develop to protect all citizens equally, although the state's institutions were oriented toward enhancing the biological security of the majority and the elite against those elements of the population felt to be threatening or "problematic." When the contradiction became apparent, the American state preserved its legitimacy by reasserting its actual commitment to security, largely through dramatic acts of opposition to communism and terrorism in Latin America and the Arab world during the 1980s, followed by the domestic "war on drugs."

But other Foucauldian explanations can be given. For example, one might argue that the postwar welfare state undertook a self-critical limitation with respect to the cultural domain similar to that which it undertook where the economy was concerned in seventeenth-century Europe—and met resistance when it did so. Like Kantian reason, liberal government is legitimated by its ability to remain within boundaries—those that it sets or those set for it by political economy as a specific field of knowledge (*NB* 21–22). It should not "govern too much," but leave a certain zone of freedom alone as if it were a "thing-in-itself." Foucault notes that liberals need not appeal to *juridical/moral* principles such as the natural right to collective or individual freedom; one could just as easily criticize governments who overgoverned and thereby undermined their own goals on *pragmatic* grounds (45). This involves the state in a delicate balancing act: on the one hand, it must shape the freedom of economic and social actors, but on the other it must be sufficiently restrained that individuals believe they are expressing their own normativity in capitalist exchanges or traditional social activities.

At the end of the 1960s, perhaps under the influence of social science and psychology, the welfare state began withdrawing from the cultural domain just as earlier liberal governments had withdrawn from the economic sphere under the influence of political economy. This withdrawal gave citizens freedom as cultural and sexual actors, not merely economic actors. The decriminalization of birth control and abortion played a crucial role in public debate over the propriety of protecting cultural or subcultural choice in the same way as economic choice. Such judicial decisions improved opportunities for women and sexual minorities to live as full citizens and even changed the meaning of "male" citizenship. Under the circumstances, however, a significant number of Americans felt this to be *interference* with traditional ways of life rather than an expression of governmental restraint with respect to secular or nontraditional lives.

Finally, one might argue that neoliberalism represents an attempt to end state racism against particular groups and individuals, at the cost of dismantling the collective forms of security that were organized by state racism (even when they were essential to the survival of racialized groups). As I have mentioned, liberalism actively encourages risk-taking and tries to protect the citizen against bad outcomes, but it also uses the possibility of disaster to motivate people. All Western entrepreneurial activity through the nineteenth and twentieth centuries took place against the backdrop of a potentially apocalyptic threat from subject races or international communism. During the 1960s, publics began to demand that the welfare state purge at least some aspects of state racism, or reorient citizenship away from risk management without increasing any group's vulnerability to disease, unemployment, or environmental threats. Rather than judge the *economy* on how well it lived up to an impartial, democratic standard of risk management, however, neoliberals shocked the left by reorienting all government activity around economic growth and competition, whose risks were disregarded as "natural" rather than "intentional." Today, state racism has not disappeared, but it is considered a "side effect" of the demand that individuals sustain a constant feeling of danger in the name of competition and regard themselves only secondarily as members of a group.

In Germany, economic growth was prioritized over other indicators of popular security after World War II demolished the state's pretensions to democratic legitimacy. "The Bi-zone" of joint German and American control served as an experimental matrix for abandoning price and wage restrictions in order to demonstrate that the successor state had none of the Nazis' totalitarian impulses (but an equal ability to deliver economic success). The ideas of the Ordoliberals, as the group of German theorists behind this experiment were known, were never fully applied in West Germany, where the constitution eventually required input from consumer and labor groups on all important economic decisions (see Schwengel, in Peters 2001, 87; also Yergin and Stanislaw 2002, 19). But thanks to the Austrian émigré Friedrich von Hayek, whom Keynes granted a position at the London School of Economics, these ideas survived in Britain and the United States (Yergin and Stanislaw 2002, 123–31). Foucault referred to their descendents at the University of Chicago economics department as "anarcho-capitalists" (NB 150).

The architects of neoliberalism were joined by a philosophical conviction that the welfare state had pernicious social as well as economic effects. The Ordoliberals proposed that society should be arranged so that every individual or community could be protected against risks—not simply to continue as before, but so that they could enter into new and profitable competitive relations, or "capitalize" as effectively as possible (NB 149). Creating partnerships between government and business, facilitating investment whose fruits would remain in the private sector, turning the provision of basic services like utilities and education over to new firms, would increase efficiency and enterprise while meeting citizens' expectations of a good life; the goal was to give each individual "a sort of economic space at the interior of which they can assume and confront risks" (149–50). The versions of neoliberalism applied in Britain and the United States involved privatization of many state economic functions. These included paying private companies or religious organizations to provide job training and case management for the poor and investing public pensions in private businesses and stocks. In the United States, as it turned out, security and information firms stood to gain a great deal from such partnerships (such as Wackenhut in prisons, or H. Ross Perot's EDS in welfare case management).

These practices were complemented by efforts to make individuals feel responsible for their own economic and medical fates or dependent on family members, rather than on a government presented "phobically" as an alien force, not a public trust (Peters 2001, 91; Handler and Hasenfeld 216–17). In Foucault's words, privatization conveys that "it's up to the individual [to protect himself] by the ensemble of reserves at his disposal, either simply in his status as individual or by mutual relations" (NB 151). Thus Patrick O'Malley contrasts "discipline" with "actuarial" technologies or "prudential" mechanisms, and shows how the public has been made responsible for assuming "prudence" (1996, 197). Ideally, prudentialism "does not involve merely a privatization of risk management . . . subjects are recast as rational, responsible, knowledgeable and calculative, in control of the key aspects of their lives" (203). Risk, moreover, is recast as "a source or condition of opportunity, an avenue for enterprise and the creation of wealth, and thus an unavoidable and invaluable part of a progressive environment" (204).

Prudentialism appeals to the citizen's desire to see his or her normativity expressed in relations with government and in everyday life. But it also exaggerates the individual's ability to bear risks affecting the environment or economy as a whole. Problems of macroeconomics are repackaged in terms of "self-esteem" and "skills training" for the unemployed: "Personal fulfillment became a social obligation" (Cruikshank 1996, 233). Paolo Virno describes the traits of the new model human as "habitual mobility, the ability to keep pace with extremely rapid conversions, adaptability in every enterprise, flexibility in moving from one group of rules to another, aptitude for both banal and omnilateral linguistic navigation, command of the flow of information, and the ability to navigate among limited possible alternatives" (Virno and Hardt 1996, 13).[32] His or her psychological traits and ability to form relationships (attributed to intelligence, property, and sexual drive and attractiveness) are assessed as so many forms of "human capital" to be developed by parents, educators, and managers. The real source of risk is lack of these traits, and not the social structure in which they are developed or used.

The function of the economy in this ideal social order is not to "exchange commodities" but to maintain "mechanisms of competition." Where discipline tried to break down the individual's space and time into ever smaller segments, this art of government breaks down the social life of the population into as many surfaces for potential competition as possible. Foucault compares the "society of businesses" proposed by the Ordoliberals and the American anarcho-capitalists to the "supermarket society" of classical liberalism. "The *homo economicus* that one wants to reconstitute" in neoliberalism "is not the man of exchange, nor the consumer, it's the man of enterprise and production" (*NB* 152). This new *homo economicus* is paradigmatically an entrepreneur and investor, capable of taking advantage of spatial and temporal differences in production, rather than an owner or worker whose interests are tied to a specific place

---

32. See also the extensive discussion of the moral dilemmas posed for members of the American workforce in Richard Sennett's *The Corrosion of Character* (1998). Skills enabling workers to move flexibly between different companies and between work and family environments, Sennett finds, tend to generate detachment from particular communities of involvement and even make commitment appear a sign of undesirable rigidity or personal weakness.

and extended over a long period of time. Exemplified in American popular culture by software magnate Bill Gates, the entrepreneur takes responsibility *as an individual* for managing the temporal and geographical uncertainties that Keynesian governments had managed for the whole of their populations. To be able to risk and yet avoid disaster as an individual in a rapidly changing financial, technological, and ecological environment is an exercise of normativity that arouses immediate admiration in most Americans.

The reason why "bodies" came to play such a central role in left-wing political theory during the late 1980s and early 1990s was that the liberal security state's configuration of discourses and practices began to *individuate* rather than *group* bodies in relation to collective risks. Neoliberalism had massive effects on citizen bodies because it withdrew resources from the enforcement of worker safety and environmental legislation, encouraged sexual expression and fitness, attempted to introduce economic efficiency into the affective care and education of children, and worked to make health care a profitable enterprise. One could therefore say that neoliberalism represents the "de-population" (though not the disembodiment) of liberalism. The privatization or individualization of risk was a change in governmental technique, implemented by cutting back on many of the social insurance programs and legal protection programs of the welfare state (O'Malley 1996, 199–204). It was designed to extract a little more profit and self-care from citizens' embodied subjectivity, and to reduce the state's obligations to mediate between the rich and poor. But it did so by moralizing the act of work, by valorizing entrepreneurial risk-taking when employment was lacking, and by evaluating communities and affinities on the basis of how well they promoted such activity (Handler and Hasenfeld 1991, 208).

This development explains how anarcho-capitalism can enter into a fragile alliance with traditionalists who criticize *consumerism* for its individualistic and potentially anticommunitarian tendencies. Neoliberal *homo economicus* is not morally admirable because he or she consumes properly, but because he or she is a unique and profitable firm. Anarcho-capitalism is also compatible with a renewed commitment to the juridical model of government for two reasons. First, as Foucault points out, firms require legislation and special legal conditions to differentiate and survive (*NB* 155). These conditions, which are often confused with ordinary constitutionalism (of the Kantian variety), are the "rule of law" often promoted by the West to developing nations or those emerging from totalitarianism as necessary for democracy (177–80). Second, prisons and security technologies are a fertile domain for public-private partnership and necessary to discipline compulsively imprudent or anti-entrepreneurial actors. The fact that American corporations are legally individuals, required by their charters to provide profits to shareholders, made it easier for Americans to conceive of (noncorporate) individuals as small firms in their own right. By defending the interests of the individual-as-firm against welfare state regulation, and by defending the interests of the community-as-Christian-family against welfare state protection for minorities and secular countercultures, the New Right forged a powerful symbolic consensus.

In practice, the consensus on neoliberalism came about in the United States by circulating images of the poor mother and minority criminal that exemplified failed or abnormal entrepreneurship as well as the threat to traditional communities posed by sexual and cultural experimentation. The "welfare mother" was the object of discussion in a number of public spheres that enabled workers and the middle class to adapt to changing global economic realities such as deindustrialization and computerization, as well as government reluctance to enforce the New Deal over time. Welfare "reform" during the 1990s sent a clear message to workers and the able-bodied poor: full employment is no longer a government goal, and every citizen is required to accept the wages available on the market; there will be no public assistance for nonunionized workers who wish to improve their bargaining strength in the low-wage sector, or who refuse to accept jobs without adequate provision for health care and day care. In previous applications of the Kantian model, Foucault's controlled, artificial spaces were always inhabited by a figure in whom the *limits* of that particular form of rationality were embodied as a threat or palpable disorder. Thus the asylum helped stabilize postrevolutionary legal and medical practice in society at large by forcing reason on the mad, and the prison helped solidify the logic of property by forcing work on those who lived from illegality.

Who plays the role of "problematic object" for the liberal practice of government that draws its legitimacy from the space of the market? Ruth Smith argues that the figure of the pauper, specifically the female pauper, was an avatar of disorder for liberalism in a way it had not been for previous governmental rationalities (1991, 320, 322). Paradoxically, the poor were considered corrupt *by nature* in an economic framework where nature was orderly. By exhibiting *need*, which implied an inability to exercise liberty and interests, and by reminding the employed and independent of the possibility of *scarcity*, the pauper embodied twin threats to liberalism as a form of *rationality* rather than a mere practice of domination. As Nancy Fraser and Linda Gordon argue, the category of the poor was also combined with the category of the colonized insofar as both were "inside and outside" liberalism and both were dependent in a society where the norm was presumed to be wage-earning independence, if not independent capital (1996, 243). "Normal" economic behavior consisted in habits that reduced or denied any possible resemblance to the poor (245). These authors "fill in" space between Foucault's stories of the prison, the deployment of sexuality, normalizing state racism, and the rise of liberal governmentality. They show that femaleness, for example, played an important role in bridging and naturalizing the various discourses of biopolitics.[33]

Of course, the poor mother and the prisoner had long been paradigmatic threats of urban disorder. In the United States, moreover, these were associated with racial degeneracy as well as economic danger. Often they were linked in the public mind as pathological mother and delinquent son. Indirectly, these images reflected the fact that African-Americans were denied many protections of economic success demanded and won by white Americans after the New Deal (Quadagno 1994, 21). Although they were often excluded from trade unions, great numbers of African-Americans traveled North during the first half of the twentieth century to escape Jim Crow and contribute to wartime industrial production. Some joined settled communities of middle-class blacks; others lived like migrants and built unconventional urban communities in which women worked, both by choice and of necessity. These communities were wellsprings of African-American activism during the late 1960s, when many whites moved to the suburbs in order to avoid school busing and took their tax dollars with them. They were also among the first communities to be hit by unemployment at the start of the 1970s, when returning Vietnam veterans and rising drug use began to fray local systems of social support. Suburban whites linked these crises of the inner city to two controversial

---

33. Two important historical works on the subject using analytical methods similar to Foucault's are Jacques Donzelot's *Policing of Families* (1979) and Mitchell Dean's *Constitution of Poverty* (1991).

acts of the regulatory welfare state: the extension of welfare eligibility to black mothers, who had previously been excluded from benefits in many states, and increased court protections for the civil rights of black suspects in criminal cases.[34] Drug use and single motherhood were certainly not limited to urban African-American communities; but the stigma of motherhood had different consequences for single white women than for single black women, because their babies were highly valued for adoption (Solinger 2000, 26–34).[35] They were heavily psychologized, however, and allowed psychological explanations for poverty to become prominent at a moment of economic instability—instability that would eventually be resolved by abandoning full employment as a goal of national policy.

How did this situation come about? Southern legislators concerned to maintain a ready supply of cheap African-American agricultural labor allowed Aid to Dependent Children (ADC) to be included in the Social Security Act of 1936 on condition that it would not be available to single black mothers, or available only at the discretion of the states, who often imposed morality tests and "man-in-the-house" surveillance (Quadagno 1994, 20, 119–20). As Nancy Fraser has pointed out, the assumption that American families should survive on a single male income allowed social insurance for (male) workers to be conceived as entitlements they deserved as citizens, while women were allowed to benefit from government programs only as wives or as the accidental heads of "defective" households without a male provider (Fraser 1989, 149–53). "Women's" public assistance was not only tied to their maternity (since nonworking women without children were supposed to be supported by their families) but also their moral character, which was oppressively surveyed by agency officials.

During the 1960s, as the divorce rate left increasing numbers of women

34. African-American poverty was largely ignored in the American media until the mid-1960s, when northern migration, urban political unrest (especially in response to police brutality), and feminist or civil rights demands for greater access to social services attracted the attention of whites (Gilens 2003, 103–8). At that point, media images began to portray American poverty as a problem for blacks and a problem for the United States posed *by* blacks (112–22).

35. During the 1950s, very few single African-American mothers applied or were accepted for ADC funding. Most kept their children, but were regarded as biologically deviant by policymakers. Single white mothers, by contrast, were eligible for ADC, but considered psychologically deviant if they failed to put their children up for adoption. By the early 1970s, many middle-class white women could keep their children without psychological stigma, which African-American women had more difficulty fending off because of their real or presumed financial precariousness (Solinger 2000, 22–26; 213–23).

with minor children to support, these programs were expanded to support single mothers who remained home with their children. But never-married mothers and divorced black mothers also demanded access to the program as a civil right, and were eventually awarded welfare benefits alongside their supposedly more "moral" white counterparts (Nadasen 2002, 275–76, 284–86). Many welfare recipients, white and black, continued to work covertly, because welfare rarely brought families to the poverty level (especially after real wages began falling in the late 1970s). But public assistance did improve their bargaining position and forced employers to offer higher wages, flexible hours, and/or benefits if they wished to attract women workers (most clustered at the lower end of the market). At the same time, more affluent married white women were beginning to work outside the home, and struggling to change the public perception that paid work was incompatible with motherhood. Where previously the affluent married woman debilitated by full-time maternity was "hystericized," the normalizing gaze increasingly turned on poor women who bore and raised children at home without a male provider—who rejected, in other words, the pre–civil rights assumption that they should pay for parenthood by making themselves available as cheap agricultural or domestic labor.

The language of "dependency," which drew an analogy between these citizens' "defective" relation to the state and the drug user's "defective" relation to his or her consumer habits, allowed both to be tainted with mental illness as well as immorality (Fraser and Gordon 1996, 252–54; Cruikshank 1996, 238–39; Kingfisher 1996, 21–37). Like "degeneracy" in the nineteenth century, "dependency" may be transmitted to children through a "culture of poverty," which involves, among other things, refusing work at overly low wages (Handler and Hasenfeld 1991, 205; Fraser and Gordon 1996, 254). In both cases, dependency was mediated by perceived bodily excess. In both cases, dependency also legitimated much older power relations involving racial anxiety and disciplinary control, imprisonment for drug offences, and eventually mandatory work at below-market wages for poor mothers.

Apart from psychological analogies, economists analyzed both types of dependency according to a logic of self-interest that bore little relationship to the actual reasoning of mothers or addicts. What this logic did, however, was elaborate an artificial "rationality" for entrepreneurial *homo economicus* in which "efficiency" was described as the highest good. Legislators and judges began using this logic as a norm against which a range of social and corporate actions could be explained or condemned for promoting or inhibiting the functioning of markets. Within this logic, welfare was interpreted as a disincentive to work, rather than as a compensation for the previously sexist and racist structure of New Deal social protections or a way for women to maintain families despite their own poverty and that of the men in their communities. Penalties for drug use, likewise, were increased in an effort to reduce "demand" for controlled substances (see NB 262–65). In the logic of the Chicago school and its political offspring, all forms of community were translated into more or less profitable relationships in which self-interested individual (or corporate) actors might invest, rather than the bed in which such individuals and businesses were born.[36] Education, political participation, and marriage involved returns and opportunity costs, and since it was assumed that rationality involved competition, any noncompetitive behavior was either pathologized or reinterpreted in terms of a hidden individual interest. As Cutrofello has noted, the strategies used to produce such calculations amounted to "techniques of mutual betrayal" (1994, 65–70).

The logic of enlightened self-interest permitted apparent contradictions in right-wing policy to be resolved. Republicans judged religious community and marriage able to produce a better "return" on educational, employment, and parental investments than nontraditional forms of community. Mothers who chose to stay home with children, however, were regarded as making bad or inefficient investments of time; they needed to be either lured or driven into competitive market relationships. In vain did feminists argue that the jobs available to poor women could not return a sufficient "profit" to justify leaving children in uncertain day care. As the real value of wages fell during the 1970s and 1980s, affluent mothers who had begun working outside the home for spiritual fulfillment found their income necessary for a middle-class standard of living. Why not poor mothers as well?

Perhaps the logic of family as "enterprise" was acceptable to Americans because it helped justify existing beliefs about the moral and psychological pathology of the urban poor. Perhaps, given American individualism, it made a virtue of the necessity of managing alone when corporations began to jettison large segments of the workforce. When

---

36. See, for example, Becker 1995. African-American women's maternity had been described (disparagingly) as economically self-interested from the origins of public assistance (Solinger 2000, 29–31).

sociologists claimed that the urban poor were "victims" of deindustrialization and racism, or politicians claimed that they were driven by inappropriate pleasures and lacked a sense for the value of work, the middle class examined its own pleasures, searched for examples of victimization, and attempted to improve its self-esteem and work ethic in hopes of avoiding the fate of their poorer brethren (Cruikshank 1996, 247). They closed down on their own imagination in the name of the same "realism" that was being forced on the poor.

American conservatives and liberals could agree on little except the debilitating effects of welfare dependency, although they debated the reasons for these effects. Leftists like Fraser and Habermas regarded the "clientalization" of welfare and other social service recipients as proof of the welfare state's potential illegitimacy or need for serious rethinking. But the New Right used very similar arguments that the welfare state "sapped" the citizen's will to self-sufficiency. In other words, they used purely economic language to interpret the disciplines and resistances that leftists who wanted to legitimate the state juridically would have condemned as excessive biopower. In the context, of course, to *defend* welfare was to place oneself outside the bounds of sane public discourse, to engage in patent *Schwärmerei*. And yet once regulations on industry were lifted and monetarist policies implemented, leftists and feminists were increasingly driven to defend the welfare state as the only possible noncommunist and therefore communicable version of their aesthetic vision. This made them appear neglectful and condescending to the poor, and rendered them all the more irrelevant according to the emerging moral consensus. Wendy Brown marvels at the counterproductivity of defending the state at the moment when it was, in fact, busy building prisons and laying the legal groundwork to restructure American industry (1995, 15–18).

Modern citizens are asked to live up to or reconcile the forms of individuality presented by many discourses and institutions. If a citizen fails in this task, his or her normativity will not be recognized, and all that will be noticeable is his or her conformity or lack of conformity with an externally imposed norm. Mothers, for example, receive a myriad of instructions from the medical and therapeutic professions. Advertising appeals both to their desire to satisfy these instructions and sympathetically and ironically to their sense that these instructions can never be entirely satisfied. Employers expect a different set of behaviors than doctors: for example, willingness to work flexible shifts or to work with hazardous chemicals. When employers recognize that these expectations are incompatible with other expectations placed on mothers, they usually insist that the women work out the difference or quit, or simply wear them out until they "slip up" in a way that justifies their dismissal. The entrepreneurial ethos of neoliberalism encourages women to make this precariousness into an opportunity for profit. If possible, others should be able to resolve their own tensions and anxieties by buying something you have to offer (child care in your own home, perhaps). The existence of tension should always be a motivation to subordinate oneself dutifully to external demands as if they were self-given, to make oneself the employer of one's own inevitable exploitation. To construe such a situation as heteronomy pure and simple offends the mother/worker's dignity, bringing her close to madness. This hidden link between rationality and responsibility is made possible and plausible by Kant's philosophy, but neoliberalism has adapted it into a practice of government unthinkable in the eighteenth century (when mothers, of course, were subjected to no less exhausting moral prescriptions) (O'Malley 1996, 199–201).

As we have seen, one way to satisfy this expectation of unity among experiential and discursive domains is against the backdrop of a foreclosed problematic object. Successful or striving individuals avoid any resemblance to those who seem to be failing in the use of property, intelligence, and sexuality: the poor, the mentally ill, the sexually exploited or undesirable. Contemplating or studying phenomenal forms of this object give rise to communicable states of mind, allowing people to make sense to themselves and others, to phenomenal forms of freedom. In her study of a Michigan welfare office, Catherine Kingfisher found that the relative economic fragility and enormous work demands on welfare agency *workers* exacerbated their desire to differentiate themselves morally or psychologically from clients who shared similar class and racial origins (1996, chap. 9). The constitution of the pauper, madman, criminal, or pervert allows societies in political or economic ferment to retain and enhance the understanding of "proper" European or American individuality they regard as a moral ideal, hope to embody, and attempt to foster in the population at large.

Often the alternative to distancing oneself from the problematic object is to regard oneself as the source of risk, a carrier of the problematic object that must be expelled or controlled for the sake of sanity, in the same way that Kant felt he must expel the *Schwärmer* (Leibniz or Swedenborg) and then the whole empirical body in order to hope for reason's unity. Those who opt for this "internalizing" strategy turn their struggle into an inner ethical substance—rather than one in which many discourses and social practices are implicated, and which many others are needed to resolve. They interpret the struggle as if it were a confused expression of an inner truth, psychologizing and moralizing it, rather than regarding it as one among many similar products—*énoncé*/events or body/events—made possible by the regimes of power and knowledge at that given point in time. This temptation is doubly forceful for groups whose bodies already represent a threat in the eyes of others, but it can haunt people with privilege as well. To engage in this potentially infinite work of interpretation ensures that one may approach the norm more or less asymptotically, and one's moral standing may be affirmed as one comes closer, but never set or participate in setting new norms.

Rejecting internalization may involve the construction of a conflicted historical terrain and the discovery of empirical knowledge about past events of individualization in the genealogical mode. Rejecting internalization does not, however, mean that someone struggling with this incongruity of expectations and norms should find an *external agent* at whose feet the blame can be laid or from whom restitution can be demanded. It means that one must seek out other bodies and *énoncés* in whom one sees the effects of the same processes of normalization one recognizes in oneself. These people and statements are "doubles" in a very different sense from the doubles of anthropology such as the empirical and the transcendental, the cogito and the unthought, and the retreat and return of the origin. These doubles—the ones one must be like and the ones one must not be like—indicate the structure of the *sensus communis* that both brings one together with others and consigns one to privacy and incommunicability.

Negative Anthropology

In the introduction, I described two ways of thinking about political imagination. On the one hand, people imagine that discourses, institutions, and practices do fit together somehow as "things-in-themselves," although they recognize that working out the details of this fit is a difficult (and according to Kant, fruitless) task. However, political imagination" can also mean the range of actions people believe are possible in a given historical situation. Institutional beliefs, problems, and habits together comprise a "problematic" in which bodies are caught, or whose coherence they must support. Eleanor Bumpurs was unable to defend her living situation without appearing to be an "immediate threat and endangerment" to the life of an armed police officer. These ways of thinking about political imagination are factual and practical, respectively. They also focus on the contents of individuals' beliefs or intentions.

A third way to understand political imagination has less to do with individual consciousness or responsiveness and more to do with normativity. Here, political imagination indicates the individual's ability to envision a way of life or quality of life that can only be achieved with an indefinite number of others. Like an aesthetic form, which gives pleasure apart from its empirical existence or moral universality, the object of such imagination is a *situation* of being-with and a source of *interest* in one's fellow citizens that involves hope and curiosity. This imagination does not result from an act of will, since its content depends upon the changing character of other human beings. Like aesthetic sensibility, moreover, it may ultimately differentiate each individual from others by suggesting alternate or competing pleasures or qualities *in* the forms they collectively find pleasing and expect others to enjoy.

This third kind of political imagination is the one that would allow someone from a secure economic and cultural position to see, think, and act without having to deny the possibility of an obscure resemblance to Mrs. Bumpurs (or the Iranian demonstrator mentioned in the introduction). Although it might seem absurd to imagine similarities at the ordinary level of conversation or the everyday scale of geographical or cultural intuitions, this kind of political imagination is attentive to the way in which class, race, and nation or religion turns humans into objects of cognition or *anthropological* understanding rather than into subjects of a shared aesthetic sensibility or forms whose associations and variability inspire a lawful play among the powers of thought. To achieve any recognition, as the following pages will explain in more detail, this kind of imagination must be embedded in institutions and physical or discursive media, since we are finite, discursive beings. These institutions and media must comprise a structure that reduces the occasions on which one is likely to perceive one's fellow human beings as a life-threatening "risk" rather than a morally ennobling challenge to existing strategies for social recognition. Finally, this structure would leave room for each of us to vary the trajectories by which we differ from each other over time, collectively and individually—without defining individuality *merely* in terms of the occasions when it conflicts with collectivity.

Rights and public spheres are mechanisms by which modern societies have managed to integrate freedom and this third form of imagination into the governmental practice of states. This integration fosters powerful identifications between the form of the "free" individual and the "lawful" or "security-minded" state. Such identifications appeal to citizens' fear of risking their physical safety as well as the conviction of individuality and agency associated with, or represented by, anthropological categories that divide humans into communicative groups. They also appeal to citizens' hope to communicate indefinitely and add the power of other bodies to their own, a hope associated with "publicity" in its commercial as well as its political sense. States and civil society institutions use the materiality and discursivity of bodies and imagination (the inevitability of limitation and the possibility of limited expansion from any given level) to exert power over citizens. Citizens are promised health and prosperity if they relate to themselves and to one another as beings of life, labor, and language—or, in the version I have presented, sexuality, property, and intelligence. The "body" whose possibilities are articulated by law or public discourse is more than brute materiality or phenomenological self-evidence.

People are only freed from such entanglements in a particular anthropological scheme of axes or categories of individualization through the act of problematization. In problematization, the incongruities between discourses and the plurality of imaginative spaces one inhabits actually become occasions for realizing new actions—actions that become plausible to others, therefore communicable and contagious. I have suggested that the state of mind required for problematization is very similar to the one required for Kant's *aesthetic reflective judgment*, because it identifies a conflict or form for which no concepts yet exist and because it is indefinitely communicable. The demonstrator confronting the shah's police seems to have had an intuitive sense that the relations of power in a given situation were changing, perhaps because of nonverbal cues from the officers or the tone of the crowd that day; he took the risk of moving forward, and was lucky to have been supported by a sufficient number of others. Although the empirical body is at stake in the act of problematization, as well as when an individual is caught by a problematic that exceeds her capacities, this body has an aspect that is nonempirical and remains open to further transformation. Bodies are most affective and creative at the moments when they cannot explain their pleasures empirically and therefore have a positive interest in reassessing what counts for them as "real." In the act of aesthetic judgment, subject and object are animated by the promise of form and by the excitement of being able to live without closing or naming that form, or even the subject's own form.

If one interprets his or her body according to the categories of positivist medicine, advertising, and the labor market, one risks thinking that the body is "merely empirical." This does not mean that the body's pain or suffering are negligible; without that body, the subject loses his or her most important vehicle for acting on him- or herself, or risking present abilities in the hope of gaining new ones. Identifying an experience as "violence," moreover, is an attempt to surmount and act on one's *incapacity* to participate, however minimally, in power relations. One's body is a point of application for state and economic power (sometimes violence, as well). But it is also a sign of our *normativity*, or the ability to reassess the form of the individual and the value of intelligence, sexuality, or property. Bodies, finally, are forms in which one recognizes another person's normativity as similar to his or her own, even though this normativity may be expressed through very different abilities and risks. If Western societies incarnated the limits of knowledge and order in bodies in order to study and administer the exercise of power more effectively, then any attempt to problematize the legacy of post-Kantian thought must begin by taking a critical perspective on the body—stepping back and asking about the discursivity and materiality of thought which position bodies at the interstices of more and less realized imaginative spaces.

Drawing on the lecture "What Is Enlightenment?" we can identify three strategies for resisting or exercising power in a historical situation. One strategy, as we have seen, is to show that a certain element of the phenomenal world or scientific discourse represents a properly "unthinkable" limit condition for that world's coherence and rationality, rather than being a simple empirical appearance. Doing so reveals another standpoint from which it has already become "thinkable" in a different manner, if not grasped in the manner of an essence. Madness, death, crime, and sex are phenomenal events whose observation and codification gave shape to sanity, health, citizenship, and biological normalcy—the problems of the post-Kantian era. Foucault contends that unlike Kantian critique, whose goal was "knowing [*savoir*] what limits knowledge [*connaissance*] must renounce exceeding," contemporary critique must reverse its orientation and ask, "In what is given to us as universal, necessary, obligatory, what place is occupied by whatever is singular, contingent, and the product of arbitrary constraints?" (*EST* 315).

In practical terms, this form of critique means breaking down concepts that make us feel that our freedom or security are at stake in an expert discourse into the *multiple practices* these concepts allow us to group together. In this second, but related, nominalist strategy, sexuality, property, and intelligence become "bundles" of social functions that can be reorganized or played against one another, and no longer appear as "powers" buried in the soul to which and for which we are answerable. The state, power, capitalism, and the individual are likewise mapped onto multiple axes from which some elements can be selected and others resisted. The practices that cohere around a social or intellectual "problem," and ensure that people with traits such as poverty or mental illness will always be sought out to embody that problem can then dissolve (perhaps, of course, to be replaced by new problems) (*EST* 318). A final strategy, associated with the care of the self, is to recognize that freedom is a relationship *internal* to action and to the body's capacity for activity that can be actualized at subindividual and transindividual levels, rather than at the level of the body as self-evidently individuated organism.

The common thread to struggles from the New Left, according to Foucault, was the challenge they posed to modes of *individualization* rather than to modes of domination or exploitation (*PWR* 329–31). Foucault did not deny the continued existence of domination or exploitation, but suggested that activists increasingly grasped such phenomena from the side of their effects on individualization. Perhaps this is because modernity already confronts societies ("traditional" or not) with inescapable self-consciousness about the processes by which individuals are distinguished from and maintained at the heart of one or more collectives. Such self-consciousness is a change in the mode of individualization, not the creation of individuals where previously there were only "communities," "masses," or "races." Such an attitude is necessary to prevent certain kinds of individuals from seeming to be the "universal, necessary, and obligatory" products of a world-historical process of emancipation or a world-historical process of moral decline.

During the 1970s, Foucault referred repeatedly to the body as a multiplicity of forces that can be articulated and managed differently by different historical technologies. The "individual" is made possible by the administrative identification of "subindividuals" within a population—discrete units of behavior and skill—which can be combined and recombined to suit the needs of its biological cultivation and preservation (Foucault *P/K* 208; see also Ransom 1997, 46). "Both disciplines and the individuals they create are plural," John Ransom explains, noting that what counts as an "individual" for one practice or discourse may not count as an "individual" for another (49). Ransom draws a link between the "subindividual," the *limit* of a society's or discourse's ability to break down the body and its movements any further in an intelligible manner, and what Foucault once referred to as the *plebian aspect* of social phenomena. In a 1977 interview, Foucault hypothesized that the "target for apparatuses of power" was not "the plebs" (the people as thing-in-itself) but the "plebian aspect" of their practice. "There is certainly no such thing as "the" plebs; rather there is, as it were, a certain plebeian quality or aspect (*'de la' plèbe*). . . . This measure of plebs is not so much what stands outside relations of power as their limit, their underside, their counterstroke, that which responds to every advance of power by a movement of disengagement. Hence it forms the motivation for every new development of networks of power" (P/K 138).[37] Because power is a relationship between actions that makes *use* of the body in order to affect the actions of another person, usually by the way it *individuates* the one who is subjected to its exercise, individualization is *never complete*; it never determines us "all the way down" because there is no end to the body's differentiability (Ransom 1997, 120; Daddabo 1999, 110). Since the body is rendered intelligible in different ways and for different purposes by different practices and systems of knowledge, it is capable of resistance along multiple axes and one discourse may enable a body to resist an otherwise compatible discourse (Ransom 1997, 126).

The collective end of human activity is equally incomplete, contrary to the claims of most positive communitarians.

> The problem is, precisely, to decide if it is actually suitable to place oneself within a "we" in order to assert the principles one recognizes and the values one accepts; or if it is not, rather, necessary to make the future formation of a "we" possible, by elabo-

---

37. The "plebian aspect" seems to correspond to what Foucault called "indiscipline" in *Discipline and Punish*, namely, behavior that refuses the question of legality and illegality. In the above interview with *Les révoltes logiques* from 1977 he is more cautious not to "naturalize" the occasion for revolt or conflict as a real kernel or "prime matter" of resistance than he was in *Discipline and Punish*; for example, "an indiscipline that is the indiscipline of native, immediate, liberty" (DP 291–92). Daddabbo observes that only in the case of popular revolt against the *supplice* did the "sociological" plebs coincide with the "ontological" plebs (see Daddabbo 1999, 104–9). Copjec argues forcefully against any idea of an "ontological" plebs (1994, 1–3).

rating the question. Because it seems to me that the "we" must not be previous to the question; it can only be the result(and the necessarily temporary result(of the question as it is posed in the new terms in which one formulates it. (Foucault 1984, 385)

Asking about the critical "we" at any point in a struggle is one way of detaching from states or communities that claim to have prior knowledge of the individual's capacities and ends. For Foucault, these questions go hand in hand with the foundational question of Kant's "What Is Enlightenment?": "What's going on just now? What's happening to us?" Being attentive to the changing character of this "we" is another way of detaching from political technologies that produce the individual and install her as the ontological point of reference for most modern political thought.

Han traces the word "problematization" to the Greek *problema* or "shield"; problematization therefore "forms the theoretical 'shield' that allows the subject to define himself through his actions in and relationships with the world" (Han 2002, 164). This observation is interesting because it suggests that problematization is a "self-interpretive movement" that permits the subject to resist her subjection to interpretive categories imposed from outside by introducing a *division* into the time of action, sifting what she does from what she does *by* doing it.[38] Dividing the action may reveal other ways in which one could do or desire that same action—creating room for play within a seemingly determined present. Thus problematization is an exercise of freedom consistent with the *inventiveness* every subject of constraining power must use. As I mentioned earlier, Foucault defines power as "a mode of action that does not act directly and immediately on others," but "upon their actions: an action upon an action, on possible or actual future or present actions." Relations of power can be established with oneself as well as with others. By putting myself in a new situation or entertaining a new thought, I risk no longer being able to make sense of earlier situations and thoughts, but I can usually be counted upon to respond in some way to the new constraints I discover, constraints that may or may not strike me as problematic. This in fact is how Foucault described his own creative work in *The Archaeology of Knowledge* and *The Use of Pleasure:* "The 'essay'—which should be understood as the assay or test by which, in the game of truth, one

---

38. "People know what they do; they frequently know why they do what they do; but what they don't know is what what they do does" (personal communication, in Dreyfus and Rabinow 1983, 187).

undergoes changes . . . is the living substance of philosophy" (Foucault 1985, 7–10; *AK* 17). Problematization enables subjects to enter into relations of power with their own activity apart from the external relations in which they are caught, to *act on their own actions without doing themselves violence*. Because it alters the actor's understanding of his or her own individuality, problematization cannot be considered an "individualist" act in any simple sense, or free in a radically autonomous sense, though it may be liberating.

Law and public spaces, insofar as they are genuine occasions for the exercise of power, can be thought as very complex modes of acting on one's own actions and those of others. Rights, most significantly, are patterns of power, actions that *limit* the actions of others and give them a zone for expression and invention. Thus rights have an immediate affective impact and can be considered as interpretations of a polity's real and ideal affective habits. As discussed earlier in this section, it is important that citizens see the full range of their power relations represented in rights.

Renata Salecl offers a psychoanalytic account of rights that tries to separate their function in relations of empirical need from their function in relations that are intersubjective, creative, and indeterminate. Although needs may vary in relation to an organism's environment, they are biological and finite. Humans demand goods and services as tokens of recognition apart from needs, but goods and services are also finite, however personal the recognition they were intended to convey. Since goods can represent another person's indifference as well as her attentiveness, demand is potentially infinite, giving rise to the problem Marxists and cultural critics call the "creation of false needs." *Desire*, on the other hand, enables people to differentiate themselves *as* they recognize one another, and thereby to change the shape of the community and what it considers normal. Desire responds to an obscure aspect of the other person, a capacity for invention that individuates him or her from other apparently interchangeable members of a species, class, race, economic status, education, or religion. When modern democracies and movements invent new rights, they do so to ensure that needs are sufficiently met and to reveal the indeterminacy of desire, but also because existing rights are always capable of being interpreted as mere "filler" for demand and need. According to Salecl, rights in modern democratic societies assure citizens that the state will protect their individuality by satisfying only *some* needs or demands—not because remaining needs or demands lack merit, but because they represent the level on which citizens are easily rendered interchangeable—normal, rather than normative (1994, 124–25).

Kant's critical turn prevents us from talking about humans as interchangeable or unique *in themselves*. They only appear to be interchangeable at particular levels of perception, within narrower or wider zones of expression, and more or less detailed conceptual vocabularies. Intuitions and concepts may apprehend the other person at a level that reveals more or less of her individuality and inventiveness, in other words, her freedom. They *also* place the other person in relation to a collective—those against whom and with whom she is defined as different. According to Kant, individuality could truly be expressed and recognized only insofar as one acted on one's own actions as a subject of law and was also recognized as a subject of law by others. But he also recognized that institutions and movements obscure or enhance this individuality, just as they can obscure or enhance her relation to the collective against whose backdrop her individuality appears at all. One task of political imagination, whether focused by law, the organization of geographical space, or public media, is to reveal lines and aspects along which the individuality of others comes once again into view. Such attention cannot respond to the other person as already "recognizable," schematized; it involves a use of imagination that can only be found in aesthetic reflective judgment.

Salecl's distinction, to which I have given an aesthetic rather than a formalistic spin, helps us make sense of left- as well as right-wing fears that the welfare state will limit the meaning of rights to *needs* and degrade citizenship into mere normality or "dependency." When she argues that welfare recipients do not just require the satisfaction of their needs, but also have to participate in the *definition* of those needs, Fraser is trying to rearticulate the relation of *desire* to *need* in the power relations and institutions bundled together as "rights" (1989, 154–55). Interpreting one's own needs is a minimal condition for being a normative being who can experiment with bodily requirements and possibilities; as Amartya Sen puts it, fasting and starving are very different ways to live the same physiological state: "It is *choosing to starve when one does have other options*" (1992, 52). This is also why Arendt insists that the properly political dimension does not involve concerns about economic distribution (e.g., Arendt 1979, 315–19). Whether needs are satisfied directly by the state or delegated to the market, neither socialists nor liberals can presume that the process of distribution itself will maintain the perpetual state of *dissatisfaction* and invention—freedom—that finds expression in politics. (Nor, I would add, should the process of competition for profit substitute for other competitive uses of freedom, especially to the extent of degrading needs.)

Although rights *do* allow for the interpretation of needs, they do not enable citizens to differentiate themselves in relation to indefinite others (Arendt 1982, 74). This is the function of public spaces, which are also associated with affective states. Foucault suggested that concerns about dependency and criticisms regarding the quality of citizenship in welfare states during the 1980s were linked to the atrophy of specific public spaces (*PWR* 370–71). The ability to act on one's actions and those of others only comes with practice; and one does not learn what a body can *do*, Spinoza might say, until one knows what bodies it can combine with. Like rights, public spaces create zones within which people can act on their own actions and the actions of others by means of sensory and imaginative situations in which they feel themselves to be at stake or "at risk." As everyone knows, very few statements or gestures have a *genuine* universal reach, and yet the belief that "the whole world is watching" gives actors an irreplaceable feeling of normativity or ability to invent themselves as creatures who can do more than need and be satisfied.

If public spaces are to facilitate the individual's reflection on his or her similarity and resistance to others, they must be material, taking time to explore and absorb, foreclosing some encounters in order to make others available. They must involve relations of actual and anticipated power with avenues for reversal; spectatorship alone does not allow for a genuine public space. Although *individuals* should not have to bear too much heterotopia in their own bodies, finally, *public spaces* should not be closed against antagonisms. Discourse, the nonhuman environment, and some collective economic resources should be protected by the public spaces that use them in order to ensure that individuals do not *need* to protect *themselves* against the unsettling resemblances to fellow citizens.

From these reflections on the way that rights and public spaces function as occasions for the exercise of individualizing and resistant power relations, we can begin to answer two questions frequently aimed at Foucault's work. First, some ask, is Foucault guilty of "aestheticizing" politics? In other words, does he risk subordinating needs (including the need for privacy) to a way of seeing and acting that is actively hostile to reason? Second, is Foucault tacitly conservative if he fails to encourage action and thought aimed at changing the totality of social relations? Is his work "aesthetic" in the sense of "impotent," leaving the infrastructure of communication and production in the hands of those who use them for exploitation?

Benjamin raised the charge of "aesthetization" against the Italian futurists, who appreciated Fascism because of its investment in speed and modern technology (1969, 241–42). Contemporary theorists usually mean something like "relativism," "individualism," or "amoralism" by the charge, especially when any of these attitudes are combined with a willingness to employ violence. According to Kant, the relationship between the aesthetic and the moral is more complex. Genuine aesthetic judgment refers to a state of mind that is universally communicable, whether or not individual movements or critics are able to recognize its presence. Beauty and sublimity can only symbolize or awaken moral judgment or action, not substitute for it. Kant also believed that genuine moral action required the subject to identify with a standpoint that could be held by anyone and that could pertain to any action, much like the subject position of modern, law-governed natural science (EST 279–80).

By distinguishing between this universal standpoint and his empirical, interested situation, one might say that Kant divided the time of his actions and thereby acted on them. In this respect, his understanding of moral judgment closely resembles the act of problematization. Like problematization, moreover, the habits by which Kant acted on his own actions reaffirmed or reawakened consciousness of his own *normativity*, as opposed to his status as a more or less "normal" anthropological being. Here, the aesthetic is not opposed to the moral or the rational, but restores the individual's power to judge what is worth valuing or subjecting to reasons. Kant refers to the division of time in the sublime as "violence," and many critics have argued that Kant's way of acting on his own actions can easily amount to doing violence to one's sensual capacities. Others, including Foucault, have simply pointed out that to experiment effectively with the scope of one's own actions, one need not identify a priori with the subject of law or knowledge (FL 330).

Modeling political or moral judgment on logic means identifying good actions with the actions *required* from a rational subject (rather than those which are merely possible, or those required for some other kind of subject) (EST 264). This may be a good way for logicians to translate moral capacities into the terms of their most highly developed talent, but it can also represent a desire to escape the unpredictability of human action (Arendt 1968, 109–14). Although, according to Arendt, this desire appears in the West as early as Plato, it has become especially urgent in the modern world because of the obvious plurality of traditions and opportunities for experimentation with life practices. Every community suspects other communities and dissidents of pursuing the agreeable rather than what is genuinely beautiful or moral. When alien communities or dissidents are regarded as sources of risk, reasoning from a universal standpoint is attractive because it allows the majority to describe troublesome differences as "irrational" rather than merely "aggravating," while neutralizing defensive inclinations to engage in persecutory violence. As we saw earlier with respect to the self-limitation of state reason in liberal government, the style of moral reasoning enjoined by Kant in the name of autonomy can also make use of the actor's interest in her own identity as a rational being to reduce risk or control others (*EST* 300). While universalistic morality may seem necessary in a world ordered by biopolitics, this kind of *world* may not be necessary.

Arguably, individuals are more likely to be *insensible* to the effects of their own aggression if they believe that such anesthesia demonstrates their normativity and moral superiority. Moral norms are required to enforce concern for others in a world where normativity is associated with *lack of feeling* toward fellow citizens, and where those who are attentive and caring are regarded as merely "normal." These are beliefs that many cultures use to differentiate men from women and elite from subordinate classes, whose individualization ordinarily involves care and emotional perceptiveness. Many also believe that indifference to the suffering of others is a side effect of competitive instincts, inevitable or desirable for enterprise or innovation. But this also seems implausible, for people *allowed or required to care* often compete both personally and professionally to provide better care or to cultivate more subtle patterns of responsiveness and appreciation in one another. There is no reason to believe that absent moral norms requiring actors to suppress their individuality for the sake of a universal human ideal or concrete human community, these actors would cease to be interested in community *or* individuality.

Consider the amount of time it takes the amateur to distinguish artistic styles from one another and to elaborate a knowledgeable taste; that is, a sense for what makes her resonate with her own mental capacities and those of others, putting her in a communicable mood. Even with the diversity of human cultures, we have made very little progress in grasping the different *ways* in which it is possible to live or to experience the sensory and social environment (*EST* 158–60, 163–64). Because of their discursivity and materiality (the fact that some options must be restricted for others to be realized), individuals and societies have different limitations in their capacity for experimentation over a given historical period. The materials for these trajectories can remain very stable as long as the resulting forms are received as signs of normativity. Most easel art, for example, uses the same pigments to produce widely varying forms. One need not be especially visibly distinctive in order to create a very satisfying and distinctive aesthetic experience; the practitioner and spectator must simply be trained to intensify the meaning of small variations. For these reasons, some countercultures prefer the aesthetic forms of premodern societies to those of modernity. But this minimalist sensibility should not deteriorate into nostalgia or a fear of novelty. In our current culture, the potentials for individualization and variation are set by consumer goods and the industries that distribute them. Those who want to engage in anticommercial practices of individualization (and therefore collectivization) often employ religious discourses and are therefore vulnerable to the nostalgia and fear of change associated with many religions. But there is no reason why a materially stable culture cannot engage in profound and constant variation in its aesthetic forms, some of which are, as always, more inspiring and communicable than others.

A second objection to the foregoing account of political imagination might be that no social movement or art of government can exercise effective power if, in the name of attentiveness to the present, it fails to demand globally decisive events. Must political change be conceived as a *revolution* in order to actualize the normativity of oppressed peoples? In "What Is Enlightenment?" Foucault regards such claims as metaphysical: "We have to give up hope of ever acceding to a point of view that could give us access to any complete and definitive knowledge [*connaissance*] of what may constitute our historical limits" (*EST* 316). The Marxist tradition opposes a metaphysics of revolution to the critical self-limitation of the bourgeois public sphere. But belief in the historical beneficence of an infinitely open market is no less metaphysical than belief in the inevitable goodness of revolution. Neither Idea of reason does justice to the complexity of materiality and discursivity within the social world. Foucault's nominalist analyses of the "historical dimension," the state, capitalism, and Enlightenment acknowledge discursivity without allowing the present to be opposed to the past or future as totalities in *absolute* conflict.[39]

39. See "For an Ethic of Discomfort," for an expression of profound skepticism about the

Many global projects only grasp the present by destroying it, and therefore prevent the critic from having any benchmark against which to identify, much less measure the preferability of new phenomena. Consequently, they fall back upon unconscious patterns of social organization that relieve participants of responsibility for their own mimetic and aversive tendencies. Such global transformations are also traumatic, in the sense that they require the activist to abandon the *conditions* for risking and altering herself without having constructed new and reliable forms of pleasure. Thus they damage the individuality and collectivity whose powers they hoped to enhance at an affective level. Finally, they tend to ignore the materiality, finitude, and multiplicity of spaces required to make their "utopia" a real place, producing an experience of domination that is either suffocating or boring.

But there may also be reasons not to shrink from the idea of revolution, as long as it is not conceived as a moral obligation or the destiny of history as "thing-in-itself." Kant warns that "one age cannot enter into an alliance on oath to put the next age in a position where it would be impossible for it to extend and correct its knowledge . . . or to make any progress whatsoever in enlightenment. . . . Later generations are thus perfectly entitled to dismiss these agreements as unauthorized and criminal" (*KPW* 57–58). This would prevent any society from taking steps to *rule out* the possibility of radical transformations on the basis of a genuine change in the present. Nor can a social movement be fully normative without *hoping* that its way of seeing, feeling, and enacting political life is universally communicable—even if it remains relatively local in its actual impact.

No historical situation is open to *every* imaginable change, and every change that does take place creates irrevocable obstacles to transformation in the future. As Harvey argues,

> Social, institutional, and material structures (walls, highways, territorial subdivisions, institutions of governance, social ine-

---

"empty form of a world revolution," which he regarded as conservatism on the left (*PWR* 446–47). Many readers of Foucault, such as Ransom (1997) and Simons (1995) are heartened by his focus on the transformative potential of local interventions that could be characterized as "revolt," rather than attempts at sweeping transformation more obviously recognizable as "revolution" (*PWR* 291). Such a reading can make Foucault into the champion of a de facto liberalism that seems to accept the inevitability of capitalist practices, but other readers, including Balibar (1994), Marsden (1999), Poster (1984), and Smart (1983), find his work useful for analyzing the intersection of economic and cultural practices resulting in structural social inequality from a distinctly nonliberal, anticapitalist perspective. *Naissance de la biopolitique* clarifies Foucault's views on liberalism.

> qualities) are either made or not made.... Once such structures are built they are often hard to change (nuclear power stations commit us for thousands of years and institutions of law gather more and more weight of precedent as time goes on).... A total reorganization of materialized organizational forms like New York City or Los Angeles is much harder to envisage let alone accomplish now than a century ago. (Harvey 2000, 185)

This follows necessarily from Kant's insight that human understanding is discursive; not only do public events unfold in time, but our ability to apprehend them is always partial or dependent on the scale of apprehension. Moreover, it is not always easy or possible to reverse earlier ways of perceiving and conceiving a situation, for these ways will be embedded in the discourse and infrastructure of law and public space, if not in our self-understanding. Foucault warns against totalizing forms of political imagination on the grounds that they promote "the empty dream of freedom." Harvey interprets Foucault's pluralism as unwillingness to face the reality of this materiality and conflict. After all, Kant also claimed that the *reality* of experience requires phenomena to be capable of real conflict, changing one another and sometimes annulling the bases for one another's existence. But neither Harvey nor Foucault imply that historical actors have no options other than affirming revolution or affirming neoliberalism.

Nor does opposition to a "metaphysical" notion of revolution mean that political work can only be done "in disorder and contingency" (*EST* 317). There are many ways in which political practices can be unified apart from the final form of a state or society; for example, at the level of method. Foucault believes that many recent social movements were unified by their stakes: disconnecting the "growth of capabilities" from the "intensification of power relations," relations that seem to involve a discrete number of practices and maintain a specific set of behaviors or identities as problematic objects (madness, disease, criminality, sexuality, or poverty). There is no reason resistance to these "individualizing and totalizing" forms of power *cannot* come from traditional left-wing or even liberal institutions, movements, and ideologies. If these institutions, movements, and ideologies are to remain politically viable, it will be despite their traditional understanding of social change, or because they participate in a "new problematic" in which they may not recognize themselves.

In other words, it is important to recognize that political imagination *does* require conviction regarding the possibility of alternate forms of life, but that fantasies regarding alternatives are *part of the process* of identifying and acting on the present that is lived simultaneously by many bodies, rather than the *telos* toward which that process must aim. Political imagination also requires that material security enable individuals to explore their resemblances to one another without losing the right or ability to engage in further self-differentiation and self-alteration with respect to the constantly changing social body and historical horizon.

Kant's egalitarianism is rooted in his conviction that all humans are capable, in principle, of identifying themselves as rational subjects and acting on their actions with the unique "lawlike" style of rational subjects. Indeed, we are obliged to identify with this standpoint, both in our personal actions and in our efforts to create a public (*EST* 305–6). Kant's understanding of the rational subject is not *just* logical but also relies upon an anthropology describing the actual problems of human knowledge and moral experience. This anthropology takes empirical psychological and social differences into consideration, and as we have already seen, allows Kant to countenance a fair degree of inequality in moral practice. But one can conceive other grounds for egalitarianism according to Kantian principles: namely, the finite actor's *inability to know* the extent of his or her resemblance to other human and even nonhuman beings as "things-in-themselves." The analysis of power relations teaches us how others can be made to sustain our empirical pleasures and abilities. But we cannot know how close or how far we are to others in the communicable state of mind that makes it possible to *vary imagination and understanding freely and lawfully*. Such a state of mind is required if a familiar world is to be shaped with new concepts, or if a group is to *rise above the level of empirical need* and declare itself equal to the forces of nature, biological or political.

In the account of political imagination I have offered, actors are haunted by their inability to predict what they and others will want from their fellow human beings "after the revolution," local or global. Like Republicans who hope that the future will resemble the form of life in their ideal 1950s suburb, all positive, schematized images of the future are based on what we have already known, the ways we imagine we were linked to others in the past and would like to remain linked or improve in a better world. This is why Kant refers to Enlightenment as an *Ausgang* rather than a specific social telos: to prevent the empirical from being calcified as a transcendental for future generations. Nor, like the libertarians who eventually gained economic authority because of coalitions with radical Christians, or the civil rights movement whose offspring identified with the Viet Cong, can one know whom problematization will lead one to resemble over time.

No anthropology can anticipate what kind of bodies and psychology we will have in the kingdom of ends. In the absence of a specific historical context or set of schematizable affinities, we do not even know how people differ—which does not mean that they can be presumed not to differ. All we can know, as Simone de Beauvoir has eloquently stated in *The Ethics of Ambiguity*, is that the exclusion of some group or trait may petrify our own possibilities as well as theirs (1994, 71–73). No movement or state can know *a priori* whose being will *fail to benefit* from a changed society. Nor should it presume to fix some group in the role of a "problematic object," "collective risk," or racial enemy so that this society can be built around his or her prison, asylum, or uncommunicating body. Because our knowledge regarding the extent of resemblances is limited and only accessible through genealogy, rather than through empirical observation or logical deduction, I refer to this form of egalitarianism as "negative anthropology" by analogy to the "negative theology" left after Kant's Copernican turn. Leaving zones like the economy alone so that citizens could self-consciously exercise their freedom allowed liberal governments to balance one set of risks against others. In practice, the state's reluctance to interfere with the activities marked off as economic favored the legislating and productive activities of certain elements in civil society, such as men and property owners, over the freedom of other elements. I would argue that the political imagination embodied in rights and publics may suffer as much from measures to reinforce *inequality* as populations may have suffered because of early modern or socialist efforts to enforce *normality* at the level of "population."

AFTERWORD: NOT SIMILAR TO SOMETHING, JUST SIMILAR

One of the most conflicted legacies of the New Left has been the perception that radical politics requires advocates to maintain a perpetual state of outrage and sadness. In the foregoing section, I have given some reasons why I think melancholy came to dominate social movements that originally had a diverse affective composition. It does seem true that political imagination involves (or springs from) perpetual dissatisfaction with the present, a constant need to be dividing the present and producing events that affect the form of individuality and collectivity. On the other hand, this dissatisfaction need not be the result of sadness, guilt, or hatred. If Arendt is right that the ultimate purpose of the state and public spheres is not just to coordinate satisfactions, but to give an open-ended, critical image of the future, then there is a kind of dissatisfaction that has nothing to do with need, but with the way we are individuated in relation to one another.

One way of introducing division into the time of the present, releasing possible actions or one's own actions and the actions of others, is to recover and remember historical struggles. The effect of problematization on everyday practices at the aggregate level is rarely as dramatic as people hope or fear—even the ways we suffer and harm each other are fairly stable. The *materiality* of people's bodies and their public spaces consists in this inertial quality, which can be reflected upon in different ways and lends itself to different schematizations or perceptions of identity. When historical struggles are suppressed, however, this reflective and reconstitutive work is impossible to carry out, and historical struggles themselves can be dead-ended by desires to avenge the loss of a particular organization of social space. Every body that acknowledges this tension, including Kant's (I imagine) is stretched through many spaces at once and feels crazy half the time. Like Roger Caillois's legendary psychasthenic, the mad critic "tries to look at *himself from* any point whatever in space. He feels himself becoming space, *dark space where things cannot be put*. He is similar, not similar to something, but just 'similar.' And he invents spaces of which he is the 'convulsive possession'" (Caillois 1984, 30). He (or she) inhabits a melancholy body that is caught *in* and *between* a multiplicitous imagination, rather than actively differentiating and thereby producing a body that would be adequate to a better society. Our bodies should always be better than the societies we currently have.

One can construe this process of transformation as risking or rejecting who one is, but it can also be thought as risking or rejecting previous understandings of our "natural" differences and affinities with other thinkers. John Brown's madness, like that of Eleanor Bumpurs, could not be easily dissociated from the expectations and governing technologies of the state in which he lived. What sanity would correspond to a feeling of kinship with single women on public assistance? What kind of nonempirical pleasure arises between Americans who see the prison as a paradigmatic product of contemporary capitalism as well as racism and public fear of drug users? Such a point of view enables the "body politic" to be reconceived in terms of its "carceral" management. Likewise, the thoughts and acts identified as "homosexual" became the basis for rethinking "heterosexuality" as a norm, not to mention (in the age of AIDS) the political and economic function of the family and health care services across national borders. These are forms of aesthetic apprehension like the one recognizing factories and bodies destroyed by capitalist exploitation as *problems* demanding reorganization of the liberal political field as well as immediate resistance to work—through strikes and machine-breaking. But they are also styles of perception that allow us to recognize Marx's revolution in thought as one of many possible acts of resistance, one of many possible deployments of "critique," and thus one which can be repeated, set free amid "similar" deployments, rather than considered the inevitable horizon of all oppositional political thought.

So when I say that the body matters for political imagination, it is not the empirical body taken as an object by medicine, but the body as a matrix enabling us to establish affinities and differences from others in a range of public spaces, as well as to develop a notion of privacy. Understood in this way, the body remains material and discursive. Although it can be made visible or provoked to speak by the right circumstances, it is probably more "invisible" than visible. This body is stronger, able to support physical risks as well as the emotional risks represented by the existence of fellow human beings, if it is supported by minimal political and economic guarantees. The body is also stronger when the community in relation to which it is individuated feels that its habits of sensibility can survive and innovate in the face of new cultural phenomena (Cornell 1998, 156, 167–69). But these guarantees, including human rights legislation, only take us to the threshold where the body learns capacities that have not yet been realized in the social landscape.

Attention to the body is only "depoliticizing" and "privatizing" if one forgets that the senses are involved in a body that is never completely empirical, never fully schematized, and never fully individuated. Indeed, it is amusing to think that "privatizing" could mean "depoliticizing," given the immense political changes that have been effected through "privatization" of business and state functions over the last twenty years. What we want to resist are particular *ways* of privatization (and collectivization), not processes associated with these words *tout court*. Although the state may be involved in coordinating satisfactions, it should not be looked at too rapturously. Like Medusa's head, it can give the impression of being all-giving and all-denying and all-defining, so that people forget about their capacity to *feel* with non-state organs (or even to invent some). The feeling of normativity that the state ought to *protect* is not born in the state but on the borderline where one's sense of being visible alternates with one's sense of being invisible, public and private by turns.

REFERENCES

LIST OF ABBREVIATIONS FOR WORKS BY KANT

AN *Anthropology from a Pragmatic Point of View.* 1798. Translated and edited by Robert B. Louden. Cambridge: Cambridge University Press, 2006.
CJ *Critique of Judgment.* 1790–93. Translated by Werner S. Pluhar. Indianapolis: Hackett, 1987.
CPR *Critique of Pure Reason.* 1781–87. Translated by Werner S. Pluhar. Indianapolis: Hackett, 1996.
CPrR *Critique of Practical Reason.* 1788. Translated and edited by Mary Gregor. Cambridge: Cambridge University Press, 1997.
GR *Groundwork of the Metaphysics of Morals.* 1785. Translated and edited by Mary Gregor. Cambridge: Cambridge University Press, 1997.
KPW *Kant's Political Writings.* Translated by H. B. Nisbet. Edited by Hans Reiss. Second edition. Cambridge: Cambridge University Press, 1991.
MM *The Metaphysics of Morals.* 1797. Translated and edited by Mary Gregor. Cambridge: Cambridge University Press, 1996.
PR *Prolegomena to Any Future Metaphysics That Will Be Able to Come Forward as a Science.* 1783. Translated and edited by Gary Hatfield. Cambridge: Cambridge University Press, 1997.

LIST OF ABBREVIATIONS FOR WORKS BY FOUCAULT

AK *The Archaeology of Knowledge and the Discourse on Language.* 1969. Translated by A. M. Sheridan Smith. New York: Pantheon Books, 1972.
AME *Aesthetics, Method, and Epistemology.* Edited by James D. Faubion. Volume 2 of *Essential Works of Foucault, 1954–1984.* New York: New Press, 1998.
BC *The Birth of the Clinic: An Archaeology of Medical Perception.* 1963. Translated by A. M. Sheridan Smith. New York: Vintage Books, 1994.
DE *Dits et écrits, 1954–1988.* Four volumes. Edited by Daniel Defert and François Ewald. Paris: Éditions Gallimard, 1994.
DP *Discipline and Punish: The Birth of the Prison.* 1975. Translated by Alan Sheridan. New York: Vintage Books, 1979.
EST *Ethics: Subjectivity and Truth.* Edited by Paul Rabinow. Translated by Robert Hurley et al. Volume 1 of *Essential Works of Foucault, 1954–1984.* New York: New Press, 1997.
FL *Foucault Live (Interviews, 1966–84).* Translated by John Johnston. Edited by Sylvère Lotringer. New York: Semiotext(e), 1989.
HM *History of Madness.* 1961. Translated by Jonathan Murphy and Jean Khalfa. Edited by Jean Khalfa. London: Routledge, 2006.

HS1   *The History of Sexuality. Vol. I: An Introduction.* 1976. Translated by Robert Hurley. New York: Pantheon Books, 1978.
NB   *Naissance de la biopolitique. Cours au Collège de France, 1978–1979.* Edited by Michel Senellart, under the direction of François Ewald and Alessandro Fontana. Paris: Gallimard/Seuil, 2004.
OT   *The Order of Things: An Archaeology of the Human Sciences.* 1966. Translated by Alan Sheridan. New York: Vintage Books, 1973.
P/K   *Power/Knowledge: Selected Interviews and Other Writings, 1972–1977.* Edited by Colin Gordon. Translated by Colin Gordon, Leo Marshall, John Mepham, and Kate Soper. New York: Pantheon Books, 1980.
PWR   *Power.* Edited by James Faubion. Translated by Robert Hurley et al. Volume 3 of *Essential Works of Foucault, 1954–1984.* New York: New Press, 2000.
SMBD *"Society Must Be Defended": Lectures at the Collège de France, 1975–1976,* 1997. Translated by David Macey. New York: Picador, 2003.
STP   *Sécurité, térritoire, population. Cours au Collège de France, 1977–1978.* Edited by Michel Senellart, under the direction of François Ewald and Alessandro Fontana. Paris: Gallimard/Seuil, 2004.

OTHER WORKS CITED
(INCLUDING OTHER WORKS BY KANT AND FOUCAULT)

Adair, Vivyan C. 2003. "Disciplined and Punished: Poor Women, Bodily Inscription, and Resistance Through Education." In *Reclaiming Class: Women, Poverty, and the Promise of Higher Education in America,* ed. Vivyan C. Adair and Sandra L. Dahlberg, 25–52. Philadelphia: Temple University Press.
Althusser, Louis. 1971. "Ideology and Ideological State Apparatuses (Notes Towards an Investigation)." In *Lenin and Philosophy, and Other Essays,* trans. Ben Brewster, 127–86. New York: Monthly Review Press.
———. 1979. "On the Young Marx." In *For Marx,* trans. Ben Brewster, 49–86. London: Verso. (Orig. pub. 1965.)
Althusser, Louis, and Etienne Balibar. 1977. *Reading Capital.* London: NLB. (Orig. pub. 1968.)
Anderson, Benedict. 1983. *Imagined Communities: Reflections on the Origin and Spread of Nationalism.* London: Verso and NLB.
Appadurai, Arjun. 1996. *Modernity at Large: Cultural Dimensions of Globalization.* Minneapolis: University of Minnesota Press.
Arendt, Hannah. 1958. *The Human Condition.* Chicago: University of Chicago Press.
———. 1968. *Between Past and Future: Eight Exercises in Political Thought.* New York: Penguin Books.
———. 1979. "On Hannah Arendt." In *Hannah Arendt: The Recovery of the Public World,* ed. Melvyn A. Hill, 301–39. New York: St. Martin's Press.
———. 1982. *Lectures on Kant's Political Philosophy.* Edited by Ronald Beiner. Chicago: University of Chicago Press.
Aristotle. 1941. *Physics.* In *The Basic Works of Aristotle,* ed. Richard McKeon. New York: Random House.

Armstrong, Karen. 2001. *The Battle for God: A History of Fundamentalism*. New York: Ballantine Books.
Aronowitz, Stanley. 2000. "Unions as Counter-Public Spheres." In *Masses, Classes, and the Public Sphere*, ed. Mike Hill and Warren Montag, 83–101. London: Verso.
Balibar, Etienne. 1992. "Foucault and Marx: The Question of Nominalism." In *Michel Foucault, Philosopher*, ed. and trans. Timothy J. Armstrong, 38–57. London: Routledge.
———. 1994. *Masses, Classes, Ideas: Studies on Politics and Philosophy Before and After Marx*. Translated by James Swenson. New York: Routledge.
Bartky, Sandra Lee. 1990. *Femininity and Domination: Studies in the Phenomenology of Oppression*. New York: Routledge.
Barthes, Roland. 1972. *Mythologies*. Translated by Annette Lavers. New York: Hill and Wang. (Orig. pub. 1957.)
Bartolovich, Crystal. 2000. "Inventing London." In *Masses, Classes, and the Public Sphere*, ed. Mike Hill and Warren Montag, 13–40. London: Verso.
Beauvoir, Simone de. 1994. *The Ethics of Ambiguity*. Translated by Bernard Frechtman. New York: Citadel Press. (Orig. pub. 1947.)
Becker, Gary. 1995. "Nobel Lecture: The Economic Way of Looking at Behavior." In *The Essence of Becker*, ed. Ramón Febrero and Pedro S. Schwartz. Stanford: Hoover Institution Press.
Beiser, Frederick C. 1987. *The Fate of Reason: German Philosophy from Kant to Fichte*. Cambridge: Harvard University Press.
Benjamin, Walter. 1969. "The Work of Art in the Age of Mechanical Reproduction." In *Illuminations: Essays and Reflections*, ed. Hannah Arendt, 217–51. New York: Schocken Books.
———. 1978. "On the Mimetic Faculty." In *Reflections*, trans. Edmund Jephcott, ed. Peter Demetz, 333–36. New York: Harcourt Brace Jovanovich.Bernauer, James W. 1990. *Michel Foucault's Force of Flight: Toward an Ethics for Thought*. Atlantic Highlands, N.J: Humanities Press.
Bhabha, Homi. 1994. "Of Mimicry and Man: The Ambivalence of Colonial Discourse." In *The Location of Culture*, 85–92. London: Routledge.
Borges, Jorge Luis. 1964. "The Analytical Language of John Wilkins." In *Other Inquisitions: 1937–1952*, trans. Ruth L.C. Simms, 101–5. Austin: University of Texas Press. (Orig. pub. 1952.)
Braidotti, Rosi. 1991. *Patterns of Dissonance: A Study of Women in Contemporary Philosophy*. Translated by Elizabeth Guild. New York: Routledge.
Brennan, Teresa. 2004. *The Transmission of Affect*. Ithaca: Cornell University Press.
Brown, Wendy. 1995. *States of Injury: Power and Freedom in Late Modernity*. Princeton: Princeton University Press.
Brunschwig, Henri. 1974. *Enlightenment and Romanticism in Eighteenth-Century Prussia*. Translated by Frank Jellinek. Chicago: University of Chicago Press. (Orig. pub. 1947.)
Buder, Leonard. 1984. "Police Kill Woman Being Evicted, Officers Say She Wielded a Knife." *New York Times*, October 30, 1984, B3, col. 5.
Burchell, Graham, Colin Gordon, and Peter Miller, eds. 1991. *The Foucault Effect: Studies in Governmentality, with Two Lectures by and an Interview with Michel Foucault*. Chicago: University of Chicago Press.

Butler, Judith. 1990. *Gender Trouble: Feminism and the Subversion of Identity.* New York: Routledge.
——. 1993. *Bodies That Matter: On the Discursive Limits of "Sex."* New York: Routledge.
——. 1997. *The Psychic Life of Power: Theories in Subjection.* Stanford: Stanford University Press.
Calhoun, Craig, ed. 1992. *Habermas and the Public Sphere.* Cambridge: MIT Press.
Caillois, Roger. 1984. "Mimicry and Legendary Psychasthenia." Translated by John Shepley. *October* 31:17–32.
Canguilhem, Georges. 1991. *The Normal and the Pathological.* Translated by Carolyn R. Fawcett. New York: Zone Books.
Carpenter, Andrew. 1998. *Kant's Earliest Solution to the Mind-Body Problem* and *True Estimation of Living Forces* (translation of Kant). Ann Arbor: UMI.
Casey, Edward S. 1976. *Imagining: A Phenomenological Study.* Bloomington: Indiana University Press.
Castoriadis, Cornelius. 1987. *The Imaginary Institution of Society.* Translated by Kathleen Blamey. Cambridge: MIT Press. (Orig. pub. 1975.)
Caygill, Howard. 1989. *Art of Judgement.* Oxford: Basil Blackwell.
——. 2000. "Life and Aesthetic Pleasure." In *The Matter of Critique: Readings in Kant's Philosophy,* ed. Andrea Rehberg and Rachel Jones, 79–91. Manchester: Clinamen Press.
Coontz, Stephanie. 1992. *The Way We Never Were: American Families and the Nostalgia Trap.* New York: Basic Books.
Copjec, Joan. 1994. *Read My Desire: Lacan Against the Historicists.* Cambridge: MIT Press.
——. 2000. "The Body as Viewing Instrument, or the Strut of Vision." In *Lacan in America,* ed. Jean-Michel Rabaté. New York: Other Press.
Cornell, Drucilla. 1995. *The Imaginary Domain: Abortion, Pornography, and Sexual Harassment.* New York: Routledge.
——. 1998. *At the Heart of Freedom: Feminism, Sex, and Equality.* Princeton: Princeton University Press.
Crossley, Nick, and John Michael Roberts, eds. 2004. *After Habermas: New Perspectives on the Public Sphere.* Oxford: Blackwell / The Sociological Review.
Cruikshank, Barbara. 1996. "Revolutions Within: Self-government and Self-esteem." In *Foucault and Political Reason: Liberalism, Neo-liberalism and Rationalities of Government,* ed. Andrew Barry, Thomas Osborne, and Nikolas Rose, 231–51. Chicago: University of Chicago Press.
Cutrofello, Andrew. 1994. *Discipline and Critique: Kant, Poststructuralism, and the Problem of Resistance.* Albany: State University of New York Press.
——. 1997. *Imagining Otherwise: Metapsychology and the Analytic A Posteriori.* Evanston: Northwestern University Press.
Daddabbo, Leonardo. 1999. *Tempocorpo: Forme temporali in Michel Foucault.* Napoli: La Città del Sole.
Daniel, Jamie Owen. 2000. "Rituals of Disqualification: Competing Publics and Public Housing in Contemporary Chicago." In *Masses, Classes, and the Public Sphere,* ed. Mike Hill and Warren Montag, 62–82. London: Verso.
David-Ménard, Monique. 1990. *La folie dans la raison pure: Kant lecteur de Swedenborg.* Paris: Vrin.

———. "La laboratoire de l'oeuvre." In *Michel Foucault: Lire l'oeuvre*, ed. Luce Giard, 27–36. Grenoble: Jérôme Millon.

———. 1997. *Les constructions de l'universel: Psychanalyse, philosophie.* Paris: Presses Universitaires de France.

Davidson, Arnold, I., ed. 1997. *Foucault and His Interlocutors.* Chicago: University of Chicago Press.

Davoine, Françoise, and Jean-Max Gaudillière. 2004. *History Beyond Trauma.* Translated by Susan Fairfield. New York: Other Press.

Dean, Mitchell. 1991. *The Constitution of Poverty: Toward a Genealogy of Liberal Governance.* London: Routledge.

Delanda, Manuel. 1997. *A Thousand Years of Nonlinear History.* New York: Zone Books.

Deleuze, Gilles. 1983. *Nietzsche and Philosophy.* Translated by Hugh Tomlinson. New York: Columbia University Press. (Orig. pub. 1962.)

———. 1988. *Foucault.* Translated and edited by Séan Hand. Minneapolis: University of Minnesota Press. (Orig. pub. 1986.)

———. 1990. *The Logic of Sense.* Translated by Mark Lester with Charles Stivale. Edited by Constantin V. Boundas. New York: Columbia University Press. (Orig. pub. 1969.)

———. 1994. *Difference and Repetition.* Translated by Paul Patton. New York: Columbia University Press. (Orig. pub. 1968.)

Derrida, Jacques, 1982. *Margins of Philosophy.* Translated by Alan Bass. Chicago: University of Chicago Press. (Orig. pub. 1972.)

De Man, Paul. 1984. "Phenomenality and Materiality in Kant." In *Hermeneutics: Questions and Prospects,* ed. Gary Shapiro and Alan Sica, 121–44. Amherst: University of Massachusetts Press.

Dews, Peter. 1979. "The *Nouvelle Philosophie* and Foucault." *Economy and Society* 8 (2): 127–71.

———. 1987. *Logics of Disintegration: Post-Structuralist Thought and the Claims of Critical Theory.* London: Verso.

Donzelot, Jacques. 1979. *The Policing of Families.* Translated by Robert Hurley. New York: Pantheon Books. (Orig. pub. 1977.)

Dosse, François. 1994. *New History in France: The Triumph of the "Annales."* Translated by Peter V. Conroy, Jr. Urbana: University of Illinois Press. (Orig. pub. 1987.)

Dreyfus, Hubert L. 1996. "Being and Power: Heidegger and Foucault." *International Journal of Philosophical Studies* 4 (1): 1–16.

Dreyfus, Hubert, and Paul Rabinow. 1983. *Michel Foucault: Beyond Structuralism and Hermeneutics.* Second edition. Chicago: University of Chicago Press.

During, Simon. 1992. *Foucault and Literature: Towards a Genealogy of Writing.* London: Routledge.

Ehrenreich, Barbara. 2004. "The Faith Factor." *Nation,* November 29, 2004, 6–7.

Ewald, François. 1986. *L'état-providence.* Paris: B. Grasset.

———. 1993. "Two Infinities of Risk." In *The Politics of Everyday Fear,* ed. Brian Massumi, 221–28. Minneapolis: University of Minnesota Press.

Eze, Emmanuel Chukwudi. 1995. "The Color of Reason: The Idea of 'Race' in Kant's Anthropology." In *Anthropology and the German Enlightenment: Perspec-*

tives on Humanity, ed. Katherine M. Faull. Lewisburg: Bucknell University Press.

———, ed. 1997. *Race and the Enlightenment: A Reader*. Cambridge, Mass.: Blackwell.

Farge, Arlette. 1995. *Subversive Words: Public Opinion in Eighteenth-Century France*. Translated by Rosemary Morris. University Park: Pennsylvania State University Press. (Orig. pub. 1992.)

Fausto-Sterling, Anne. 2000. *Sexing the Body: Gender Politics and the Construction of Sexuality*. New York: Basic Books.

Fenves, Peter D. 1991. *A Peculiar Fate: Metaphysics and World-History in Kant*. Ithaca: Cornell University Press.

———, ed. 1993. *Raising the Tone of Philosophy: Late Essays by Immanuel Kant, Transformative Critique by Jacques Derrida*. Baltimore: Johns Hopkins University Press.

Ferry, Luc, and Alain Renaut. 1990. *French Philosophy of the Sixties: An Essay on Antihumanism*. Translated by Mary H.S. Cattani. Amherst: University of Massachusetts Press. (Orig. pub. 1985.)

Fimiani, Mariapaola. 1998. *Foucault et Kant: Critique clinique éthique*. Translated from the Italian by Nadine Le Lirzan. Paris: L'Harmattan.

Flynn, Thomas R. 2005. *Sartre, Foucault, and Historical Reason*. Vol. 2, *A Poststructuralist Mapping of History*. Chicago: University of Chicago Press.

Focillon, Henri. 1989. *The Life of Forms in Art*. Translated by Charles B. Hogan and George Kubler. New York: Zone Books. (Orig. pub. 1934.)

Foucault, Michel. 1961. *Introduction à l'anthropologie de Kant: Thèse complementaire pour le doctorat ès lettres*. Ph.D. diss, Centre Michel Foucault, L'IMEC, Paris.

———. 1983. *This Is Not a Pipe*. Translated and edited by James Harkness. Berkeley and Los Angeles: University of California Press. (Orig. pub. 1973.)

———. 1984. *The Foucault Reader*. Edited by Paul Rabinow. New York: Pantheon Books.

———. 1985. *The Use of Pleasure*. Vol. 2 of *The History of Sexuality*. Translated by Robert Hurley. New York: Vintage Books. (Orig. pub. 1984.)

———. 1986a. *Death and the Labyrinth: The World of Raymond Roussel*. Translated by Charles Ruas. Garden City, N.Y.: Doubleday. (Orig. pub. 1963.)

———. 1986b. "Dream, Imagination, and Existence." Translated by Forrest Williams. *Review of Existential Psychology and Psychiatry* 19 (1): 29–78. (Orig. pub. 1954.)

———. 1997. *The Politics of Truth*. Edited by Sylvère Lotringer and Lysa Hochroth. New York: Semiotext(e).

Fraser, Nancy. 1989. *Unruly Practices: Power, Discourse, and Gender in Contemporary Social Theory*. Minneapolis: University of Minnesota Press.

———. 1992. "Rethinking the Public Sphere: A Contribution to the Critique of Actually Existing Democracy." In *Habermas and the Public Sphere*, ed. Craig Calhoun, 109–42. Cambridge: MIT Press.

Fraser, Nancy, and Linda Gordon. 1996. "The Genealogy of Dependency: Tracing a Keyword of the U.S. Welfare State." In *For Crying Out Loud: Women's Poverty in the United States*, ed. Diane Dujon and Ann Withorn, 235–67. Boston: South End Press.

Freud, Sigmund. 1961. "The Ego and the Id." In *The Ego and the Id and Other Works*, vol. 19 of *The Standard Edition of the Complete Psychological Works of Sigmund Freud*, ed. and trans. James Strachey. London: Hogarth Press. (Orig. pub. 1923.)

Freund, Peter E.S., and Miriam Fisher. 1982. *The Civilized Body: Social Domination, Control, and Health*. Philadelphia: Temple University Press.

Gardiner, Michael E. 2004. "Wild Publics and Grotesque Symposiums: Habermas and Bakhtin on Dialogue, Everyday Life and the Public Sphere." In *After Habermas: New Perspectives on the Public Sphere*, ed. Nick Crossley and John Michael Roberts. Oxford: Blackwell / The Sociological Review.

Gatens, Moira. 1996. *Imaginary Bodies: Ethics, Power, and Corporeality*. London: Routledge.

Gennochio, Benjamin. 1995. "Discourse, Discontinuity, Difference: The Question of 'Other' Spaces." In *Postmodern Cities and Spaces*, ed. Sophie Watson and Katherine Gibson. Cambridge: Blackwell.

Gibbons, Sarah L. 1994. *Kant's Theory of Imagination: Bridging Gaps in Judgement and Experience*. Oxford: Clarendon Press.

Gil, José. 1985. "L'espace du corps." *Le Cahier du Collège International de Philosophie* 1:94–97.

———. 1998. *Metamorphoses of the Body*. Translated by Stephen Muecke. Minneapolis: University of Minnesota Press. (Orig. pub. 1995.)

Gilens, Martin. 2003. "How the Poor Became Black: The Racialization of American Poverty in the Mass Media." In *Race and the Politics of Welfare Reform*, ed. Sanford F. Schram, Joe Soss, and Richard C. Fording, 101–30. Ann Arbor: University of Michigan Press.

Gitlin, Todd. 1989. *The Sixties: Years of Hope, Days of Rage*. New York: Bantam Books.

Goetschel, Willi. 1994. *Constituting Critique: Kant's Writing as Critical Praxis*. Translated by Eric Schwab. Durham: Duke University Press.

Gould, Stephen Jay. 1981. *The Mismeasure of Man*. New York: W. W. Norton.

Greider, William. 1992. *Who Will Tell the People: The Betrayal of American Democracy*. New York: Touchstone.

Grosz, Elizabeth. 1994. *Volatile Bodies: Toward a Corporeal Feminism*. Bloomington: Indiana University Press.

———. 1995. *Space, Time, and Perversion: Essays on the Politics of Bodies*. New York: Routledge.

Gutting, Gary. 1989. *Michel Foucault's Archeology of Scientific Reason*. Cambridge: Cambridge University Press.

Guyer, Paul. 1997. *Kant and the Claims of Taste*. Second edition. Cambridge: Cambridge University Press.

Habermas, Jürgen. 1987. *The Theory of Communicative Action*. Vol. 2 of *Lifeworld and System: A Critique of Functionalist Reason*. Translated by Thomas McCarthy. Boston: Beacon Press.

———. 1990. *The Philosophical Discourse of Modernity: Twelve Lectures*. Translated by Frederick Lawrence. Cambridge: MIT Press. (Orig. pub. 1985.)

———. 1996. *Between Facts and Norms: Contributions to a Discourse Theory of Law and Democracy*. Translated by William Rehg. Cambridge: MIT Press. (Orig. pub. 1992.)

———. 1999. *The Structural Transformation of the Public Sphere: An Inquiry into a Category of Bourgeois Society*. Translated by Thomas Burger with the assistance of Frederick Lawrence. Cambridge: MIT Press. (Orig. pub. 1962.)

Hacking, Ian. 1975. *Why Does Language Matter to Philosophy?* Cambridge: Cambridge University Press.

Hahn, Lewis Edwin, ed. 2000. *Perspectives on Habermas*. Chicago: Open Court.

Han, Béatrice. 2002. *Foucault's Critical Project: Between the Transcendental and the Historical*. Translated by Edward Pile. Stanford: Stanford University Press. (Orig. pub. 1998.)

Handler, Joel F., and Yeheskel Hasenfeld. 1991. *The Moral Construction of Poverty: Welfare Reform in America*. Newbury Park, Calif.: Sage Publications.

Hartman, Saidiya V. 1997. *Scenes of Subjection: Terror, Slavery, and Self-Making in Nineteenth-Century America*. New York: Oxford University Press.

Hartsock, Nancy. 1990. "Foucault on Power: A Theory for Women?" In *Feminism/Postmodernism*, ed. Linda J. Nicholson, 157–75. New York: Routledge.

Harvey, David. 1990. *The Condition of Postmodernity: An Enquiry into the Origins of Cultural Change*. Oxford: Blackwell.

———. 2000. *Spaces of Hope*. Berkeley and Los Angeles: University of California Press.

———. 2005. *A Brief History of Neoliberalism*. Oxford: Oxford University Press.

Heidegger, Martin. 1977a. *Basic Writings*. Translated and edited by David Farrell Krell. San Francisco: Harper & Row.

———. 1977b. *The Question Concerning Technology and Other Essays*. Translated by William Lovitt. New York: Harper & Row.

———. 1990. *Kant and the Problem of Metaphysics*. Fourth edition, enlarged. Translated by Richard Taft. Bloomington: Indiana University Press. (Orig. pub. 1929.)

Henrich, Dieter. 1994. "On the Unity of Subjectivity." In *The Unity of Reason: Essays on Kant's Philosophy*, ed. Richard L. Velkley, trans. Guenter Zoeller, 17–54. Cambridge: Harvard University Press.

Hetherington, Kevin. 1997. *The Badlands of Modernity: Heterotopia and Social Ordering*. London: Routledge.

Hill, Mike, and Warren Montag, eds. 2000. *Masses, Classes, and the Public Sphere*. London: Verso.

Holland, Catherine A. 2001. *The Body Politic: Foundings, Citizenship, and Difference in the American Political Imagination*. New York: Routledge.

Horkheimer, Max, and Theodor W. Adorno. 1995. *Dialectic of Enlightenment*. Translated by John Cumming. New York: Continuum. (Orig. pub. 1944.)

Hume, David. 1973. *A Treatise of Human Nature*. Edited by L. A. Selby-Bigge. Oxford: Clarendon Press. (Orig. pub. 1739.)

Hurtado, Aida. 1996. *The Color of Privilege: Three Blasphemies on Race and Feminism*. Ann Arbor: University of Michigan Press.

Husserl, Edmund. 1962. *Ideas: General Introduction to Pure Phenomenology*. Translated by W. R. Boyce Gibson. New York: Collier Books.

Hutchings, Kimberly. 1995. *Kant, Critique, and Politics*. London: Routledge.

Kant, Immanuel. 1900–1913. *Kants gesammelte Schriften*. Twenty-nine volumes. Berlin: Königlich Preußische Akademie der Wissenschaften.

———. 1965. *Observations on the Feeling of the Beautiful and Sublime*. Translated by John T. Goldthwait. Berkeley and Los Angeles: University of California Press. (Orig. pub. 1764.)
———. 1992a. *The Conflict of the Faculties*. Translated by Mary J. Gregor. Lincoln: University of Nebraska Press. (Orig. pub. 1798.)
———. 1992b. *Lectures on Logic*. Translated and edited by J. Michael Young. Cambridge: Cambridge University Press.
———. 1992c. *Theoretical Philosophy, 1755–1770*. Translated and edited by David Walford and Ralf Meerbote. Cambridge: Cambridge University Press.
Kearney, Richard. 1988. *The Wake of Imagination: Ideas of Creativity in Western Culture*. London: Hutchinson.
Kerszberg, Pierre. 1997. *Critique and Totality*. Albany: State University of New York Press.
Kingfisher, Catherine Pélissier. 1996. *Women in the American Welfare Trap*. Philadelphia: University of Pennsylvania Press.
Klossowski, Pierre. 1984. *La ressemblance*. Marseille: Éditions Ryôan-ji.
———. 1988. *The Baphomet*. Translated by Sophie Hawkes and Stephen Sartarelli. New York: Marsilio Publishers. (Orig. pub. 1965.)
———. 1997. *Nietzsche and the Vicious Circle*. Translated by Daniel W. Smith. Chicago: University of Chicago Press. (Orig. pub. 1969.)
Kristol, Irving. 2004. "A Conservative Welfare State." In *The Neocon Reader*, ed. Irwin Stelzer. New York: Grove Press.
Kubler, George. 1962. *The Shape of Time: Remarks on the History of Things*. New Haven: Yale University Press.
Kuhn, Thomas. 1996. *The Structure of Scientific Revolutions*. Third edition. Chicago: University of Chicago Press.
———. 2000. *The Road Since Structure: Philosophical Essays, 1970–1993*. Edited by James Conant and John Haugeland. Chicago: University of Chicago Press.
Kusch, Martin. 1991. *Foucault's Strata and Fields: An Investigation into Archaeological and Genealogical Science Studies*. Dordrecht: Kluwer Academic Publishers.
Lacan, Jacques. 1977. *Écrits: A Selection*. Translated by Alan Sheridan. New York: W. W. Norton.
Laclau, Ernesto, and Chantal Mouffe. 1985. *Hegemony and Socialist Strategy: Towards a Radical Democratic Politics*. London: Verso.
Lawlor, Leonard. 2003. *Thinking Through French Philosophy: The Being of the Question*. Bloomington: Indiana University Press.
Lear, Jonathan. 1988. *Aristotle: The Desire to Understand*. Cambridge: Cambridge University Press.
Lebrun, Gérard. 1970. *Kant et la fin de la métaphysique: Essai sur "La critique de la faculté de juger."* Paris: Librarie Armand Colin.
Lecourt, Dominique. 1975. *Marxism and Epistemology: Bachelard, Canguilhem, and Foucault*. Translated by Ben Brewster. London: NLB.
Leder, Drew. 1990. *The Absent Body*. Chicago: University of Chicago Press.
Lefebvre, Henri. 1991. *The Production of Space*. Translated by Donald Nicholson-Smith. Oxford: Blackwell.
Leibniz, G. W. 1996. *New Essays on Human Understanding*. Translated and edited

by Peter Remnant and Jonathan Bennett. Cambridge: Cambridge University Press. (Orig. pub. 1765.)

———. 1998. *Philosophical Texts.* Translated and edited by R. S. Woolhouse and Richard Francks. Oxford: Oxford University Press.

Levin, David Michael. 1989. "The Body Politic: The Embodiment of Praxis in Foucault and Habermas." *Praxis International* 9 (1/2): 112–32.

Lévi-Strauss, Claude. 1963. "The Structural Study of Myth." In *Structural Anthropology,* trans. Claire Jacobson and Brooke Grundfest Schoepf, 206–31. New York: Basic Books. (Orig. pub. 1958.)

———. 1968. *The Savage Mind.* Translated by George Weidenfeld and Nicolson Ltd. Chicago: University of Chicago Press. (Orig. pub. 1962.)

———. 1987. *Introduction to the Work of Marcel Mauss.* Translated by Felicity Baker. London: Routledge and Kegan Paul. (Orig. pub. 1950.)

Littlefield, Daniel C. 1995. "Blacks, John Brown, and a Theory of Manhood." In *His Soul Goes Marching On: Responses to John Brown and the Harpers Ferry Raid,* ed. Paul Finkelman. Charlottesville: University Press of Virginia, 67–97.

Loewen, James W. 1995. *Lies My Teacher Told Me: Everything Your American History Textbook Got Wrong.* New York: Touchstone.

Longuenesse, Beatrice. 1998. *Kant and the Capacity to Judge: Sensibility and Discursivity in the Transcendental Analytic of "The Critique of Pure Reason."* Translated by Charles T. Wolfe. Princeton: Princeton University Press. (Orig. pub. 1993.)

Losonsky, Michael. 2001. *Enlightenment and Action from Descartes to Kant: Passionate Thought.* Cambridge: Cambridge University Press.

Lowe, Donald M. 1995. *The Body in Late-Capitalist USA.* Durham: Duke University Press.

Lugones, María. 2003. "Playfulness, 'World'-Traveling, and Loving Perception." In *Pilgrimages/Peregrinajes: Theorizing Coalition Against Multiple Oppressions.* Lanham, Md.: Rowman and Littlefield.

Lyotard, Jean-François. 1989. *The Postmodern Condition: A Report on Knowledge.* Translated by Geoff Bennington and Brian Massumi. Minneapolis: University of Minnesota Press. (Orig. pub. 1979.)

———. 1991. *The Inhuman: Reflections on Time.* Translated by Geoffrey Bennington and Rachel Bowlby. Stanford: Stanford University Press. (Orig. pub. 1988.)

———. 1994. *Lessons on the Analytic of the Sublime.* Translated by Elizabeth Rottenberg. Stanford: Stanford University Press. (Orig. pub. 1991.)

Makkreel, Rudolf A. 1987. "Hermeneutics and the Limits of Consciousness." *Noûs* 21: 7–18.

———. 1990. *Imagination and Interpretation in Kant: The Hermeneutical Import of "The Critique of Judgment."* Chicago: University of Chicago Press.

Mansbridge, Jane J. 1986. *Why We Lost the ERA.* Chicago: University of Chicago Press.

Marcus, Sharon. 1992. "Fighting Bodies, Fighting Words: A Theory and Politics of Rape Prevention." In *Feminists Theorize the Political,* ed. Judith Butler and Joan W. Scott. New York: Routledge.

Marsden, Richard. 1999. *The Nature of Capital: Marx After Foucault*. London: Routledge.
Marx, Karl. 1977. *Capital: Volume I*. Translated by Ben Fowkes. New York: Vintage Books. (Orig. pub. 1867.)
Marx, Karl, and Frederick Engels. 1986. *The German Ideology, Part I*. Edited by C. J. Arthur. New York: International Publishers.
Mbembe, Achille. 2001. *On the Postcolony*. Berkeley and Los Angeles: University of California Press.
McGlone, Robert E. 1995. "John Brown, Henry Wise, and the Politics of Insanity." In *His Soul Goes Marching On: Responses to John Brown and the Harpers Ferry Raid*, ed. Paul Finkelman, 213–52. Charlottesville: University Press of Virginia.
McLuhan, Marshall. 1964. *Understanding Media: The Extensions of Man*. New York: McGraw-Hill.
McNay, Lois. 1992. *Foucault and Feminism: Power, Gender, and the Self*. Boston: Northeastern University Press.
McWhorter, Ladelle. 1999. *Bodies and Pleasures: Foucault and the Politics of Sexual Normalization*. Bloomington: Indiana University Press.
Megill, Allan. 1985. *Prophets of Extremity: Nietzsche, Heidegger, Foucault, Derrida*. Berkeley and Los Angeles: University of California Press.
Melnick, Arthur. 1989. *Space, Time, and Thought in Kant*. Dordrecht: Kluwer Academic Publishers.
Merleau-Ponty, Maurice. 1962. *Phenomenology of Perception*. Translated by Colin Smith. London: Routledge and Kegan Paul. (Orig. pub. 1945.)
Mernissi, Fatima. 2002. *Islam and Democracy: Fear of the Modern World*. Translated by Mary Jo Lakeland. Second edition. Cambridge, Mass.: Perseus Books.
Mills, Charles. 1997. *The Racial Contract*. Ithaca: Cornell University Press.
Nadasen, Premilla. 2002. "Expanding the Boundaries of the Women's Movement: Black Feminism and the Struggle for Welfare Rights." *Feminist Studies* 28 (2): 271–301.
Nedelsky, Jennifer. 1990. *Private Property and the Limits of American Constitutionalism: The Madisonian Framework and Its Legacy*. Chicago: University of Chicago Press.
Negt, Oskar, and Alexander Kluge. 1993. *Public Sphere and Experience: Toward an Analysis of the Bourgeois and Proletarian Public Sphere*. Translated by Peter Labanyi, Jamie Owen Daniel, and Assenka Oksiloff. Minneapolis: University of Minnesota Press. (Orig. pub. 1972.)
Nerlich, Graham. 1994. *The Shape of Space*. Second edition. Cambridge: Cambridge University Press.
Nietzsche, Friedrich. 1989. *On the Genealogy of Morals* and *Ecce Homo*. Translated by Walter Kaufmann. New York: Vintage Books. (Orig. pub. 1887.)
———. 1999. *Philosophy and Truth: Selections from Nietzsche's Notebooks of the Early 1870s*. Edited and translated by Daniel Breazale. Amherst, N.Y.: Humanity Books.
O'Malley, Pat. 1996. "Risk and Responsibility." In *Foucault and Political Reason: Liberalism, Neo-Liberalism and Rationalities of Government*, ed. Andrew Barry, Thomas Osborne, and Nikolas Rose, 189–207. Chicago: University of Chicago Press.

Owen, David. 1994. *Maturity and Modernity: Nietzsche, Weber, Foucault and the Ambivalence of Reason.* London: Routledge.
Patterson, Thomas. 2003. *The Vanishing Voter: Public Involvement in an Age of Uncertainty.* New York: Vintage.
Peters, Michael A. 2001. *Poststructuralism, Marxism, and Neoliberalism: Between Theory and Politics.* Lanham, Md.: Rowman and Littlefield.
Plato. 1961. *The Collected Dialogues of Plato.* Edited by Edith Hamilton and Huntington Cairns. Bollingen Series 71. Princeton: Princeton University Press.
Polis, Dennis. 1991. "A New Reading of Aristotle's *Hyle.*" *Modern Schoolman* 68:225–44.
Poster, Mark. 1984. *Foucault, Marxism, and History: Mode of Production Versus Mode of Information.* Cambridge: Polity Press.
Price, Daniel. 1997. *Without a Woman to Read: Toward the Daughter in Postmodernism.* Albany: State University of New York Press.
Proust, Françoise. 1991. "Introduction." In *Vers la paix perpétuelle, que signifie s'orienter dans la pensée? Qu'est-ce que les lumières?* trans. Jean-François Poirier and Françoise Proust. Paris: Flammarion.
Pryor, Ben. 1998. "Counter-Remembering the Enlightenment." In *Conflicts and Consequences: Selected Studies in Phenomenology and Existential Philosophy* 24, ed. Linda Martin Atcoff and Merold Westphal. SPEP Special Supplement. *Philosophy Today* 42:147–59.
Quadagno, Jill. 1994. *The Color of Welfare: How Racism Undermined the War on Poverty.* New York: Oxford University Press.
Quine, W. V. 1969. *Ontological Relativity and Other Essays.* New York: Columbia University Press.
Raab, Selwyn. 1984a. "Eviction Death Leads the City to Demote Two." *New York Times,* November 21, 1984, B3, col. 1.
———. 1984b. "Police and Victim's Daughter Clash on Shooting." *New York Times,* November 1, 1984, B5, col. 1.
———. 1985. "State Judge Dismisses Indictment of Officer in the Bumpurs Killing." *New York Times,* April 13, 2006, A1, col. 2.
Racevskis, Karlis. 1983. *Michel Foucault and the Subversion of Intellect.* Ithaca: Cornell University Press.
Rajchman, John. 1985. *Michel Foucault: The Freedom of Philosophy.* New York: Columbia University Press.
Ransom, John S. 1997. *Foucault's Discipline: The Politics of Subjectivity.* Durham: Duke University Press.
Rawls, John. 1971. *A Theory of Justice.* Cambridge: Belknap Press of Harvard University Press.
Redding, Paul. 1999. *The Logic of Affect.* Ithaca: Cornell University Press.
Rehberg, Andrea, and Rachel Jones, eds. 2000. *The Matter of Critique: Readings in Kant's Philosophy.* Manchester: Clinamen Press.
Renaut, Alain. 1993. "Kant, penseur de la modernité." Interview with François Ewald. *Magazine Littéraire* 309:18–22.
Ricken, Ulrich. 1994. *Linguistics, Anthropology, and Philosophy in the French Enlightenment.* Translated by Robert E. Norton. London: Routledge. (Orig. pub. 1984.)

Roberts, John Michael. 2003. *The Aesthetics of Free Speech: Rethinking the Public Sphere*. Basingstoke, U.K.: Palgrave.
Rogozinski, Jacob. 1996. *Kanten: Esquisses Kantiennes*. Paris: Éditions Kimé.
Ross, Kristin. 2002. *May '68 and Its Afterlives*. Chicago: University of Chicago Press.
Sacks, Oliver. 1987. "The Possessed." In *The Man Who Mistook His Wife for a Hat and Other Clinical Tales*, 120–25. New York: Harper and Row.
Salecl, Renata. 1994. *The Spoils of Freedom: Psychoanalysis and Feminism After the Fall of Socialism*. London: Routledge.
———. 2004. *On Anxiety*. London: Routledge.
Sallis, John. 1987. *Spacings—of Reason and Imagination: In Texts of Kant, Fichte, Hegel*. Chicago: Chicago University Press.
Scarry, Elaine. 1985. *The Body in Pain: The Making and Unmaking of the World*. New York: Oxford University Press.
Schott, Robin May. 1988. *Cognition and Eros: A Critique of the Kantian Paradigm*. Boston: Beacon Press.
Schrader, George. 1976. "The Status of Feeling in Kant's Philosophy." In *Proceedings of the Ottowa Congress on Kant in the Anglo-American and Continental Traditions, Held October 10–14, 1974*, ed. Pierre Laberge, François Duchesneau, and Bryan E. Morrisey. Ottowa: University of Ottowa Press.
Scott, James C. 1998. *Seeing Like a State: How Certain Schemes to Improve the Human Condition Have Failed*. New Haven: Yale University Press.
Sen, Amartya. 1992. *Inequality Reexamined*. New York: Russell Sage Foundation; Cambridge: Harvard University Press.
Sennett, Richard. 1998. *The Corrosion of Character: The Personal Consequences of Work in the New Capitalism*. New York: W. W. Norton.
Sennett, Richard, and Jonathan Cobb. 1972. *The Hidden Injuries of Class*. New York: W. W. Norton.
Shell, Susan Meld. 1996. *The Embodiment of Reason: Kant on Spirit, Generation, and Community*. Chicago: Chicago University Press.
Simons, Jon. 1995. *Foucault and the Political*. London: Routledge.
Singer, Daniel. 1999. *Whose Millennium? Theirs or Ours?* New York: Monthly Review Press.
Smart, Barry. 1983. *Foucault, Marxism, and Critique*. London: Routledge and Kegan Paul.
Smith, Ruth. 1991. "Order and Disorder: The Naturalization of Poverty." In *The American Constitutional Experiment*, ed. David M. Speak and Creighton Peden, 317–4. Studies in Social and Political Theory, vol. 12, no. 5. Lewiston, N.Y.: Edwin Mellen Press.
Solinger, Rickie. 2000. *Wake Up Little Susie: Single Pregnancy and Race Before Roe v. Wade*. New York: Routledge.
Spinoza, Baruch. 1982. *The Ethics and Selected Letters*. Translated by Samuel Shirley. Edited by Seymour Feldman. Indianapolis: Hackett Publishing. (Orig. pub. 1677.)
Stoler, Ann Laura. 1995. *Race and the Education of Desire: Foucault's "History of Sexuality" and the Colonial Order of Things*. Durham: Duke University Press.

Strathern, Marilyn. 1991. *Partial Connections*. Savage, Md.: Rowman and Littlefield.
Sunstein, Cass. 2001. *Republic.com*. Princeton: Princeton University Press.
Taussig, Michael. 1992. *The Nervous System*. New York: Routledge.
———. 1993. *Mimesis and Alterity: A Particular History of the Senses*. New York: Routledge.
Terada, Rei. 2001. *Feeling in Theory: Emotion After the "Death of the Subject."* Cambridge: Harvard University Press.
Tonda, Joseph. 2002. "Économie des miracles et dynamiques de 'subjectivation/civilization' en Afrique centrale." *Politique Africaine* 87:22–44.
Tronto, Joan. 1993. *Moral Boundaries: A Political Argument for an Ethic of Care*. New York: Routledge.
Unger, Roberto Mangabeira. 1986. *The Critical Legal Studies Movement*. Cambridge: Harvard University Press.
Virno, Paolo, and Michael Hardt, eds. 1996. *Radical Thought in Italy: A Potential Politics*. Minneapolis: University of Minnesota Press.
Von Dirke, Sabine. 1997. *"All Power to the Imagination!" The West German Counterculture from the Student Movement to the Greens*. Lincoln: University of Nebraska Press.
Wallenstein, Peter. 1995. "Incendiaries All: Southern Politics and the Harpers Ferry Raid." In *His Soul Goes Marching On: Responses to John Brown and the Harpers Ferry Raid*, ed. Paul Finkelman, 149–73. Charlottesville: University Press of Virginia.
Warner, Michael. 2002. *Publics and Counterpublics*. New York: Zone Books.
Wiegel, Sigrid. 1996. *Body- and Image-Space: Re-reading Walter Benjamin*. London: Routledge.
Wilkinson, Richard G. 1996. *Unhealthy Societies: The Afflictions of Inequality*. London: Routledge.
Williams, Patricia J. 1991. *The Alchemy of Race and Rights: Diary of a Mad Law Professor*. Cambridge: Harvard University Press.
Wilson, James Q., and George L. Kelling. 2004. "Broken Windows: The Police and Neighborhood Safety." In *The Neocon Reader*, ed. Irwin Stelzer, 149–66. New York: Grove Press.
Wolin, Richard. 1986. "Foucault's Aesthetic Decisionism." *Telos* 67:71–86.
Woodmansee, Martha. 1994. *The Author, Art, and the Market: Rereading the History of Aesthetics*. New York: Columbia University Press.
Yergin, Daniel, and Joseph Stanislaw. 2002. *The Commanding Heights: The Battle for the World Economy*. New York: Touchstone.
Young, Iris Marion. 1990. *Justice and the Politics of Difference*. Princeton: Princeton University Press.
Zammito, John H. 1992. *The Genesis of Kant's "Critique of Judgment."* Chicago: University of Chicago Press.
Žižek, Slavoj. 1989. *The Sublime Object of Ideology*. London: Verso.
———. 1993. *Tarrying with the Negative: Kant, Hegel, and the Critique of Ideology*. Durham: Duke University Press.

INDEX

advertising, 108, 245–46, 259, 280, 285
aesthetic, transcendental
  body and, 44–45, 89–92, 94, 112–13
  space, 44–45, 68, 89–92, 112–13
  time, 48, 94, 109–10
  See also body; scale of apprehension; sensibility; space; temporality
aestheticization of politics, 291–94
African-Americans, 4–5, 102–4, 185, 217–18, 239, 241–42, 258, 276–79
  slavery, 4–5, 239, 241–42, 258
Althusser, Louis, 159–60
Anderson, Benedict, 19, 219
Annales, 121–23, 160
anthropology, philosophical, 34, 260, 282, 297–98
  human and inhuman, 132, 211, 224, 228–31, 234, 260, 266, 283, 297
  Kant's views, 28, 47, 73, 89, 101, 120–21, 124, 155–56, 159, 175, 197
  See also anti-humanism; "Man"
anthropology, structural, 27, 42, 60, 159, 173, 197
anti-humanism, 57 n. 25, 158–60, 162–63
  critical, 16, 118, 298
  See also humanism
anti-Semitism, 185. *See also* racism, state
Appadurai, Arjun, 51, 53–54, 100, 170
a priori, historical, 162, 178–79
archaeology, 163, 174, 175–81
  and genealogy, 161–62, 168, 180, 211
  and power, 177–78
  See also a priori; historical; archive
archive, 153, 162–66
Arendt, Hannah, 65 n. 33, 82 n. 42, 100–101, 237, 256, 290–92, 301
Aristotle, 17, 69, 72 n. 35, 177, 187
asylum, 7, 20, 131–34, 151–54, 179, 231, 275
attention, 61, 111–12, 169, 244–46, 267
author, 126–27
  unity of oeuvre, 27–28, 127

Balibar, Etienne, 171
Barthes, Roland, 159–60

Bataille, Georges, 207
beauty, judgments of, 8, 200, 205
  and bodies, 99, 102, 211
  compared to empirical pleasure, 105
  and discursivity, 68, 169
  disinterestedness and universality of, 59–66
  and morality, 65 n. 32, 292–93
  relation to form, 76–77, 97–98
Beauvoir, Simone de, 298
Benjamin, Walter, 17, 184 n. 28
Bergson, Henri, 17
biopolitics, 222, 230–31, 235, 264–65, 276–81, 293
body in Kant's thought
  and bounds of reason, 14–15, 35–37, 90–93, 105, 110, 113–14
  and imagination, 106–7, 114, 117, 192–94
  individuating subjects, 47, 89–90, 95–96, 99–101, 106, 109, 150
  interaction with mind, 45, 89
  interaction with other bodies, 90, 93–95, 110, 113
  as object of experience, 7, 45, 49, 89–90, 95, 113–14
  and transcendental aesthetic, 44–45, 89–92, 94, 110, 112–13
  See also community of substances; individualization; real conflict
body
  and communicability, 90, 102–4, 106–8, 238, 240–41, 245–47, 257–58, 284
  in early modern philosophy, 68–69
  and *énoncés*, 180–81, 191–92, 196, 198
  and finitude, 260
  Foucault and Kant, 15
  and imagination, 5–6, 22–24
  and materiality, 68, 108–9, 113, 181, 186–89, 196, 208
  as measure, 95, 102–4, 253
  as post-Kantian problematic object, 6, 9, 17, 29, 90, 151
  and power, 6, 143–45, 181, 190–91, 235–36

body (*continued*)
    singular and plural, 2, 15, 95, 144, 150, 166
    in social critique, 117
    See also communicability; feeling; materiality
Brown, John, 4–5, 300
Bumpurs, Eleanor, 1–3, 5–6, 9, 11–13, 16, 24, 283, 300
Butler, Judith, 185, 187–89, 236

Caillois, Roger, 192 n. 29, 299
Canguilhem, Georges, 206, 249–53, 262
Caygill, Howard, 76
citizenship, 4, 24, 133, 217–18, 236, 271
    compared to clientelism, 10, 247, 249, 280, 290–91
class, socioeconomic, 141, 149, 186, 222, 239–41, 243, 264, 265, 261, 280
colonialism, 16, 102–3, 114, 123, 147–48, 185, 216–17, 222, 231 n. 9, 238, 240, 256, 260, 264
commentary. *See* interpretation
communicability, 218, 237, 243–47, 265–66
    and body, 90, 102, 104, 106–8, 188, 198, 204, 224, 238, 240–41, 245–47, 257–58
    of concepts, 49, 61–63, 99, 156
    of feeling, 65–67, 80, 108, 207, 253, 292
    and imagination, 81–83, 284, 297
    and language, 66, 69–70, 148, 175, 243
    and madness, 82–85
    and power, 102–5, 108, 126, 128, 242, 244, 253
    and publicity, 83, 85, 87–89, 100–102, 108, 110, 114, 158, 245
    and tone/music, 81–82, 85
    See also aesthetic ideas; feeling; genius
community
    imagined, 219, 241
    political, 19–21, 66, 87, 93–95, 99–100, 108–9
    of substances, 93–95, 97, 109, 112, 121
    See also nationalism; real conflict
concepts
    communicability of, 60–66
    generation and application of, 43–44, 61–62, 192
concepts, of reflection, 72–77
    inner and outer, 95–97
    matter and form, 72–75
    See also differences, inner; Kant, Immanuel, critique of Leibniz; materiality
concepts, problematic
    noumenon, 6–7, 40–44, 55, 59, 92, 129, 130–31, 229
    supersensible, 58–60, 62, 89, 99–100, 200
    See also object, problematic
confession, 148, 150, 152–53, 197–98, 204, 229, 281. *See also* interpretation
conflict, real, 35, 119, 127, 296. *See also* community
Constitution, American, 232, 243
contract, social, 5, 224, 262–63
    in Foucault, 149–50
    in Kant, 66–67, 81, 87–88
countercultures, 217, 219, 221, 294
critique
    as ethos, 58, 79, 88–89, 285–86
    Foucault and Kant, 8 n. 1, 31 n. 7, 34–35, 39, 47, 58, 67, 131, 155–56, 162, 197, 285–86, 288
    social criticism and body, 117
    social criticism and madness, 3–6, 11, 16, 300
Cutrofello, Andrew, 107, 279

David-Ménard, Monique, 7, 34 n. 8, 42, 82
death, 128, 135–40, 235, 251
    historicality, 139–40
    and limits of power, 145–46
    as problematic object, 154, 229
    and temporality, 137–38
    See also disease; life; medicine, quasi-transcendentals
Deleuze, Gilles, 160–63, 184, 186, 196, 205, 208, 209 n. 42, 210
deliberation, transcendental, 45–55, 59, 65 n. 33, 73, 96, 104, 153, 165
    and introspection, 47–49
delinquency, 128–29, 140–46
    and liberalism, 143, 152
    as problematic object, 154, 229
    and property, 141–42
    and revolution, 141
De Man, Paul, 200
Derrida, Jacques, 78 n. 41, 177 n. 10
Descartes, René, 68, 82, 132, 135
difference, inner, 102, 106, 109–10, 192, 253–54
    and aesthetic judgment, 95–100
    and body, 36, 90–91, 95–96, 113, 251
    and inner sense, 95–97, 104
    See also concepts of reflection; feeling

difference, sexual, 172, 182–83, 187–88, 190, 257, 277
    Kant's views, 101, 194–95
discipline
    Kant's views on skill and, 28–29, 67 n. 34, 81, 230–31
    See also power, disciplinary
discipline, scholarly, 12, 17, 22, 24, 28–29, 129–30, 154, 173
    and enlightenment, 221–22
    human sciences, 159, 163
    and power, 229–30
    and subjugated knowledges, 258
discursivity, 14, 114, 198, 294
    and body, 167–68, 215, 285, 299
    of communications media, 158, 169, 243–45, 296
    and finitude, 38, 197
    of institutions, 296
    and language, 68–70, 124, 213
    and materiality, 299
    of understanding, 41–42, 67–68, 70–72, 81, 83–84, 112, 127
    See also language, materiality
disease, 135–38, 246–47, 250
doublet, transcendental-empirical, 118, 127, 155–64, 227, 255, 282. *See also* anthropology, philosophical; "Man"
Dreyfus, Hubert, 155, 180, 198
drugs, American "war on," 270, 277–79

economics
    Chicago School, 274, 278–79
    Ordoliberals, 271–73
economy, political, 225, 227–29
    Foucault and Kant, 230–31, 270
education, 148, 256, 257, 259, 261
emotion, 96 n. 49, 198–210, 223–24, 299
    moral significance, 199, 205
    See also feeling
enlightenment, 33, 261, 133
    in Foucault, 221, 266–67, 288, 297
    in Kant, 52–53, 85, 88, 151, 238–39, 263
    and *Schwärmerei*, 33–35, 89
énoncé, 167, 174–81
    and bodies, 118, 180–81, 191–92, 196–98
    interpretability and materiality of, 169, 173–74, 177, 179, 180
entrepreneurialism, 13, 271, 273–75, 278, 281
ethics, 8, 182, 202, 209. *See also* morality

events, 118–19, 166, 196–97, 282, 237
    bodily, 179–84
    in language, 175–76
exteriorization, psychic, 190–91

feeling
    associated with representations, 96–97
    and body in Kant, 90–91, 99–104, 106–8, 110, 113, 253
    communicable, in aesthetic judgment, 16, 62–67, 77, 81–82, 100–103, 224
    moral significance of, 199, 209–10, 253, 293, 301
    and power, 14, 16, 189–90, 200, 205, 212, 252
    pure contrasted with sensibility, 8, 95–99, 107–8, 211, 284, 293
    and reflection, 46 n. 13
    of resistance, 202–3, 206, 146, 181, 209
    of respect, 37 n. 13, 77 n. 40, 106
    See also differences, inner; emotion; pleasure
feminism, 217–18, 275–80
    feminist theorists, 118, 172, 186–92, 205 n. 39
    See also difference, sexual
finitude
    and body, 15, 139, 146
    fundamental, 57–58, 70, 130, 139–40, 156–59
    historical, 124–25, 127–29, 132–33, 179
    of Kantian understanding, 16, 38–39, 55, 67–68, 71, 110, 114–15, 161, 197
    of public, 87, 89
form
    in aesthetic judgment, 60–61, 76, 283
    as concept of reflection, 72–75
    formlessness, 6, 51, 77, 200, 204, 206, 212
    materialization of, 169–70, 177, 187, 189, 195
    See also materiality, sublime
Foucault, Michel
    on archaeology, 161–63, 168, 174, 180, 175–81, 211
    on body, 15, 162–63
    Copernican Revolution, 129, 138, 228, 230–32
    on critique, 8 n. 1, 34–35, 39, 47, 58, 67, 131, 155–56, 162, 197, 285–86, 288
    on enlightenment, 221, 266–67, 288, 297
    on freedom, 162, 182, 224, 236–38, 249–50, 258, 286, 288, 290–92, 296, 298

Foucault, Michel (*continued*)
   on genealogy, 142, 161–62, 176, 180, 211, 221–24, 262–65, 268, 282, 298
   on imagination, 120–22, 128–29, 162–63, 167
   on Kant, 27, 28 n. 2, 73 n. 36, 119–20, 126–27, 155–58, 210, 224, 227, 267
   Kant's influence on, 8, 15, 120, 124
   on Las Meninas, 49–51
   on political economy, 230–31, 270
   on temporality, 124, 165
   on transcendental philosophy, 157–60, 162, 164, 178
   See also government, art of; heterotopia; power; rates of change, historical
freedom, 200, 205–6, 208–9, 218, 258, 273, 284
   civil, Kantian, 230, 238–39
   in Foucault, 162, 182, 224, 236–38, 249–50, 258, 286, 288, 290–92, 296, 298
   and limits of state reason, 215–16, 232–33, 270–71
   moral, Kantian, 55, 58, 233
Freud, Sigmund, 81, 173, 240

Gatens, Moira, 95, 187, 189–91
genealogy, 142, 221–24, 262–65, 268, 282, 298
   and archaeology, 161–62, 176, 180, 211
genius, 69, 80–81, 155, 257
God, 31
   and critical philosophy, 92–93
   as idea of reason, 43–44, 59, 119, 197
   and intellectual intuition, 34, 43, 71, 110, 114–15, 130, 201–2
   and language, 69–70, 124
   See also religion
government, art of, 15, 23–24, 225–26, 228, 269
   juridical, 215, 222, 231, 236–37, 255, 270, 274
   liberal. *See* liberalism
   neoliberal. *See* neoliberalism
   normalizing, 215, 237, 247, 249, 253–55
   pastoral, 215, 225
   socialist, 264–65
Grosz, Elizabeth, 186–87, 189–90

Habermas, Jürgen, 33, 163, 198, 240, 247–50, 253, 258, 266, 280
Hacking, Ian, 48
Hamann, Johann, 34

Han, Beatrice, 129, 139, 155 n. 13, 156 n. 14, 162–63, 288
Harvey, David, 22, 32, 117, 170, 220–21, 250, 295–96
Heidegger, Martin, 52 n. 22, 56–57, 68, 75, 88, 115, 123–24, 155–56, 159, 172, 248
Herder, Johann, 33–34, 80, 85
heterochrony, 22–23, 154. *See also* rates of change, historical; temporality
heterotopia, 217, 243, 264, 291, 300
   in Foucault, 17, 20–25, 127, 146, 153–54, 181, 191, 243
   in Kant, 32
   See also spaces, conceptual; world
historical dimension, 122–23, 130, 154, 157, 160, 166, 176, 263–65, 294–95
   of individual, 212–14, 239, 266, 274, 280, 283
   in Kant, 87–88, 110
   of state, 212–14, 218, 225
   See also rates of change, historical
Hobbes, Thomas, 43, 213
humanism, 126, 143, 199 n. 34, 267. *See also* anthropology, philosophical; "Man"
Hume, David, 30, 33–34, 38–39, 48, 101, 172, 197
Husserl, Edmund, 17, 47, 121–23, 153, 159
hylozoism, 68, 79–80
hypochondria, 7, 45, 106–7

ideas, aesthetic, 81–82
ideas of reason, 43–44, 77, 79, 92, 121–22, 157, 197, 228, 294
   God, 44
   soul, 42–43, 48–49, 51, 154, 197
   world, 111, 121, 127, 157
   See also concept, problematic; reason; soul; heterotopia
identification, psychic, 213, 284. *See also* psychoanalysis
illness, mental, 1, 2, 4, 7, 278
   Kant's views on, 82–84
   See also madness
imagination, 2–6, 17–23, 32, 257, 283
   and body, 10–11, 23, 68–69, 107–8
   Foucault's views on, 120–22, 128–29, 162–63, 167
   in history of philosophy, 17–18, 68–70, 123, 172
   and language, 18, 69–70, 173
   materializing, 169–71

INDEX   321

and political community, 10–11, 19
and power, 24, 181–83, 186–87, 189, 201, 208
See also heterotopia, phenomenology
imagination, Kantian
in aesthetic judgment, 63–64, 76–78
and body, 7–8, 89, 92, 106–7
communicability and, 81–83, 88, 202
and enthusiasm, 35–36 n. 10, 78–79, 82–83, 205
finitude and, 67–68
Foucault and, 162–63
Heidegger's views on, 75, 88, 115, 172–73
mimetic, 75, 192–94
schematizing, 19, 22, 71–72, 74–78, 109, 118, 153, 162, 168, 173–74, 178, 212
transcendental, 18, 28–30, 56–57, 75, 172
imagination, political, 13–14, 16, 25, 28, 212, 214, 282–84, 290, 300
and American New Left, 223–24
and body, 9–12, 126, 236
and revolution, 294–97
individual, dangerous, 154, 237, 247, 264, 267
and psychiatry, 152, 233
individuality, 47, 202, 239–40, 261–62, 234, 283–84, 293
and body, 6, 13, 54, 150, 274, 281
empowering or disempowering, 6, 202, 206, 209
as historical form, 139–40, 143–45, 150, 166–67, 212–14, 239, 266, 280
and rights, 289–90
and risk, 65, 234, 236, 255–56, 259, 274
See also individualization; sphere, public; subindividuals
individualization, 215, 254–55, 282–84, 287, 289–91, 294–95
and bodily practices, 106, 143, 146, 189, 220
and economics, 231
and feeling or desire, 99, 183, 209–10, 261
incomplete, 287, 301
and population, 149, 189, 225
and totalization, 219, 261–62, 286, 296
See also differences, inner; individuality; interiorization, psychic
inequality, social, 108, 249, 220, 297–98. See also poverty
intelligence, 255–57, 260–62, 273, 286. See also genius; power relations

interiorization, psychic, 51, 110, 190–91, 197–99, 256, 261, 280–82
and disciplinary power, 141, 143–44, 186
interiority of consciousness, 51, 110, 204–9
See also confession; exteriorization, psychic
interpretation
of énoncés, 125, 173–76
of needs, 290–91
and power, 161, 181–83
and problematization, 288–89
and subjectification, 6, 101, 197–98, 282
introspection, 45, 47–49, 147, 197–98
intuition, intellectual, 34, 41–43, 85 n. 45, 105, 114, 157, 201 n. 35. See also understanding, intuitive

judgment, aesthetic, 8–9, 61–67, 164–66
dialectic of, 60, 99, 105–6
and embodiment, 8, 15, 89–90, 99–105, 110
empirical aspects, 62–66, 99–100, 105–6, 293
in modern medicine and psychiatry, 152
moral implications, 104, 292
political implications, 100–104, 192, 237
and powers of thought, 59, 76–78, 100
and problematization, 8–9, 290
and pure pleasure, 8–9, 62–64, 76, 97–99
See also beauty, judgment of; taste, judgment of
judgment, infinite, 41, 120, 130
judgment, reflective, 59, 62
contrasted with determinate judgment, 43–44, 46, 76, 153, 169, 192, 194, 290
in Foucault, 163–66
and problematization, 8, 284, 290
See also concepts, of reflection; deliberation; imagination, schematizing
judgment, teleological, 47, 58–61, 89

Kant, Immanuel
critique of Leibniz, 33–34, 35, 42, 71–73, 91, 96
critique of Schwärmerei, 7, 15, 28, 33–35, 42, 78, 202, 221
on medicine, 82–84, 106–7
on philosophical anthropology, 28, 47, 73, 89, 101, 120–21, 124, 155–56, 159, 175, 197
precritical essays, 35–36, 44, 60, 90–92, 101, 105, 203

Kant, Immanuel (*continued*)
   on sexual difference, 101, 194–95
   on real conflict, 35, 44, 119, 127, 296
   on revolution, 85–88, 101, 231
Kapuściński, Ryszard, 11
Klossowski, Pierre, 184–85, 196–97
Kuhn, Thomas, 75, 173

labor
   division of labor, 67 n. 34, 183
   and finitude, 137, 140, 143
   and labor power, 171–72, 213, 245, 257–58
   organized, 215, 219, 220, 223, 270
   and property, 256, 259
   See also quasi-transcendentals; unions, trade
Lacan, Jacques, 50 n. 21
language, 34, 48, 123–26, 183
   being of, 155, 172, 174–75, 183, 208
   communicability of, 66, 69–70, 148, 175, 243
   and imagination, 18, 69–70, 173
   in Kant, 68–70
   and materiality, 81, 125–26, 159, 172, 187, 198
   and power, 126, 177, 243
   and representations, 69, 174, 183
   and resemblance, 18, 172–75, 184–85
   See also communicability; *énoncés*; quasi-transcendentals
law, 208, 215, 224–25, 236–37, 270
   and delinquency, 141–43
   as discourse, 20, 133–34, 150, 290, 296
   moral, Kantian, 29, 78, 106, 203, 205, 212, 233, 252–53, 297
   and neoliberalism, 221, 274
   as political/moral norm, 231, 247–49, 254–55, 262–63, 289–90
Lefebvre, Henri, 22, 51–55, 75, 95, 97, 104, 123
Leibniz, G. W., 30, 33–35, 42–43, 48, 52, 69, 71, 73, 91, 96, 105, 201–2, 205, 281
Lévi-Strauss, Claude, 27, 42, 61, 75 n. 38, 159
liberalism, 225, 229–30, 234, 247, 269–70, 273–76
   and delinquency, 143, 152
   and madness, 133–34
   and sexuality, 149
   See also government, art of; markets
life, 209, 129, 130, 138–40, 157
   and biopower, 147, 148, 209, 222, 235–36, 250–51
   Kant's views, 79, 99, 193
   See death; quasi-transcendentals
literature, 21–22, 103, 124, 159, 174–75, 184–85, 207
Longuenesse, Beatrice, 94
Lyotard, Jean-François, 32, 37 n. 13, 46 nn. 18, 19, 96, 104–5

madness
   and communicability, 82–85
   in Foucault, 128, 131–35, 140, 147, 152, 229
   "Great Confinement," 148, 179, 196
   historicality, 133–35, 154, 178 n. 21
   in Kant, 34 n. 8, 82–84, 88
   as problematic object, 229
   and public space, 131
   and social criticism, 4–5, 16, 88–89, 299–300
   See also illness, mental; psychiatry
magic realism, 21–22
Makkreel, Rudolf, 43, 46, 76, 174 n. 18
"Man," 17–18, 115, 118, 124–27, 154, 159–62, 172, 175, 195, 260
   being of, 159, 161, 165, 175, 208, 227
   See also anthropology, philosophical; anti-humanism
markets, 135, 149, 171, 260, 269, 290
   currency, 220–21
   labor, 275, 278–79
   as phenomenal field, 215, 224, 228–31, 245, 294
   slave, 241–42
Marx, Karl, 171, 173, 183, 231, 234, 248, 258, 265, 300
materiality
   and affect, 189, 208
   and body, 113–14, 180, 181, 183, 185, 187–89, 196, 208, 284–85, 300
   and communicability, 103
   of *énoncés*, 176–77, 179, 180
   and finitude, 79–80, 127, 139, 197
   of forms, 73–77, 167, 169–71, 200
   and imagination, 67–68
   of language, 81, 125–26, 159, 172, 187, 198
   and materializing form, 170, 177, 187, 195, 171–72
   and public space, 294, 296, 300
   and state's power, 215, 229, 284

See also concepts, of reflection; hylozoism; rates of change, historical; scale of apprehension; understanding
materialism, 171–72
measure, 132, 140, 183, 245
   body as, 103, 111, 113, 186
   in sublime, 109–12
media, communication, 19, 22, 100 n. 50, 108, 242–46, 290
   and materiality, 102–4, 158, 169, 246
   See also sphere, public
medicine, 135–40, 152, 166, 229, 246–47
   and madness, 131–34
   and normativity, 205, 249–51
Merleau-Ponty, Maurice, 23, 111 n. 54, 123, 180, 254 n. 23
modernity, 123, 159, 210, 286, 288, 292–93
   Kantian/Foucauldian, 266, 294–95
   See also Enlightenment
morality, 153, 166–67, 292–93
   Kantian, 18, 78, 85, 106, 204–5
   and poverty, 258–59, 277–78
   and power, 212, 215, 254
   and the unthought, 161, 260
   See also ethics; norms
mothers, 12, 19, 275–79, 280–81
movements, political/social
   left-wing, 214–19, 222–23, 265, 286, 298–99
   right-wing, 221–23, 274–75, 279, 298
   See also citizenship; countercultures; feminism
music
   African-American, 103
   and communicability, 81–82, 100 n. 50

nationalism, 19, 85, 149, 226, 240, 264–65
   and transnationalism, 218–19, 221, 235
needs, 99, 234, 249, 253, 260, 289–91
Negt, Oskar and Hans Kluge, 242–44
neoliberalism, 12–13, 15–16, 210, 215, 220–21, 223, 255, 267, 269–81, 296
Nerlich, Graham, 112–13
Nietzsche, Friedrich, 165, 172–73, 175, 183–84, 186–87, 190–91, 201, 222, 254
nominalism, 195–96, 212–14, 228, 232, 260, 286, 294
normal, 147, 251, 261–62, 298
   and pathological/abnormal, 147, 150, 204, 236, 249–54
normativity, 216, 249–55, 261–62, 264, 267, 291
   and individuality, 260–62, 280, 285
   in medicine, 107, 249–51
   and moral norms, 53–55, 59–62, 249, 293
   and public sphere, 243, 246
   and risk, 255, 267, 273, 301
   of social movements, 283, 291, 294–95
   and the state, 270, 290, 301
   in the sublime, 205–6, 207, 252–53
norms
   anthropological, 107–8, 234, 243, 260–61
   associated with body, 102–5, 108, 114, 145–46, 188, 240, 249
   in medicine, 206, 249–53, 254
   moral, 153, 198–99, 248, 292–93
   phenomenological, 24–25, 32, 123
   psychological, 6, 147–48, 300–301
   and resistance, 146, 148, 187–88
   See also law; morality
noumenon. *See* concept, problematic
Nouveau Philosophes, 248

object, transcendental = x, 42, 195
object, problematic, 51, 236, 275, 281, 298
   and aesthetic judgment, 152, 164–66
   body as, 6–7, 9, 17, 90, 118, 148
   in Foucault, 118–19, 129–30, 135, 139, 151, 154–61, 164, 167, 172
   in Kant, 28–29, 31, 43, 49, 58, 90, 122, 129
   See also death, delinquency, madness, sexuality
ontology, 2, 123, 151, 154–57, 159, 166–67, 175, 213–14, 266. *See also* language, being of; "Man"

paralogism, 37 n. 14, 42, 51, 89, 144 n. 10, 154. *See also* soul
parties, political, 13, 242, 244
pathological. *See* normal
phenomenology
   and body, 23, 126, 139, 186, 190–91, 198–99
   and imagination, 17–18, 120–21, 153
   transcendental and anthropological reflection, 122–23, 125, 159–62
   and "world," 118–19, 120–21, 153
Plato, 135, 202, 204, 292
pleasure
   in aesthetic judgment, 15, 62–64, 66–67, 76, 97–100
   as axis of individualization or communication, 103–4, 107–8, 115, 148, 204, 244, 283

pleasure (*continued*)
  and bodies in Foucault, 117, 118, 127, 155, 219, 284
  and normativity, 253
  pure vs. empirical, 9, 96, 98, 102, 106, 166, 203
  of resistance, 146, 204, 210
  sexual, 147, 151
  and sublime, 206–7
  and transcendental deliberation, 61, 96
police, early modern, 226–28, 233
  violence, 1–2, 7, 10–11, 217, 277 n. 34
  See also *raison d'état*
population
  and health, 31, 136, 140, 149, 150
  and liberal individuals, 231–33, 234
  and privatization, 16, 273, 274
  and public, 237–38, 241, 263
  and *raison d'état*, 222, 225, 227, 296
  and risk, 234–36, 247, 274
  See also *raison d'état*
positivity. *See* a priori, historical
post-coloniality, 17, 19, 22, 223, 243, 268
postmodernity, 32
poverty, 139, 250 n. 20, 275–81
power, 3, 14, 16, 155, 195–96
  and bodies, 181–82, 189, 249, 253, 274, 284–85, 287, 291
  and communicability, 102–4, 108, 253
  and communication media, 102–4, 242–47
  disciplinary, 143–45, 150, 225, 227–29, 255, 258
  and feeling, 10, 114, 189–90, 200, 205–6, 209, 210, 212
  and imagination, 14, 181–83, 201, 208, 232
  and knowledge, 120, 126, 162, 197, 213, 229–30, 258, 265
  and language, 126, 177, 243
  and law, 237, 249, 270, 280, 289–90
  limits of state, 141, 145, 149, 232 n. 11, 239, 287
  and morality, 212, 215
  normalizing, 150, 209, 215, 249, 254–55
  pastoral, 215, 225–26
  power relations, 149, 181–82, 200, 208–10, 214–15, 256, 287–88
  repressive and generative, 146, 148, 151, 178, 212–13
  and resemblance, 185–87, 191, 197
  and resistance, 146, 153, 181, 204–5, 209, 219, 296

  and risk, 234, 296–97
  in self-formation, 162, 197–98, 261–62, 289
  and violence, 181–82, 252
powers, balance of, in *raison d'état*, 86, 225, 230
powers of thought, 27–29, 45–47, 49, 61–63, 96–98, 114, 193–94, 257. *See also* imagination; reason; understanding
prison, 142–45, 208, 231, 272, 275–76, 280, 300
privatization, 12–13, 220, 270, 272, 274, 301. *See also* neoliberalism
problematic, 12, 92 n. 48, 282, 296
problematization, 9, 12, 105, 153, 155, 162, 208
  and action on actions, 288–89, 292, 300
  and public sphere, 237–38, 266 n. 31, 284, 286,
property
  and delinquency, 141–42
  and liberalism, 232, 298
  as principle of interiorization, 214, 255–56, 258–60, 262, 284
  See also power relations
prudentialism, 272–73. *See also* neoliberalism
psychiatry, 2, 19–20, 24, 131–35, 152, 233
psychoanalysis, 23, 166, 289–90
  in deployment of sexuality, 148, 152, 197, 204
  and narcissistic identification, 65, 213, 284
  on resemblance and anxiety, 185, 186, 191
punishment, 141–45, 241, 258
  divine, 131
  in Nietzsche, 183, 186–87, 190

quasi-transcendentals, 124–25, 155–58, 255, 259–60
  and historicity, 15, 130, 157
  life, labor, and language, 234, 284
  and "Man," 125, 227
Quine, Willard V. O., 24 n. 3

racism, 185, 188–89, 217, 268, 278
  in Kant, 101, 257
  state, 222, 235, 237, 247, 264–65, 271, 276
raison d'état
  discipline, 225, 273
  and liberalism, 228–30, 233

pastoral. *See* power, pastoral
sécurité, 224–27, 247, 252
  See also liberalism
Ransom, John, 286–87
rates of change, historical, 122–23, 125, 154, 165–66
  breakup of classical episteme, 158–60, 263
  dividing the "present," 266–69
  heterochrony, 22–23, 154
  See also events; history; scale of apprehension; temporality
Rawls, John, 67, 247–49
Reagan, Ronald, 1, 3, 6, 13, 24, 214, 220
reason, Kantian, 29
  bounds of, 38–39
  division of, 56–58, 104, 110, 119, 151
  practical, 43, 55, 203
  unity of, 14–15, 43–44, 55–59, 105
  See also ideas of reason
reflection
  anthropological, 28 n. 2, 156
  transcendental. *See* deliberation
  See also judgment, reflective
religion, 19, 34, 190, 221, 240, 259, 269, 294
  in American culture, 241, 242, 275, 279
  in modern Europe, 33–34, 156
representations, 30–32, 48–53, 123–24, 142, 172–74, 183
  of body, 113, 186–87
  feeling associated with, 96–97, 203
  Kantian and Classical, in Foucault, 123–24, 157–59
  and language, 69, 174, 183
  sorted by deliberation, 45–46
  of space, 51–52, 54, 113, 123
resemblance, 297–98, 300
  between bodies, 149–50, 167, 181, 190–92, 195
  between social positions, 235, 268–69, 276, 281, 283, 291
  between subject and object, 75–76, 170, 200
  in Foucault, 118, 125, 128, 167, 172–75, 196–98
  and imagination, 169, 192–95
  and imitation, 185, 187–89, 196
  in Kant, 68, 99–100, 192–95
  and language, 18, 172–75, 184–85
  and pornography, 184, 192–93
  and power, 181–82, 185–87, 197, 251
resistance, 89, 107, 296, 300
  body as point of, 117, 166, 182, 249, 287
  in Foucault, 146, 148, 153, 209, 258
  in Kant, 202–6
  to resemblance, 3, 7, 42, 181, 185–86, 189, 276, 281, 291
  See also power
Revolution, Copernican
  Kantian, 30–32, 37–39, 92–93, 114, 158–59, 168, 202
  and liberalism, 228, 230–32
  and madness, 129
  and medicine, 138
Revolution, 85–87, 141, 234, 243, 294–97
  English, 263
  French, 86, 128, 131, 133, 141–143, 264–65, 268
  Iranian, 10–11, 283–84
  Kant's views on, 85–88, 101, 231
rights, 3, 95, 141–43, 284, 289–91
  civil, for African-Americans, 217–18, 232, 277–78
  See also citizenship
risk, 214, 225–26, 231–36, 238, 247, 251–53, 258, 272–74, 283, 298
  environmental, 245–46
  and freedom, 232, 233. *See also* freedom
  and individuality, 9, 234, 236, 255, 259, 281
  and neoliberalism, 13, 16, 271–74
  and power relations, 232
Rogozinski, Jacob, 77, 78 n. 41, 85 n. 46, 111–12
Rousseau, 34, 36, 69, 133

Sade, Donatien Alphonse François, Marquis de, 183–84, 192–93, 197, 207
Salecl, Renata, 259, 289–90
Sartre, Jean-Paul, 17
scale of apprehension, 14, 16, 53
  and materiality, 68, 72–77
  and normativity, 252–54, 287
  of public communication, 103, 237, 245, 257, 290, 296
  of state reason, 226, 229
  in sublime, 109–12, 160, 169–70
  See also discursivity, materiality, sublime
Scarry, Elaine, 10, 170, 182, 190, 211
Schwärmerei, 230, 257, 267
  and enlightenment, 33–35, 89, 267
  and Foucault, 199, 208–9, 230, 267
  and genius, 80–81
  and Kant, 7, 15, 28, 33–35, 42, 78, 202, 221

Schwärmerei (continued)
    and madness, 82–84, 281
    and revolution, 86–89
security, 222, 252–53, 255, 270, 297. See also *raison d'état*
sensibility, Kantian, 96–97
    aesthetic (in general), 254, 265, 283
sensus communis, 63–67, 81, 99, 102–6, 109, 115, 187, 191–92, 195, 211, 222, 282
sexuality, 19, 127–29, 146–51, 271
    and bodily difference or similarity, 148, 189–90, 192–93
    and bourgeois identity, 102–3, 149, 239, 258–59, 264
    and mental illness, 132, 140, 141
    and neoliberalism, 271, 274, 275–76
    as principle of interiorization, 147–48, 255–56, 259 262, 264, 284
    as problematic object, 117–18, 147–48, 229, 281, 286, 296
    queer or "perverse," 6, 147–48, 150–51, 181, 188–89, 205, 218, 271, 300
    resistance to, 117–18, 148, 152–53, 208
    and risk, 235–36, 259, 264
    See also power relations
Shell, Susan Meld, 83, 91, 93
slavery, 4–5, 218, 222, 232 n. 11, 239, 241–42, 258
    Kant's views, 102
soul, 23, 143–45, 204, 286
    in Kant, 36, 41–44, 48–49, 51, 74, 154, 197
    in Leibniz, 35–36
    and simulacra, 184–85
    See also ideas, of reason; introspection
sovereignty, 150, 215, 224, 228–31, 236, 263
    Kant's views, 31, 87
    and contemporary power, 254–55
space
    form of outer sense, 91–97, 110–13
    Kant's views, 30, 36, 42, 44–45, 49–55, 91–97, 110–14
    materiality of objects in, 68, 72–73, 109–10, 139
    non-Euclidean, 112–13
    as product of social practice, 51–55, 113, 134, 256, 244
    relation to time, 89–90, 104, 127
    representations of, 51–52, 54, 113, 123
    transcendental topic, 47, 54, 97
space, conceptual, 32, 47–59, 97, 104, 110, 113, 123, 153, 179, 191, 300

    and death, 135–37, 140
    and delinquency, 143, 152
    in Foucault, 19–22, 134, 155, 163
    market as, 224, 231, 272–73
    and sexuality, 147
    See also deliberation; heterotopia; market; sphere, public
sphere, public, 19, 236–38, 284
    bourgeois, 33, 35, 238–40, 242–43, 294
    as check on government, 83, 85–88, 238–39
    and communicability of aesthetic judgment, 99–101, 102, 110, 267–69
    and embodiment, 238, 240–42, 247
    plebian/proletarian, 239–43
    and population, 238, 263
    and press/media, 242, 244–46, 296
    and problematization, 105, 266, 289, 291, 300
Spinoza, Baruch, 30, 33, 36, 69, 79, 189–90, 201–2, 205, 291
state, 214–15, 301
    as historical form, 212–14, 218, 225
    as idea of political reason, 228, 231
    legitimation, 215, 231, 247, 269
    phobie d'état, 4–6, 214–15, 248, 272
    welfare, 214, 216–21, 223–24, 248, 255, 270–72, 274, 280, 291
statement. See *énoncé*
Strathern, Marilyn, 75, 112
subindividuals, 286–87. *See also* individuality
subject
    and emotion, 199–200, 208–9
    of knowledge, 6, 132, 199, 213, 292
    of law, 134, 147, 292
    of power and action, 162, 181–82, 205–6
    transformation of, 167, 170, 285, 288
sublime, judgments of, 64, 200, 206, 207, 210
    anthropological aspects, 101, 105
    and fanaticism, 78–79
    in Foucault's work, 207–8, 210
    moral implications, 78, 212, 252, 262
    restriction of imagination, 76–79, 83, 109–11
    and temporality, 165, 206, 292
supersensible. *See* concept, problematic
Surrealism, 21, 174–75, 207
surveillance
    of individuals, 6, 9, 141–43,
    of populations, 226, 238, 241
Swedenborg, Emmanuel, 35 n. 10, 82, 281

taste, judgment of
    anthropological aspects, 62–67, 100–106
    communicability of, 60–67
    pure aesthetic pleasure in, 61–64, 97–99
    universality of, 59, 61, 63, 108
    See also feeling; judgment, aesthetic; pleasure; sensibility; sensus communis
temporality
    apprehending events, 164, 188–89, 196, 245–46, 267, 292, 296
    and criminality, 143–44
    and disciplined body, 143–44, 225
    and disease, 137–39
    division of, in problematization, 291–92, 296
    and entrepreneurialism, 273–74
    form of inner sense, 48, 52, 71, 73 n. 36, 90, 94, 130, 206
    historical, 130, 154, 157, 160
    and language, 124
    multiple, 109–10, 111–12, 121–23, 154, 165–66
    permanence and succession, 93–94, 109
    pure intuitions, 68, 72, 78, 79, 254
    in relation to space, 99, 103–4, 109
    space-time, 112–13
    and synthesis of understanding, 56, 68, 70–71, 168–69, 288, 291
    See also heterochrony; historical dimension; understanding
topic, transcendental, 46, 50, 59, 153. See also deliberation; spaces, conceptual

understanding, Kantian
    discursivity, 14, 41, 67–68, 70–72, 74–75, 81, 114, 127, 167–68, 183
    and imagination, 67–68, 71–76, 76–77, 83, 153, 192, 194, 297
    intuitive, 41–43, 70, 110
    and language, 68–70
    materiality, 67–68, 79, 81, 103, 127, 167–68, 183
    temporality, 70–71, 110, 168–69
    See also powers of thought
unions, trade, 220, 242, 244, 276. See also labor, organized
unthought, 119, 161, 163–67, 260, 282
unthinkable, 285
USSR, 12, 14, 86, 216, 218, 223, 248, 271

Velázquez, Diego, 49–51
Veyne, Paul, 38, 164
violence, 253, 269, 285, 289, 292, 293
    and power, 181–82

war, 10
    Algerian-French, 217
    civil war, 4, 85–89, 263
    Cold (Soviet-American), 268
    race, 150, 154
    Vietnam-American, 217, 276
    See also colonialism; drugs, American "War on;" racism, state; revolution; USSR
welfare (ADC/AFDC), 13, 275, 277–81, 290
whiteness, 103–4
Williams, Patricia, 1–2
Wolff, Christian, 33
world
    in Kant, 120–21, 153, 157
    in phenomenology, 121–22, 168
    See also heterotopia; ideas of reason

Žižek, Slavoj, 11, 42 n. 16, 92 n. 48, 100 n. 50

www.ingramcontent.com/pod-product-compliance
Lightning Source LLC
Chambersburg PA
CBHW031544300426
44111CB00006BA/173